W9-BKY-197

FLIGHT THEORY AND AERODYNAMICS

FLIGHT THEORY AND AERODYNAMICS

A Practical Guide for Operational Safety

SECOND EDITION

Charles E. Dole

James E. Lewis

A WILEY-INTERSCIENCE PUBLICATION

JOHN WILEY & SONS, INC.

New York • Chichester • Weinheim • Brisbane • Singapore • Toronto

This book is printed on acid-free paper. ♾

Published simultaneously in Canada.

Library of Congress Cataloging-in-Publication Data:

Dole, Charles E. (Charles Edward), 1916–
Flight theory and aerodynamics : a practical guide for operational safety / by Charles E. Dole. -- 2nd ed.
p. cm.
Includes index.
ISBN 0-471-37006-1 (alk. paper)
1. Aerodynamics--Handbooks, manuals, etc. 2. Airplanes--Piloting--Handbooks, manuals, etc. 3. Aeronautics--Safety measures--Handbooks, manuals, etc. I. Title.
TL570.D56 2000
629.132--dc21 99-38912
CIP

Printed in the United States of America

10 9 8 7 6 5 4

PREFACE

The first edition of *Flight Theory and Aerodynamics* was written as a text for the USAF Flying Safety Officer course and other similar courses presented by the Institute of Safety and Systems Management of the University of Southern California. The students of these courses were mature pilots of the ranks of captain and lieutenant colonel. Many of these students had had little or no engineering education and lacked the mathematical skills necessary for a complete study of aerodynamics. With this in mind, the text was written using only basic algebra.

The book has been well received by the general aviation public. Now, 18 years later, it seems appropriate to revise the text. This revision is slanted toward the book's use as an introductory and familiarization course text in colleges and universities offering aeronautics curricula. It is not intended to replace any of the many fine advanced aerodynamics texts now on the market. Instead, it presents a broad coverage of the theory of flight.

The first edition contained eight chapters, two of which were too long to be easily digested by students. In this edition, these two chapters have been revised into eight "bite-sized" chapters. A new chapter on helicopter flight has been added, and the entire book has been brought up to date.

A summary of the symbols and equations used in each chapter is found at the end of the chapter, along with problems to assist the students in evaluating their understanding of the material. The answers to the problems are given at the end of the book, together with selected references.

The authors would like to gratefully acknowledge the previous works done by H. H. Hurt, Jr., *Aerodynamics for Naval Aviators*, and John R. Montgomery, *Sikorsky Helicopter Flight Theory for Pilots and Mechanics*, Sikorsky Aircraft, Division of United Technologies, from which many illustrations in this book were obtained.

CHARLES E. DOLE
JAMES E. LEWIS

CONTENTS

FLIGHT THEORY AND AERODYNAMICS

Boeing 757-3000 (Courtesy the Boeing Company).

1 Introduction

A basic understanding of the physical laws of nature that affect aircraft in flight and on the ground is a prerequisite for the study of aerodynamics. A brief summary of the branch of physics called *mechanics* is presented here. Concepts of work, energy, power, and friction are discussed in this chapter. For a more detailed explanation of these subjects, the reader should consult any basic physics text.

BASIC QUANTITIES

Because the metric system of measurement has not yet been widely accepted in the United States, the English system of measurement is used in this book. The fundamental units are

Force	pounds (lb)
Distance	feet (ft)
Time	seconds (sec)

From the fundamental units, other quantities can be derived:

Velocity (distance/time)	ft/sec (fps)
Area (distance squared)	square ft (ft^2)
Pressure (force/unit area)	lb/ft^2 (psf)
Acceleration (change in velocity)	ft/sec/sec (fps^2)

Aircraft measure airspeed in knots (nautical miles per hour) or in Mach number (the ratio of true airspeed to the speed of sound). Rates of climb and descent are measured in feet per minute, so quantities other than those above are used in some cases. Some useful conversion factors are listed below:

Multiply	by	to get
knots	1.69	feet per second (fps)
fps	0.5925	knots
miles per hour (mph)	1.47	fps
fps	0.6818	mph
nautical miles (nmi)	6076	feet (ft)
nmi	1.15	statute miles (stmi)
stmi	0.869	nmi
knots	101.3	feet per minute (fpm)

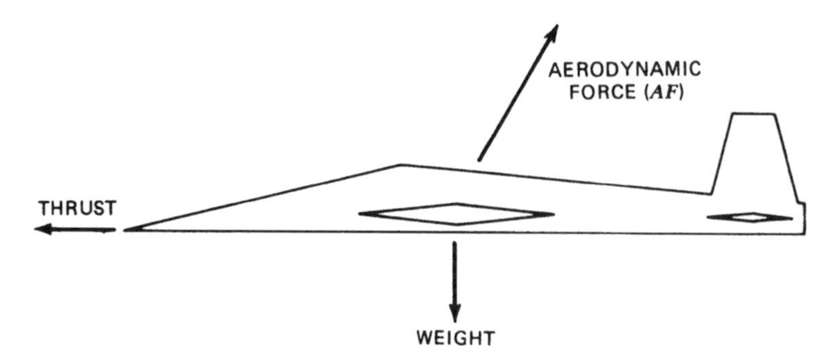

Fig. 1.1. Forces on an airplane in steady flight.

FORCES

A force is a push or a pull tending to change the state of motion of a body. Typical forces acting on an aircraft in steady flight are shown in Fig. 1.1. Figure 1.2 shows the resolution of the aerodynamic force into two components. The component that is 90° to the flight path and acts toward the top of the airplane is called *lift*. The component that is parallel to the flight path and acts toward the rear of the airplane is called *drag*.

MASS

Mass is a measure of the amount of material contained in a body. *Weight*, on the other hand, is a force caused by the gravitational attraction of the earth, moon, sun, or other heavenly bodies. Weight will vary, depending on where the body is located in space. Mass will not vary with position.

$$\text{Weight } (W) = \text{Mass } (m) \times \text{Acceleration of gravity } (g)$$

$$W = mg \tag{1.1}$$

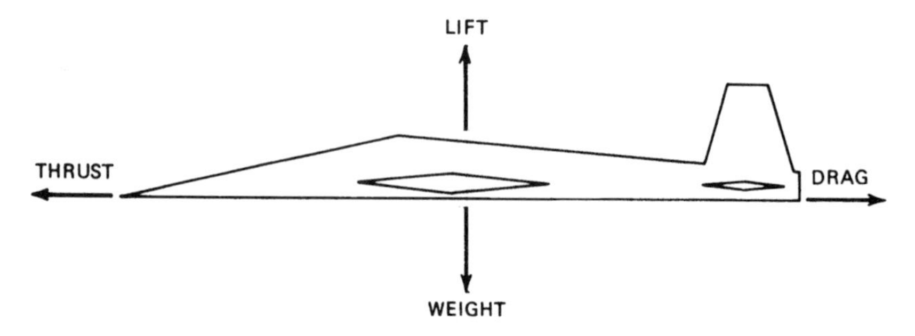

Fig. 1.2. Resolved forces on an airplane in steady flight.

Rearranging gives

$$m = \frac{W}{g} \frac{\text{lb}}{\text{ft/sec}^2} = \frac{\text{lb} \cdot \text{sec}^2}{\text{ft}}$$

This mass unit is called the *slug*.

SCALAR AND VECTOR QUANTITIES

A quantity that has size or magnitude only is called a *scalar* quantity. The quantities of mass, time, and temperature are examples of scalar quantities. A quantity that has both magnitude and direction is called a *vector* quantity. Forces, accelerations, and velocities are examples of vector quantities. Speed is a scalar, but if we consider the direction of the speed, then it is a vector quantity called *velocity*. If we say a car traveled 40 mi, the distance is a scalar. But if we say a car traveled 40 mi to the north, the distance is a vector quantity.

Scalar Addition

Scalar quantities can be added (or subtracted) by simple arithmetic. For example, if you have 5 gallons of gas in your car's tank and you stop at a gas station and top off your tank with 9 gallons more, your tank now holds 14 gallons.

Vector Addition

Vector addition is more complicated than scalar addition. Vector quantities are conveniently shown by arrows. The length of the arrow represents the magnitude of the quantity, and the orientation of the arrow represents the directional property of the quantity. For example, if we consider the top of this page as representing north and we want to show the velocity of an aircraft flying east at an airspeed of 300 knots, the velocity vector is as shown in Fig. 1.3. If there is a 30-knot wind from the north, the wind vector is as shown in Fig. 1.4.

To find the aircraft's flight path, groundspeed, and drift angle, we add these two vectors as follows. Place the tail of the wind vector at the arrow of the aircraft vector and draw a straight line from the tail of the aircraft vector to

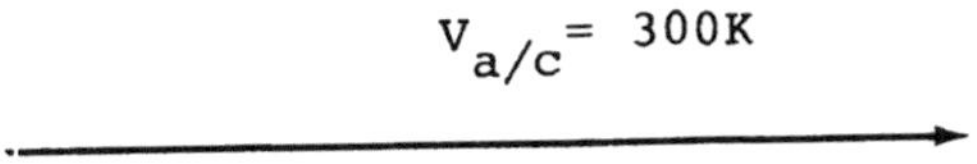

$V_{a/c} = 300\text{K}$

Fig. 1.3. Vector of an eastbound aircraft.

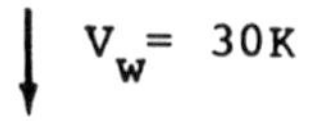

$V_w = 30K$

Fig. 1.4. Vector of a north wind.

the arrow of the wind vector. This *resultant* vector represents the path of the aircraft over the ground. The length of the resultant vector represents the groundspeed, and the angle between the aircraft vector and the resultant vector is the drift angle (Fig. 1.5).

The groundspeed is the hypotenuse of the right triangle and is found by use of the Pythagorean theorem $V_r^2 = V_{a/c}^2 + V_w^2$:

$$\text{Groundspeed} = V_r = \sqrt{(300)^2 + (30)^2} = 301.5 \text{ knots}$$

The drift angle is the angle whose tangent is $V_w/V_{a/c} = 30/300 = 0.1$, which is 5.7° to the right (south) of the aircraft heading.

Vector Resolution

It is often desirable to replace a given vector by two or more other vectors. This is called *vector resolution.* The resulting vectors are called component vectors of the original vector and, if added vectorially, they will produce the original vector. For example, if an aircraft is in a steady climb, at an airspeed of 200 knots, and the flight path makes a 30° angle with the horizontal, the groundspeed and rate of climb can be found by vector resolution. The flight path and velocity are shown by vector $V_{a/c}$ in Fig. 1.6.

In Fig. 1.7 to resolve the vector $V_{a/c}$ into a component V_h parallel to the horizontal, which will represent the groundspeed, and a vertical component, V_v, which will represent the rate of climb, we simply draw a straight line vertically upward from the horizontal to the tip of the arrow $V_{a/c}$. This vertical line represents the rate of climb and the horizontal line represents the groundspeed of the aircraft. If the airspeed $V_{a/c}$ is 200 knots and the climb angle is 30°, mathematically the values are

$$V_h = V_{a/c} \cos 30° = 200(0.866) = 173.2 \text{ knots} \qquad \text{(Groundspeed)}$$

$$V_v = V_{a/c} \sin 30° = 200(0.500) = 100 \text{ knots or } 10{,}130 \text{ fpm} \qquad \text{(Rate of climb)}$$

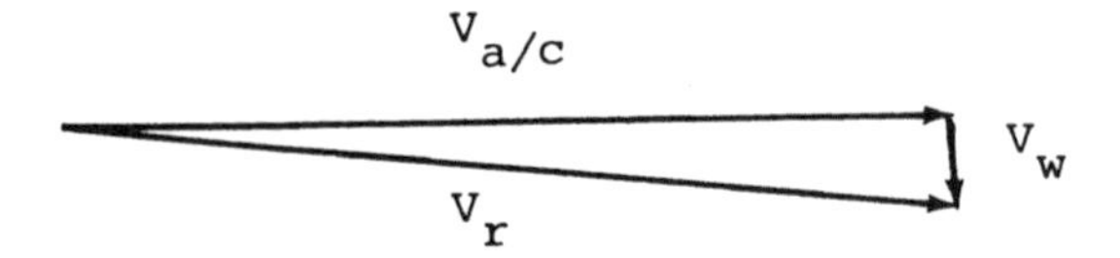

Fig. 1.5. Vector addition.

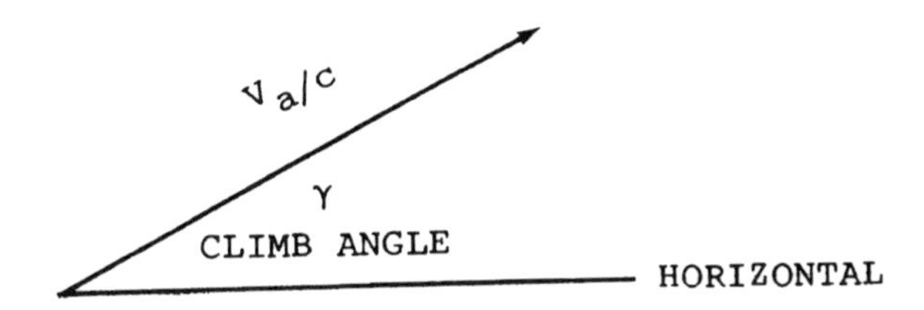

Fig. 1.6. Vector of an aircraft in a climb.

MOMENTS

If a mechanic tightens a nut by applying a force to a wrench, a twisting action, called a *moment*, is created about the center of the bolt. This particular type of moment is called *torque* (pronounced "tork"). Moments, M, are measured by multiplying the amount of the applied force, F, by the *moment arm*, L:

$$\text{Moment} = \text{force} \times \text{arm} \quad \text{or} \quad M = FL \tag{1.2}$$

The moment arm is the perpendicular distance from the line of action of the applied force to the center of rotation. Moments are measured as foot-pounds (ft-lb) or as inch-pounds (in.-lb). If a mechanic uses a 10-in.-long wrench and applies 25 lb of force, the torque on the nut is 250 in.-lb.

The aircraft moments that are of particular interest to pilots include pitching moments, yawing moments, and rolling moments. Pitching moments, for example, occur when an aircraft's elevator is moved. Air loads on the elevator, multiplied by the distance to the aircraft's center of gravity (CG), create pitching moments, which cause the nose to pitch up or down.

Several forces may act on an aircraft at the same time, and each will produce its own moment about the aircraft's CG. Some of these moments may oppose others in direction. It is therefore necessary to classify each moment, not only by its magnitude, but also by its direction of rotation. One such classification could be by *clockwise* or *counterclockwise* rotation. In the case of pitching moments, a *nose-up* or *nose-down* classification seems appropriate.

Mathematically, it is desirable that moments be classified as positive (+) or negative (−). For example, if a clockwise moment is considered to be a +

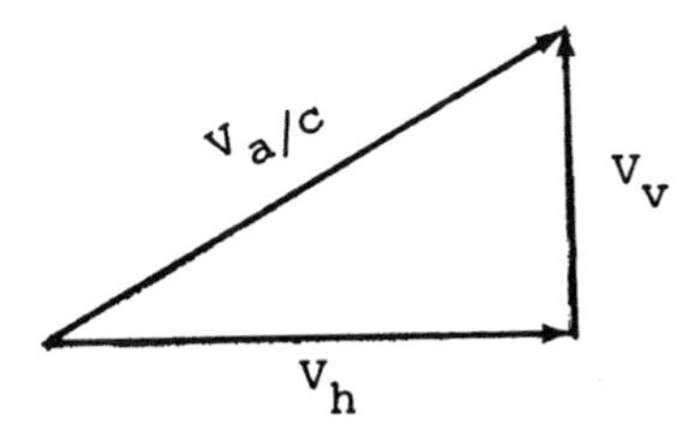

Fig. 1.7. Vectors of groundspeed and rate of climb.

moment, then a counterclockwise moment must be considered to be a − moment. By definition, aircraft nose-up pitching moments are considered to be + moments. This is discussed in more detail later in our study of stability and control.

EQUILIBRIUM CONDITIONS

Webster defines equilibrium as "a state of balance or equality between opposing forces." A body must meet two requirements to be in a state of equilibrium:

1. There must be no unbalanced forces acting on the body. This is written as the mathematical formula $\Sigma F = 0$, where Σ (cap sigma) is the Greek symbol for "sum of." Figures 1.1 and 1.2 illustrate situations where this condition is satisfied (lift = weight, thrust = drag, etc.)
2. There must be no unbalanced moments acting on the body. Mathematically, $\Sigma M = 0$ (Fig. 1.8).

Moments at the fulcrum in Fig. 1.8 are 50 ft-lb clockwise and 50 ft-lb counterclockwise. So, $\Sigma M = 0$. To satisfy the first condition of equilibrium, the fulcrum must press against the seesaw with a force of 15 lb. So, $\Sigma F = 0$.

NEWTON'S LAWS OF MOTION

Sir Isaac Newton summarized three generalizations about force and motion. These are known as the *laws of motion.*

Newton's First Law

In simple language, the first law states that *a body at rest will remain at rest and a body in motion will remain in motion, in a straight line, unless acted upon*

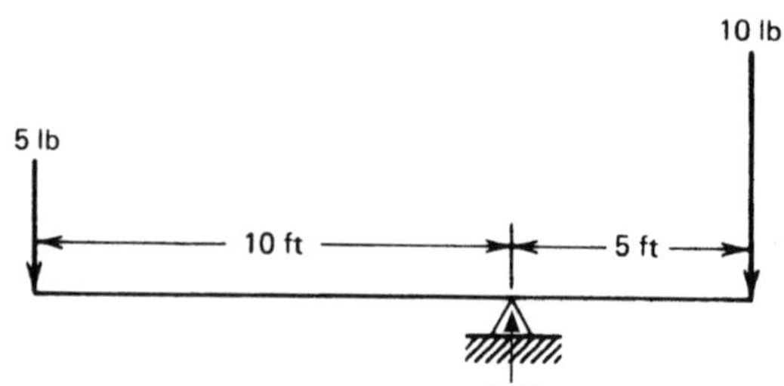

Fig. 1.8. Seesaw in equilibrium.

by an unbalanced force. The first law implies that bodies have a property called *inertia.* Inertia may be defined as the property of a body that results in its maintaining its velocity unchanged unless it interacts with an unbalanced force. The measure of inertia is what is technically known as *mass.* According to the first law, a spaceship flying in a frictionless atmosphere, and not under the gravitational influence of any heavenly body, could shut off its engines and its airspeed would remain constant.

Newton's Second Law

The second law states that *if a body is acted on by an unbalanced force, the body will accelerate in the direction of the force and the acceleration will be directly proportional to the force and inversely proportional to the mass of the body.* *Acceleration* is the change in motion (speed) of a body in a unit of time. The amount of the acceleration a, is directly proportional to the unbalanced force, F, and is inversely proportional to the mass, m, of the body. These two effects can be expressed by the simple equation

$$a = \frac{F}{m}$$

or, more commonly,

$$F = ma \tag{1.3}$$

Newton's Third Law

The third law states that *for every action force there is an equal and opposite reaction force.* Note that for this law to have any meaning, there must be an interaction between the force and a body. For example, the gases produced by burning fuel in a rocket engine are accelerated through the rocket nozzle. The equal and opposite force acts on the interior walls of the combustion chamber, and the rocket is accelerated in the opposite direction. The recoil of a gun is a clear demonstration of Newton's third law of motion.

LINEAR MOTION

Newton's laws of motion express relationships among force, mass, and acceleration, but they stop short of discussing velocity, time, and distance. These are covered here. In the interest of simplicity, we assume here that acceleration is constant. Then,

$$\text{Acceleration } a = \frac{\text{Change in velocity}}{\text{Change in time}} = \frac{\Delta V}{\Delta t} = \frac{V - V_0}{t - t_0}$$

where

Δ (cap delta) means "change in"
V = velocity at time t
V_0 = velocity at time t_0

If we start the time at $t_0 = 0$ and rearrange the above, then

$$V = V_0 + at \tag{1.4}$$

The distance s traveled in a certain time is

$$s = V_{av}t$$

The average velocity V_{av} is

$$V_{av} = \frac{V + V_0}{2}$$

Therefore,

$$s = \frac{V_0 + at + V_0}{2}t \quad \text{or} \quad s = V_0 t + \frac{1}{2}at^2 \tag{1.5}$$

Solving Eqs. 1.4 and 1.5 simultaneously and eliminating t, we can derive a third equation:

$$s = \frac{V^2 - V_0^2}{2a} \tag{1.6}$$

Equations 1.3, 1.4, and 1.5 are useful in calculating takeoff and landing factors. They are studied in some detail in Chapters 9 and 10.

ROTATIONAL MOTION

Without derivation, some of the relationships among tangential (tip) velocity, V_t; radius of rotation, r; revolutions per minute, rpm; centripetal forces, CF; weight of rotating parts, W; and acceleration of gravity, g, are shown below. The centripetal force is that force that causes an airplane to turn. The apparent force that is equal and opposite to this is called the centrifugal force.

$$V_t = \frac{r(\text{rpm})}{9.55} \quad (\text{fps}) \tag{1.7}$$

$$\text{CF} = \frac{WV_t^2}{gr} \quad (\text{lb}) \tag{1.8}$$

$$\text{CF} = \frac{W\,r(\text{rpm})^2}{2930} \tag{1.9}$$

WORK

In physics, work has a meaning different from the popular definition. You can push against a solid wall until you are exhausted but, unless the wall moves, you are not doing any work. Work requires that a force must move an object in the direction of the force. Another way of saying this is that *only the component of the force in the direction of movement does any work*:

$$\text{Work} = \text{Force} \times \text{Distance}$$

Work is measured in ft-lb.

ENERGY

Energy is the ability to do work. There are many kinds of energy: solar, chemical, heat, nuclear, and others. The type of energy that is of interest to us in aviation is *mechanical energy*.

There are two kinds of mechanical energy. The first is called *potential energy of position*, or more simply *potential energy*, PE. No movement is involved in calculating PE. A good example of this kind of energy is water stored behind a dam. If released, the water would be able to do work, such as running a generator. PE equals the weight, W, of an object multiplied by the height, h, of the object above some base plane:

$$\text{PE} = Wh \quad \text{(ft-lb)} \tag{1.10}$$

The second kind of mechanical energy is called *kinetic energy*, KE. As the name implies, kinetic energy requires movement of an object. It is a function of the mass, m, of the object and its velocity, V:

$$\text{KE} = \tfrac{1}{2}mV^2 \quad \text{(ft-lb)} \tag{1.11}$$

The total mechanical energy, TE, of an object is the sum of its PE and KE:

$$\text{TE} = \text{PE} + \text{KE} \tag{1.12}$$

The law of conservation of energy states that the total energy remains constant. Both potential and kinetic energy can change in value, but the total energy must remain the same: *Energy cannot be created or destroyed, but can change in form.*

A simple example of transferring energy is a roller coaster. The car and passengers are first moved up the ramp to the highest point on the ride. As the car is moved up the ramp it is gaining potential energy. As the car starts downward and accelerates, the potential energy is transformed into kinetic energy. When the car reaches ground level, all the potential energy has been transformed into kinetic energy, and the speed of the car is at a maximum. As the car starts up the next ramp, the kinetic energy is being transformed into potential energy, and the car slows down.

POWER

In our discussion of work and energy we have not mentioned time. *Power* is defined as "the rate of doing work" or work/time. We know:

$$\text{Work} = \text{force} \times \text{distance}$$

and

$$\text{Speed} = \text{distance/time}$$

$$\text{Power} = \frac{\text{work}}{\text{time}} = \frac{\text{force} \times \text{distance}}{\text{time}} = \text{force} \times \text{speed} \quad \text{(ft-lb/sec)}$$

James Watt defined the term *horsepower* (HP) as 550 ft-lb/sec:

$$\text{Horsepower} = \frac{\text{Force} \times \text{Speed}}{550}$$

If the speed is measured in knots, V_k, and the force is the *thrust*, T, of a jet engine, then

$$\text{HP} = \frac{\text{Thrust} \times V_k}{325} = \frac{TV_k}{325} \tag{1.13}$$

Equation 1.13 is very useful in comparing thrust-producing aircraft (turbojets) with power-producing aircraft (propeller aircraft and helicopters).

FRICTION

If two surfaces are in contact with each other, then a force develops between them when an attempt is made to move them relative to each other. This force is called *friction*. Generally, we think of friction as something to be avoided because it wastes energy and causes parts to wear. Friction is not always our enemy, however, for without it there would be no traction between an aircraft's tires and the runway.

Several factors are involved in determining friction effects on aircraft during takeoff and landing operations. Among these are runway surfacing material, condition of the runway, tire material and tread, and the amount of brake slippage. All of these variables determine a *coefficient of friction* μ (mu). The actual braking force, F_b, is the product of this coefficient μ (Greek symbol mu) and the normal force, N, between the tires and the runway:

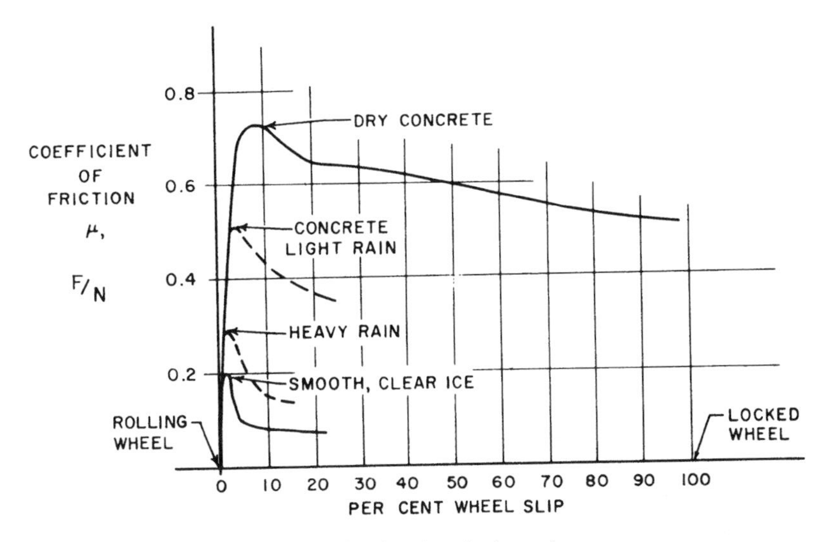

Fig. 1.9. Coefficients of friction for airplane tires on a runway.

$$F_b = \mu N \quad \text{(lb)} \tag{1.14}$$

Figure 1.9 shows typical values of the coefficient of friction for various conditions.

SYMBOLS

a	Acceleration (ft/sec^2)
CF	Centrifugal force (lb)
E	Energy (ft-lb)
KE	Kinetic energy
PE	Potential energy
TE	Total energy
F	Force (lb)
F_b	Braking force
g	Acceleration of gravity (ft/sec^2)
h	Height (ft)
HP	Horsepower
L	Moment arm (ft or in.)
m	Mass (slugs, lb-sec^2/ft)
M	Moment (ft-lb or in.-lb)
N	Normal force (lb)

r	Radius (ft)
rpm	Revolutions per minute
S	Distance (ft)
T	Thrust (lb)
t	Time (sec)
V	Speed (ft/sec)
V_k	Speed (knots)
V_0	Initial speed
V_t	Tangential (tip) speed
W	Weight (lb)
μ (mu)	Coefficient of friction (dimensionless)

EQUATIONS

1.1 $W = mg$

1.2 $M = FL$

1.3 $F = ma$

1.4 $V = V_0 + at$

1.5 $s = V_0 t + \frac{1}{2}at^2$

1.6 $s = \dfrac{V^2 - V_0^2}{2a}$

1.7 $V_t = \dfrac{r(\text{rpm})}{9.55}$

1.8 $\text{CF} = \dfrac{WV_t^2}{gr}$

1.9 $\text{CF} = \dfrac{Wr(\text{rpm})^2}{2930}$

1.10 $\text{PE} = Wh$

1.11 $\text{KE} = \frac{1}{2}mV^2$

1.12 $\text{TE} = \text{PE} + \text{KE}$

1.13 $\text{HP} = \dfrac{TV_k}{325}$

1.14 $F_b = \mu N$

PROBLEMS

Note: Answers to problems are given at the end of the book.

1. An airplane weighs 16,000 lb. The local gravitational acceleration g is 32 fps^2. What is the mass of the airplane?

2. The airplane in Problem 1 accelerates down the takeoff runway with a net force of 6000 lb. Find the acceleration of the airplane.

3. An airplane is towing a glider to altitude. The tow rope is 20° below the horizontal and has a tension force of 300 lb exerted on it by the airplane. Find the horizontal drag of the glider and the amount of lift that the rope is providing to the glider. Sin 20° = 0.342; cos 20° = 0.940.

4. A jet airplane is climbing at a constant airspeed in no-wind conditions. The plane is directly over a point on the ground that is 4 statute miles from the takeoff point and the altimeter reads 15,840 ft. Find the tangent of the plane's climb angle and the distance that it has flown through the air.

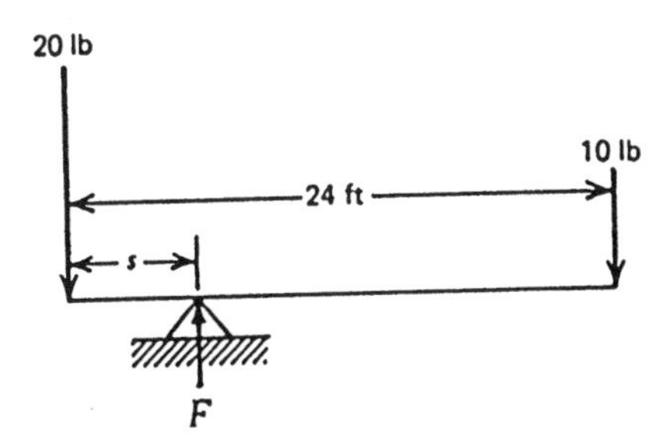

Problem 5

5. Find the distance s and the force F on the seesaw fulcrum shown in the figure. Assume that the system is in equilibrium.

6. The airplane in Problem 2 starts from a brakes-locked position on the runway. The airplane takes off at an airspeed of 200 fps. Find the time for the aircraft to reach takeoff speed.

7. Under no-wind conditions, what takeoff roll is required for the aircraft in Problem 6?

8. Upon reaching a velocity of 100 fps, the pilot of the airplane in Problem 6 decides to abort the takeoff and applies brakes and stops the airplane in 1000 ft. Find the airplane's deceleration.

9. A helicopter has a rotor diameter of 30 ft and it is being operated in a hover at 286.5 rpm. Find the tip speed V_t of the rotor.

10. An airplane weighs 16,000 lb and is flying at 5000 ft altitude and at an airspeed of 200 fps. Find (a) the potential energy, (b) the kinetic energy, and (c) the total energy. Assuming no extra drag on the airplane, if the pilot dove until the airspeed was 400 fps, what would the altitude be?

11. An aircraft's turbojet engine produces 10,000 lb of thrust at 162.5 knots true airspeed. What is the equivalent power that it is producing?

12. An aircraft weighs 24,000 lb and has 75% of its weight on the main (braking) wheels. If the coefficient of friction is 0.7, find the braking force F_b on the airplane.

Boeing 777-300 (Courtesy the Boeing Company).

2 Atmosphere and Airspeed Measurement

PROPERTIES OF THE ATMOSPHERE

The aerodynamic forces and moments acting on an aircraft in flight are due, in great part, to the properties of the air mass in which the aircraft is flying. By volume the atmosphere is composed of approximately 78% nitrogen, 21% oxygen, and 1% other gases. The most important properties of air that affect aerodynamic behavior are its static pressure, temperature, density, and viscosity.

Static Pressure

The *static pressure* of the air, P, is simply the weight per unit area of the air above the level under consideration. For instance, the weight of a column of air with a cross-sectional area of 1 ft^2 and extending upward from sea level through the atmosphere is 2116 lb. The sea level static pressure is, therefore, 2116 psf (or 14.7 psi). Static pressure is reduced as altitude is increased because there is less air weight above. At 18,000 ft altitude the static pressure is about half that at sea level. Another commonly used measure of static pressure is *inches of mercury*. On a standard sea level day the air's static pressure will support a column of mercury (Hg) that is 29.92 in. high. Weather reports use a third method of measuring static pressure called *millibars*. In addition to these rather confusing systems, there are the metric measurements in use throughout most of the world.

In aerodynamics it is convenient to use pressure ratios, rather than actual pressures. Thus the units of measurement are canceled out:

$$\text{Pressure ratio } \delta \text{ (delta)} = \frac{P}{P_0} \tag{2.1}$$

where P_0 is the sea level standard static pressure (2116 psf or 29.92 in. Hg). Thus, a pressure ratio of 0.5 means that the ambient pressure is one-half of the standard sea level value. At 18,000 ft, on a standard day, the pressure ratio is 0.4992.

Temperature

The commonly used measures of temperature are the Fahrenheit, F, and Celsius, C (formerly called centigrade) scales. Neither of these scales has absolute zero as a base, so neither can be used in calculations. Absolute temperature must be used instead. Absolute zero in the Fahrenheit system is -460° and in the Celsius system is -273°. To convert from the Fahrenheit system to the absolute system, called Rankine, R, add 460 to the °F. To convert from the Celsius system to the absolute system, called Kelvin, K, add 273 to the °C. The symbol for absolute temperature is T and the symbol for sea level standard temperature is T_0:

$$\begin{aligned} T_0 &= 519^\circ\text{R } (59^\circ\text{F}) \\ &= 288^\circ\text{K } (15^\circ\text{C}) \end{aligned}$$

By using temperature ratios, instead of actual temperatures, the units cancel. The temperature ratio is the Greek letter theta, θ:

$$\theta = \frac{T}{T_0} \tag{2.2}$$

At sea level, on a standard day, $\theta_{0^\circ} = 1.0$. Temperature decreases with an increase in altitude until the tropopause is reached (36,089 ft on a standard day). It then remains constant until an altitude of about 82,023 ft. The temperature at the tropopause is -69.7°F and $\theta = 0.7519$.

Density

Density is the most important property of air in the study of aerodynamics. Density is the mass of the air per unit of volume. The symbol for density is ρ (rho):

$$\rho = \frac{\text{Mass}}{\text{Unit volume}} \quad (\text{slugs/ft}^3)$$

Standard sea level density is $\rho_0 = 0.002377$ slugs/ft^3. Density decreases with an increase in altitude. At 22,000 ft, the density is 0.001183 slugs/ft^3 (about one-half of sea level density).

It is desirable in aerodynamics to use density ratios instead of the actual values of density. The symbol for density ratio is σ (sigma):

$$\sigma = \frac{\rho}{\rho_0} \tag{2.3}$$

The universal gas law shows that density is directly proportional to pressure and inversely proportional to absolute temperature:

$$\rho = \frac{P}{RT} \tag{2.4}$$

Forming a ratio gives

$$\frac{\rho}{\rho_0} = \frac{P/RT}{P_0/RT_0} = \frac{P/P_0}{T/T_0}$$

R is the gas constant and cancels, so

$$\sigma = \frac{\delta}{\theta} \tag{2.5}$$

Viscosity

Viscosity can be simply defined as the internal friction of a fluid caused by molecular attraction that makes it resist its tendency to flow. The viscosity of the air is important when discussing airflow in the region very close to the surface of an aircraft. This region is called the *boundary layer*. We discuss viscosity in more detail when we take up the subject of boundary layer theory.

ICAO STANDARD ATMOSPHERE

To provide a basis for comparing aircraft performance at different parts of the world and under varying atmospheric conditions, the performance data must be reduced to a set of standard conditions. These are defined by the International Civil Aviation Organization (ICAO) and are compiled in a standard atmosphere table. An abbreviated table is shown here as Table 2.1. Columns in the table show standard day density, density ratio, pressure, pressure ratio, temperature, temperature ratio, and speed of sound at various altitudes.

Two types of altitude interest the pilot: pressure altitude and density altitude. Pressure altitude is that altitude in the standard atmosphere corresponding to a certain static pressure. For instance, if the pressure at a certain altitude is 1455 psf, then the pressure ratio is

$$\delta = \frac{1455}{2116} = 0.6876$$

Entering Table 2.1 with this value, we find the corresponding pressure altitude of 10,000 ft.

Density altitude is found by correcting pressure altitude for nonstandard temperature conditions. If the air has a density ratio of 0.6292, the density ratio

Table 2.1. Standard Atmosphere Table

Altitude (ft)	Density Ratio, σ	$\sqrt{\sigma}$	Pressure Ratio, δ	Temperature (°F)	Temperature Ratio, θ	Speed of Sound (knots)	Kinematic Viscosity, ν (ft²/sec)
0	1.0000	1.0000	1.0000	59.00	1.0000	661.7	.000158
1,000	0.9711	0.9854	0.9644	55.43	0.9931	659.5	.000161
2,000	0.9428	0.9710	0.9298	51.87	0.9862	657.2	.000165
3,000	0.9151	0.9566	0.8962	48.30	0.9794	654.9	.000169
4,000	0.8881	0.9424	0.8637	44.74	0.9725	652.6	.000174
5,000	0.8617	0.9283	0.8320	41.17	0.9656	650.3	.000178
6,000	0.8359	0.9143	0.8014	37.60	0.9587	647.9	.000182
7,000	0.8106	0.9004	0.7716	34.04	0.9519	645.6	.000187
8,000	0.7860	0.8866	0.7428	30.47	0.9450	643.3	.000192
9,000	0.7620	0.8729	0.7148	26.90	0.9381	640.9	.000197
10,000	0.7385	0.8593	0.6877	23.34	0.9312	638.6	.000202
15,000	0.6292	0.7932	0.5643	5.51	0.8969	626.7	.000229
20,000	0.5328	0.7299	0.4595	−12.32	0.8625	614.6	.000262
25,000	0.4481	0.6694	0.3711	−30.15	0.8281	602.2	.000302
30,000	0.3741	0.6117	0.2970	−47.98	0.7937	589.5	.000349
35,000	0.3099	0.5567	0.2353	−65.82	0.7594	576.6	.000405
36,089[a]	0.2971	0.5450	0.2234	−69.70	0.7519	573.8	.000419
40,000	0.2462	0.4962	0.1851	−69.70	0.7519	573.8	.000506
45,000	0.1936	0.4400	0.1455	−69.70	0.7519	573.8	.000643
50,000	0.1522	0.3002	0.1145	−69.70	0.7519	573.8	.000818

[a]The tropopause.

column in Table 2.1 shows that this value corresponds to a density altitude of 15,000 ft. Density altitude influences aircraft performance. Therefore, aircraft performance charts are provided for various density altitudes.

CONTINUITY EQUATION

Consider the flow of air through a pipe of varying cross section as shown in Fig. 2.1. There is no flow through the sides of the pipe. Air flows only through the ends. The mass of air entering the pipe, in a given unit of time, equals the mass of air leaving the pipe, in the same unit of time. The mass flow through the pipe must remain constant. The mass flow at each station is equal. Constant mass flow is called *steady-state flow*. The mass airflow is equal to the volume of air multiplied by the density of the air. The volume of air, at any station, is equal to the velocity of the air multiplied by the cross-sectional area of that station.

The mass airflow symbol Q is the product of the density, the area, and the velocity:

$$Q = \rho A V \tag{2.6}$$

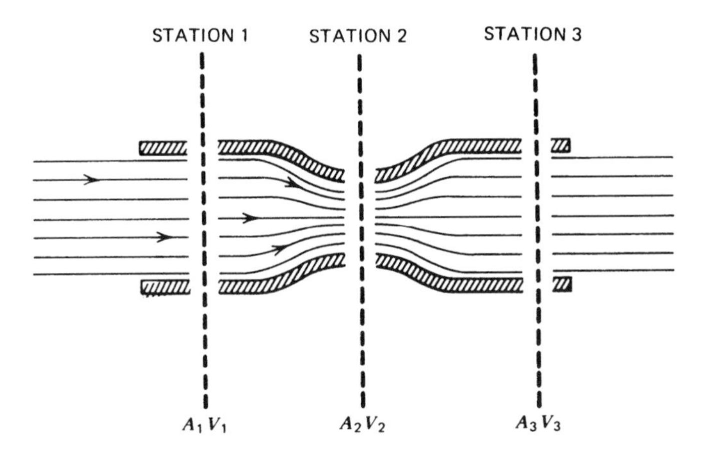

Fig. 2.1. Flow of air through a pipe.

The continuity equation states that the mass airflow is a constant:

$$\rho_1 A_1 V_1 = \rho_2 A_2 V_2 = \rho_3 A_3 V_3 = \text{constant} \tag{2.7}$$

The continuity equation is valid for steady-state flow, both in subsonic and supersonic flow. For subsonic flow the air is considered to be incompressible, and its density remains constant. The density symbols can then be eliminated; thus, for subsonic flow,

$$A_1 V_1 = A_2 V_2 = A_3 V_3 = \text{constant} \tag{2.8}$$

Velocity is inversely proportional to cross-sectional area or as cross-sectional area decreases, velocity increases.

BERNOULLI'S EQUATION

The continuity equation explains the relationship between velocity and cross-sectional area. It does not explain differences in static pressure of the air passing through a pipe of varying cross sections. Bernoulli, using the principle of conservation of energy, developed a concept that explains the behavior of pressures in gases.

Consider the flow of air through a Venturi tube as shown in Fig. 2.2. The energy of an airstream is in two forms: It has a *potential energy*, which is its static pressure, and a *kinetic energy*, which is its dynamic pressure. The total pressure of the airstream is the sum of the static pressure and the dynamic pressure. The total pressure remains constant, according to the law of conser-

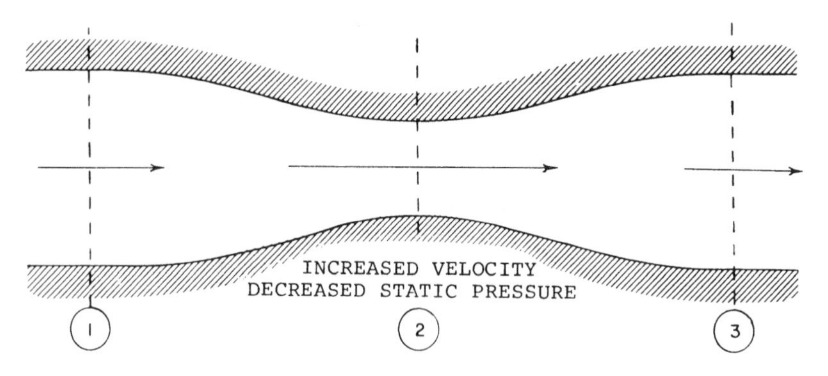

Fig. 2.2. Venturi tube.

vation of energy. Thus, an increase in one form of pressure must result in a decrease in the other.

Static pressure is an easily understood concept (see the discussion earlier in this chapter). Dynamic pressure, q, is similar to kinetic energy in mechanics and is expressed by

$$q = \tfrac{1}{2}\rho V^2 \quad \text{(psf)} \tag{2.9}$$

where V is measured in feet per second. Pilots are much more familiar with velocity measured in knots instead of in feet per second, so a new equation for dynamic pressure, q, is used in this book. Its derivation is shown here:

$$\text{Density ratio, } \sigma = \frac{\rho}{\rho_0} = \frac{\rho}{0.002377}$$

$$\text{or } \rho = 0.002377\sigma$$

$$V_{\text{fps}} = 1.69 V_{\text{k}}$$

$$(V_{\text{fps}})^2 = 2.856(V_{\text{k}})^2$$

Substituting in Eq. 2.9 yields

$$q = \frac{\sigma V_{\text{k}}^2}{295} \quad \text{(psf)} \tag{2.10}$$

Bernoulli's equation can now be expressed as Total pressure, H = Static pressure, P + Dynamic pressure, q:

$$H = P + \frac{\sigma V_{\text{k}}^2}{295} \quad \text{(psf)} \tag{2.11}$$

To visualize how lift is developed on a cambered airfoil, draw a line down

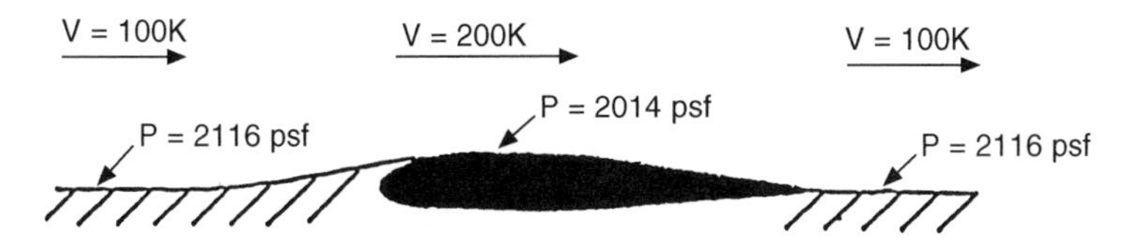

Fig. 2.3. Velocities and pressures on an airfoil superimposed on a venturi tube.

the middle of a venturi tube. Discard the upper half of the figure and superimpose an airfoil on the necked down section of the tube (Fig. 2.3). Note that the static pressure over the airfoil is less than that ahead of it.

AIRSPEED MEASUREMENT

If a symmetrically shaped object is placed in a moving airstream (Fig. 2.4), the flow pattern will be as shown. Some airflow will pass over the object and some will flow beneath it, but at the point at the nose of the object, the flow will be stopped completely. This point is called the *stagnation point*. Since the air velocity at this point is zero, the dynamic pressure is also zero. The stagnation pressure is, therefore, all static pressure and must be equal to the total pressure, H, of the airstream.

In Fig. 2.5 the free stream values of velocity and pressure are shown at station 1 and are for sea level standard day conditions. The pitot tube is shown as the *total pressure port* at station 2. The pitot tube must be pointed into the relative wind for accurate readings. The air entering the pitot tube comes to a complete stop and thus the static pressure in the tube, P_2, is equal to the total free stream pressure, H. This pressure (2150 psf) is ducted into a diaphragm.

The static pressure port can be made as a part of the point tube or, in more expensive indicators, it can be at a distance from the pitot tube. It should be located at a point where the local air velocity is exactly equal to the airplane velocity. The static port is made so that none of the velocity enters the port.

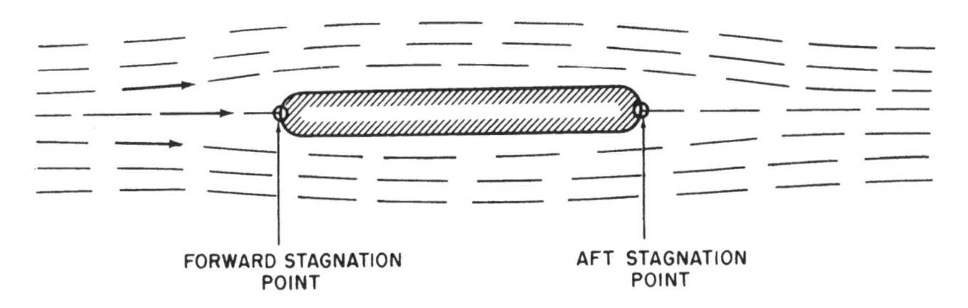

Fig. 2.4. Flow around a symmetrical object.

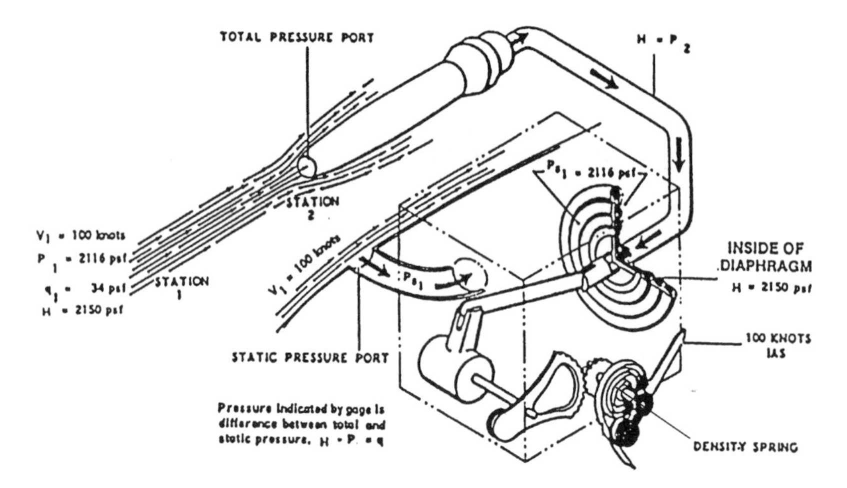

Fig. 2.5. Schematic of a pitot–static airspeed indicator.

The port measures only static pressure, P_1, and none of the dynamic pressure, q_1. The static pressure is ducted into the chamber surrounding the diaphragm.

Now we have P_2 inside the diaphragm and P_1 outside the diaphragm. Remember that $P_2 = H = P_1 + q_1 = 2150\,\text{psf}$ and that $P_1 = 2116\,\text{psf}$. The difference between these pressures is $H - P_1 = q_1$. This differential pressure deflects the flexible diaphragm that is geated to the airspeed pointer. The airspeed indicator is calibrated to read airspeed.

Indicated Airspeed

Indicated airspeed (IAS) is the reading of the airspeed indicator dial. If there are any errors in the instrument, they will be shown on an instrument error card located near the instrument. Position error results if the static pressure port is not located on the aircraft where the local air velocity is exactly equal to the free stream velocity of the airplane. If this error is present, it will be included in the instrument error chart.

Calibrated Airspeed

Calibrated airspeed (CAS) is obtained when the necessary corrections have been made to the IAS. In fast, high-altitude aircraft, the air entering the pitot tube is subjected to a ram effect, which causes the diaphragm to be deflected too far. The resulting airspeed indication is too high and must be corrected.

Figure 2.6 shows a compressibility correction chart. A rule of thumb is that, if flying above 10,000 ft and 200 knots, the compressibility correction should be made. Unlike the instrument and position error charts, which vary with different aircraft, this chart is good for any aircraft.

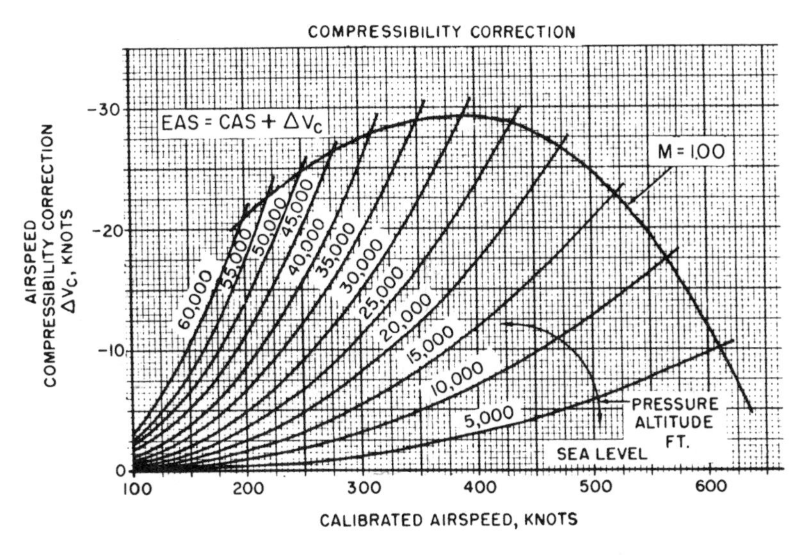

Fig. 2.6. Compressibility correction chart.

Equivalent Airspeed

Equivalent airspeed (EAS) results when the CAS has been corrected for compressibility effects. One further correction must be made to obtain true airspeed. The airspeed indicator measures dynamic pressure and is calibrated for sea level standard day density. As altitude increases, the density ratio decreases and a correction must be made. The correction factor is $\sqrt{\sigma}$.

True Airspeed

True airspeed (TAS) is obtained when EAS has been corrected for density ratio:

$$\text{TAS} = \frac{\text{EAS}}{\sqrt{\sigma}} \tag{2.12}$$

Values of $\sqrt{\sigma}$ can be found in the ICAO Standard Altitude Chart (Table 2.1); values of $1/\sqrt{\sigma}$ (called "SMOE") can be found in Fig. 2.7.

SYMBOLS

A	Area (ft^2)
CAS	Calibrated airspeed (knots)
°C	Celsius temperature (deg.)
EAS	Equivalent airspeed
°F	Fahrenheit temperature

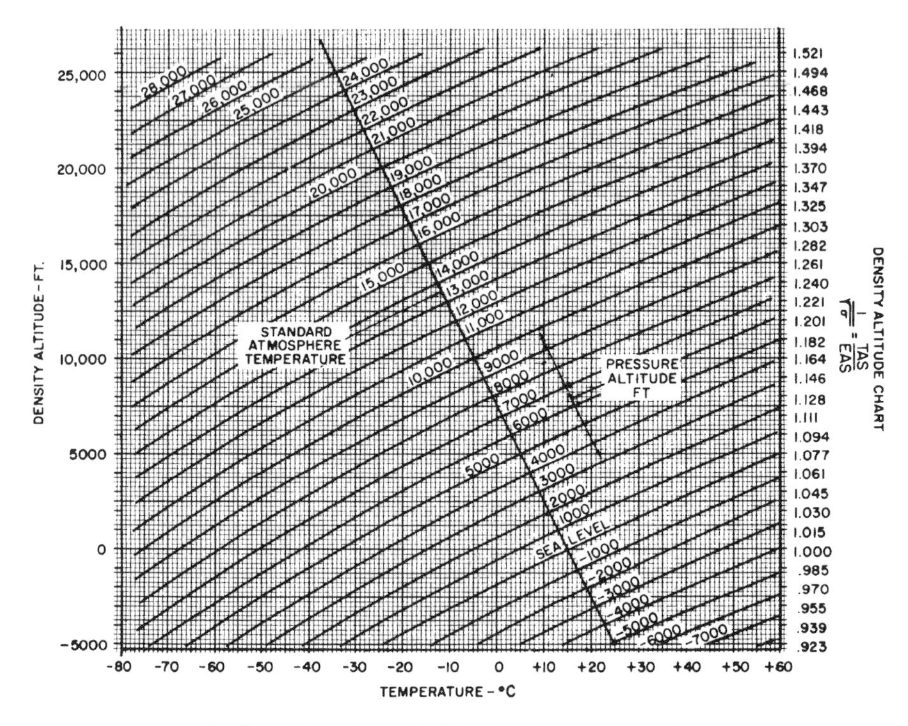

Fig. 2.7. Altitude and EAS to TAS correction chart.

H	Total pressure(head)(psf)
IAS	Indicated airspeed
°K	Kelvin temperature
P	Static pressure
P_0	Sea level standard pressure
q	Dynamic pressure
R	Universal gas constant
°R	Rankine temperature
TAS	True airspeed
T	Absolute temperature
T_0	Sea level standard temperature
V	Velocity (fps)
V_k	Velocity (knots)
δ (delta)	Pressure ratio
θ (theta)	Temperature ratio
ρ (rho)	Density
σ (sigma)	Density ratio

EQUATIONS

(2.1) $\delta = \dfrac{P}{P_0}$

(2.2) $\theta = \dfrac{T}{T_0}$

(2.3) $\sigma = \dfrac{\rho}{\rho_0}$

(2.4) $\rho = \dfrac{P}{RT}$

(2.5) $\sigma = \dfrac{\delta}{\theta}$

(2.6) $Q = \rho A V$

(2.7) $\rho_1 A_1 V_1 = \rho_2 A_2 V_2$

(2.8) $A_1 V_1 = A_2 V_2$

(2.9) $q = \frac{1}{2}\rho V^2$ (psf)

(2.10) $q = \dfrac{\sigma V_k^2}{295}$ (psf)

(2.11) $H = P + \dfrac{\sigma V_k^2}{295}$

(2.12) $\text{TAS} = \dfrac{\text{EAS}}{\sqrt{\sigma}}$

PROBLEMS

1. An increase in static air pressure
- **a.** affects air density by decreasing the density.
- **b.** does not affect air density.
- **c.** affects air density by increasing the density.

2. A decrease in temperature
- **a.** affects air density by decreasing the density.
- **b.** does not affect air density.
- **c.** affects air density by increasing the density.

3. Pressure ratio is
- **a.** ambient pressure divided by sea level standard pressure measured in the same units.

b. ambient pressure in millibars divided by 29.92.
c. ambient pressure in pounds per square inch divided by 2116.
d. sea level standard pressure in inches of mercury divided by 29.92.

4. Density ratio, σ (sigma) is
a. equal to pressure ratio divided by temperature ratio.
b. measured in slugs per cubic foot.
c. equal to the ambient density divided by sea level standard pressure.
d. None of the above

5. Bernoulli's equation for subsonic flow states that
a. if the velocity of an airstream within a tube is increased, the static pressure of the air increases.
b. if the area of a tube decreases, the static pressure of the air increases.
c. if the velocity of an airstream within a tube increases, the static pressure of the air decreases, but the sum of the static pressure and the velocity remains constant.
d. None of the above

6. Dynamic pressure of an airstream is
a. directly proportional to the square of the velocity.
b. directly proportional to the air density.
c. Neither (a) nor (b)
d. Both (a) and (b)

7. In this book, we use the formula for dynamic pressure, $q = \sigma V_k^2/295$, rather than the more conventional formula, $q = \frac{1}{2}\rho V^2$, because
a. V in our formula is measured in knots.
b. density ratio is easier to handle (mathematically) than the actual density (slugs per cubic foot).
c. Both (a) and (b)
d. Neither (a) nor (b)

8. The corrections that must be made to indicated airspeed (IAS) to obtain calibrated airspeed (CAS) are
a. position error and compressibility error.
b. instrument error and position error.
c. instrument error and density error.
d. position error and density error.

9. The correction that must be made to CAS to obtain equivalent airspeed (EAS) is called compressibility error, which
a. is always a negative value.
b. can be ignored at high altitude.

  c. can be ignored at high airspeed.
  d. can be either a positive or a negative value.

10. The correction from EAS to true airspeed (TAS) is dependent on
  a. temperature ratio alone.
  b. density ratio alone.
  c. pressure ratio alone.
  d. None of the above

11. An airplane is operating from an airfield that has a barometric pressure of 27.82 in. Hg and a runway temperature of 100°F. Calculate (or find in Table 2.1) the following:
  a. Pressure ratio
  b. Pressure altitude
  c. Temperature ratio
  d. Density ratio
  e. Density altitude

12. Fill in the values below for the stations in the drawing.

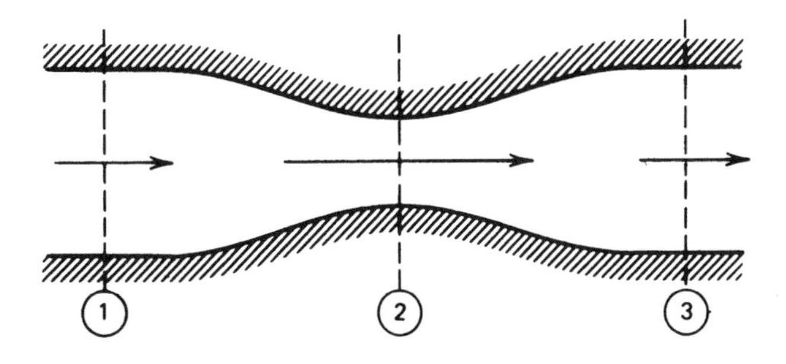

At station 1	At station 2	At station 3
$A_1 = 10\text{ ft}^2$	$A_2 = 5\text{ ft}^2$	$V_3 = 80$ knots
$V_1 = 100$ knots	$V_2 = ?$	$A_3 = ?$
$P_1 = 2030$ psf	$P_2 = ?$	$P_3 = ?$
$\sigma = 0.968$		
$H = ?$	$q_2 = ?$	$q_3 = ?$

13. Using Table 2.1, calculate the dynamic pressure, q, at 5000 ft density altitude and 200 knots TAS.

14. The airspeed indicator of an airplane reads 355 knots. There are no instrument or position errors. If the airplane is flying at a pressure altitude of 25,000 ft, find the equivalent airspeed (EAS).

15. Find the true airspeed (TAS) of the airplane in Problem 14 if the outside air temperature is −40°C.

3 Airfoils and Aerodynamic Forces

AIRFOILS

An *airfoil* or, more properly, an airfoil section, is a vertical slice of a wing (see Fig. 3.1). In discussing airfoils, the planform (or horizontal plane) of the wing is ignored. Wingtip effects, sweepback, taper, wash/out or wash/in, and other design features are not considered.

Airfoil Terminology

The terminology used to discuss an airfoil is shown in Fig. 3.2:

1. *Chord line* is a straight line connecting the leading edge and the trailing edge of the airfoil.
2. *Chord* is the length of the chord line. All airfoil dimensions are measured in terms of the chord.
3. *Mean camber line* is a line drawn halfway between the upper surface and the lower surface.
4. *Maximum camber* is the maximum distance between the mean camber line and the chord line. The location of maximum camber is important in determining the aerodynamic characteristics of the airfoil.
5. *Maximum thickness* is the maximum distance between the upper and lower surfaces. The location of maximum thickness is also important.
6. *Leading edge radius* is a measure of the sharpness of the leading edge. It may vary from zero for a knife-edge supersonic airfoil to about 2% (of the chord) for rather blunt leading-edge airfoils.

Geometry Variables of Airfoils

There are four main variables in the geometry of an airfoil:

1. Shape of the mean camber line
2. Thickness
3. Location of maximum thickness
4. Leading-edge radius

If the mean camber line coincides with the chord line, the airfoil is said to be symmetrical. In symmetrical airfoils, the upper and lower surfaces have the same shape and are equidistant from the chord line.

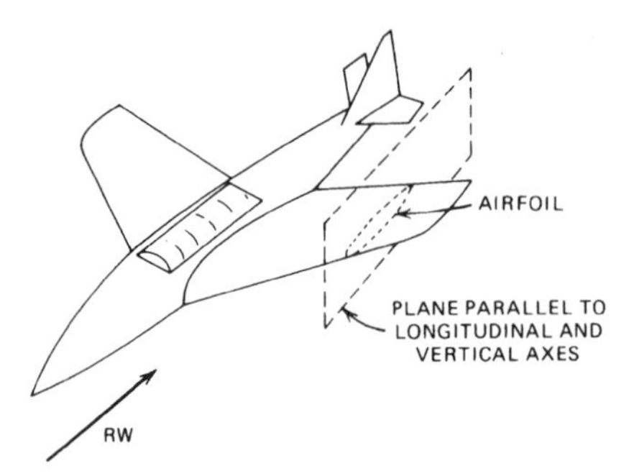

Fig. 3.1. Airfoil section.

Classification of Airfoils

A simple description of the most common types of airfoils is presented. A more complete treatment of the subject can be found in Reference 1.

Most airfoil development in the United States was done by the National Advisory Committee for Aeronautics (NACA) starting in 1929. NACA was the forerunner of the National Aeronautics and Space Administration (NASA). The first series of airfoils investigated was the "four-digit" series. The first digit gives the amount of camber, in percentage of chord. The second digit gives the position of maximum camber, in tenths of chord, and the last two give the maximum thickness, in percentage of chord. For example, a NACA 2415 airfoil has a maximum camber of 2% C, located at 40% C (measured from the leading edge), and has a maximum thickness of 15% C. A NACA 0012 airfoil is a symmetrical airfoil (has zero camber) and has a thickness of 12% C.

Further development led to the "five-digit" series, the "1-series," and, with the advent of higher speeds, to the so-called *laminar flow* airfoils. The laminar

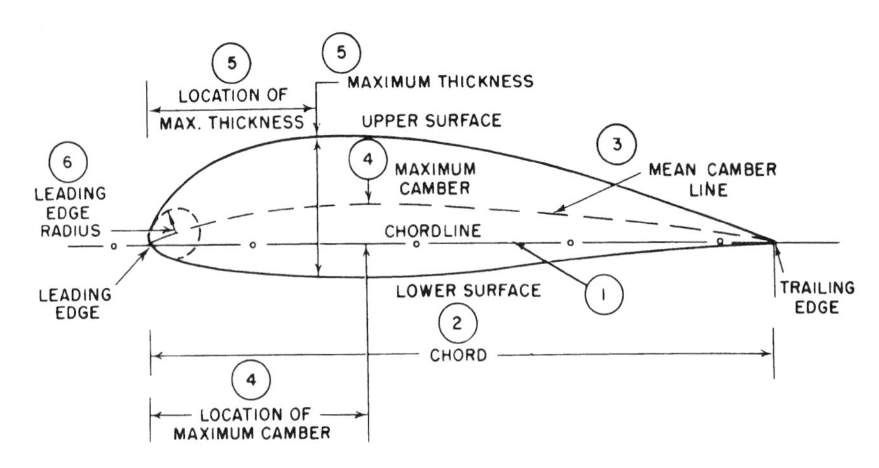

Fig. 3.2. Airfoil terminology.

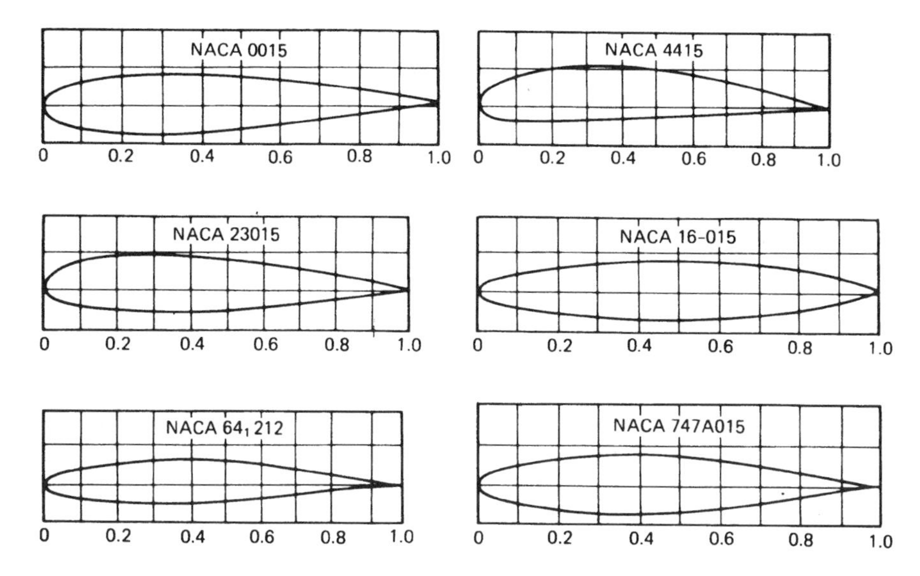

Fig. 3.3. NACA airfoils (NACA data).

flow airfoils are the "6-series" and "7-series" airfoils and result from moving the maximum thickness back and reducing the leading-edge radius. Two things happen with this treatment. First, the point of minimum pressure is moved backward, thus increasing the distance from the leading edge that laminar (smooth) airflow exists, which reduces drag. Second, the critical Mach number is increased, thus allowing the airspeed of the aircraft to be increased without encountering compressibility problems. In the 6-series, the first digit indicates the series and the second gives the location of minimum pressure in tenths of chord. The third digit represents the design lift coefficient in tenths, and the last two digits (as in all NACA airfoils) show the thickness in percentage of chord. For example, NACA 64-212 is a 6-series airfoil with minimum pressure at 40% C, a design lift coefficient of 0.2, and a thickness of 12% C. Sketches of NACA subsonic airfoil series are shown in Fig. 3.3.

A recent development is the *supercritical* airfoil and the general aviation counterpart, the NASA GA(W)-1 airfoil. These airfoils, as well as supersonic airfoils, are discussed in Chapter 15.

DEVELOPMENT OF FORCES ON AIRFOILS

Leonardo da Vinci stated the cardinal principle of wind tunnel testing nearly 400 years before the Wright brothers achieved powered flight. Near the beginning of the sixteenth century, da Vinci said: *the action of the medium upon a body is the same whether the body moves in a quiescent medium, or whether the particles of the medium impinge with the same velocity upon the quiescent body*. This principle allows us to consider only relative motion of the airfoil and the air surrounding it. We may use such terms as "airfoil passing through the air" and "air passing over the airfoil" interchangeably.

DEFINITIONS

Flight Path Velocity The speed and direction of a body passing through the air.

Relative Wind (RW) The speed and direction of the air impinging on a body passing through it. It is equal and opposite in direction to the flight path velocity.

Angle of Attack (AOA or α) The acute angle between the relative wind and the chord line of an airfoil.

Aerodynamic Force (AF) The net resulting static pressure multiplied by the planform area of an airfoil.

Lift The component L of the aerodynamic force that is perpendicular to the relative wind.

Drag The component D of the aerodynamic force that is parallel to the relative wind.

Center of Pressure (CP) The point on the chord line where the aerodynamic force acts.

Laminar Flow Smooth airflow with little transfer of momentum between parallel layers.

Streamlined Flow Same as laminar flow.

Turbulent Flow Flow where the streamlines break up and there is much mixing of the layers.

Pressure Disturbances on Airfoils

If an airfoil is subjected to a moving airflow, velocity and pressure changes take place that create pressure disturbances in the airflow surrounding it. These disturbances originate at the airfoil surface and propagate in all directions at the speed of sound. If the flight path velocity is subsonic, the pressure disturbances that are moving ahead of the airfoil affect the airflow approaching the airfoil (Fig. 3.4).

Velocity and Static Pressure Changes About an Airfoil

The air approaching the leading edge of an airfoil is first slowed down. It then speeds up again as it passes over or beneath the airfoil. Figure 3.5 compares two local velocities with the flight path velocity V_1 and with each other. As the velocity changes, so does the dynamic pressure and, according to Bernoulli's principle, so does the static pressure. Air near the stagnation point has slowed

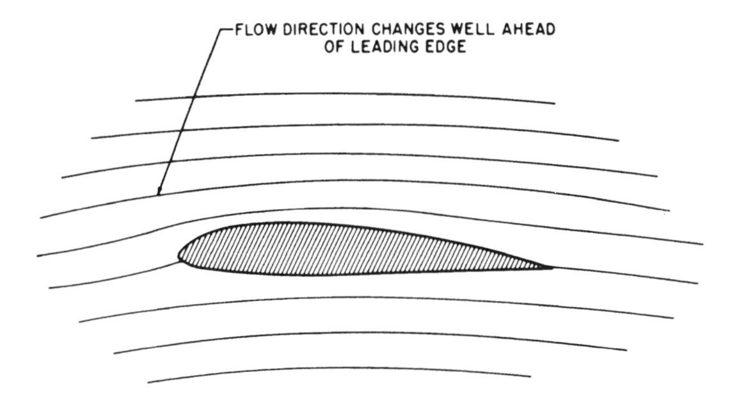

Fig. 3.4. Effect of pressure disturbances on airflow around an airfoil.

down, so the static pressure in this region is higher than the ambient static pressure. Air that is passing above and below the airfoil, and thus has speeded up to a value higher than the flight path velocity, will produce static pressures that are lower than ambient static pressure.

At a point near maximum thickness, maximum velocity and minimum static pressure will occur. Because air has viscosity, some of its energy will be lost to friction and a "wake" of low-velocity, turbulent air exists near the trailing edge, resulting in a small, high-pressure area. Figure 3.6 shows a symmetrical airfoil (a) at zero AOA and the resulting pressure distribution (b) at a positive AOA and its pressure distribution. Arrows pointing away from the airfoil indicate static pressures that are below ambient static pressure; arrows pointing toward the airfoil indicate pressures higher than ambient.

AERODYNAMIC FORCE

Aerodynamic force (AF) is the resultant of all static pressures acting on an airfoil in an airflow multiplied by the planform area that is affected by the pressure. The line of action of the AF passes through the chord line at a point

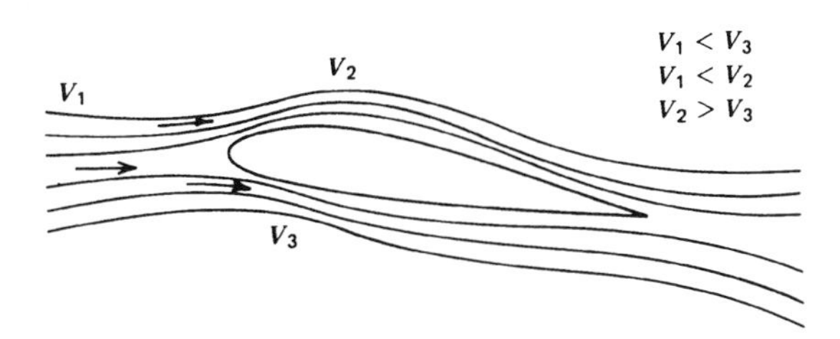

Fig. 3.5. Velocity changes around an airfoil.

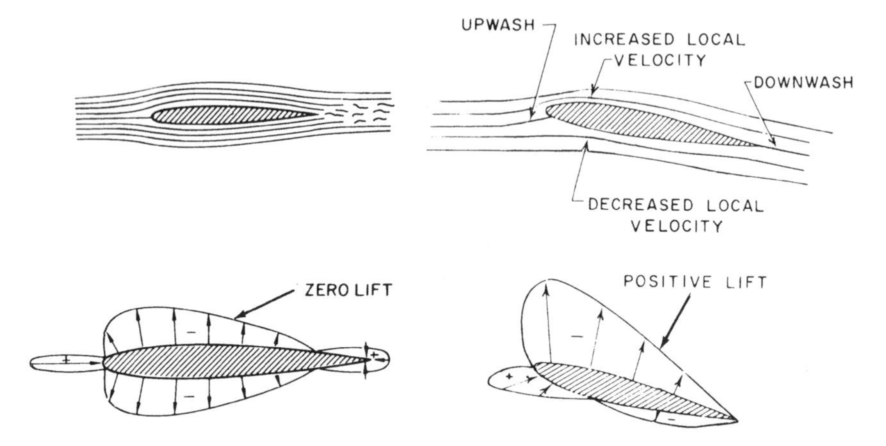

Fig. 3.6. Static pressure on an airfoil (a) at zero AOA and (b) at a positive AOA.

called the *center of pressure* (CP). It is convenient to consider that the forces acting on an aircraft, or on an airfoil, do so in some rectangular coordinate system. One such system could be defined by the longitudinal and vertical axes of an aircraft. Another could be defined by axes parallel to and perpendicular to the earth's surface. A third rectangular coordinate system is defined by the relative wind direction and an axis perpendicular to it. This last system is chosen to define lift and drag forces. Aerodynamic force (AF) is resolved into two components: one parallel to the relative wind, called *drag* (D), and the other perpendicular to the relative wind, called *lift* (L). Figure 3.7 shows the resolution of AF into its components L and D.

Pressure Distribution on a Cylinder

A stationary (nonrotating) cylinder is located in a wind tunnel as shown in Fig. 3.8a. The cylinder is equipped with static pressure taps. These measure the

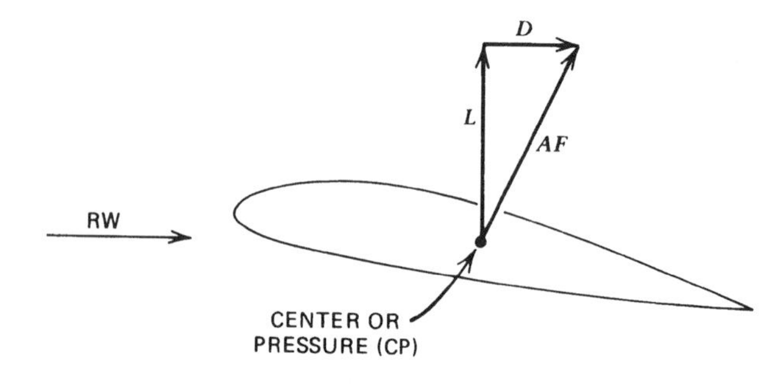

Fig. 3.7. Components of aerodynamic force.

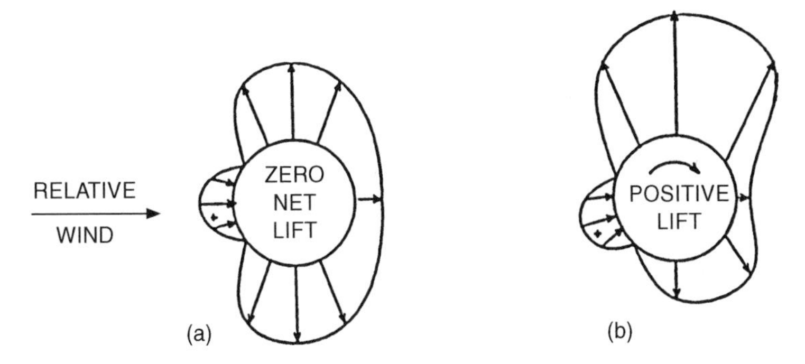

Fig. 3.8. Pressure forces on (a) nonrotating cylinder and (b) rotating cylinder.

local static pressure with respect to the ambient static pressure in the test chamber. When the tunnel is started, the airflow approaches the cylinder from the left as shown by the relative wind vector. Arrows pointing toward the cylinder show pressures that are higher (+) than ambient static pressure; arrows pointing away from the cylinder show pressures that are less (−) than ambient static pressure. Figure 3.8a shows that the upward forces are resisted by the downward forces and no net vertical force (lift) is developed by the cylinder.

Now, consider that the wind tunnel is stopped and that the cylinder is rotated in the clockwise direction. Friction effects between the cylinder's surface and the air will cause some air to stick to the cylinder and to be rotated with it. This circular movement of the air is called *circulation.* When the wind tunnel is started, the air passing over the top of the cylinder will be speeded up by circulation, while the air passing over the bottom of the cylinder will be retarded. According to Bernoulli's equation, the static pressure on the top will be reduced and the static pressure on the bottom will be increased. The new pressure distribution will be as shown in Fig. 3.8b. Lift will now be developed by the cylinder. This is called the *Magnus effect*, named after Gustav Magnus, who discovered it in 1852. It explains why you *slice* (or *hook*) your golf ball or why a good pitcher can throw a curve.

AERODYNAMIC PITCHING MOMENTS

Consider the pressure distribution about a symmetrical airfoil at zero angle of attack (AOA) (Fig. 3.9a). The large arrows show the sum of the low pressures on the top and bottom of the airfoil. They are at the *center of pressure* (CP) of their respective surfaces. The CP on the top of the airfoil and the CP on the bottom are located at the same point on the chord line. The large arrows indicate that the entire pressure on the top and bottom surfaces is acting at the CP. Because these two forces are equal and opposite in direction, no net lift is generated. Note also that the lines of action of these forces coincide, so there

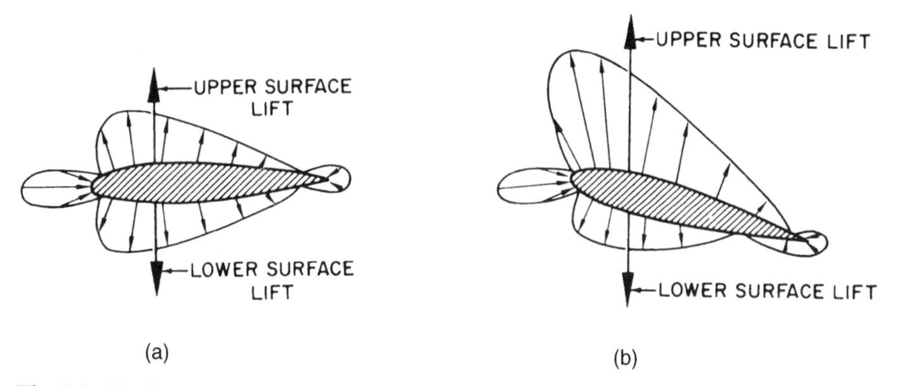

Fig. 3.9. Pitching moments on a symmetrical airfoil (a) at zero AOA and (b) at positive AOA.

is no unbalance of moments about any point on the airfoil. Figure 3.9b shows the pressure distribution about a symmetrical airfoil at a positive angle of attack (AOA). There is now an unbalance in the upper surface and lower surface lift vectors, and positive lift is being developed. However, the two lift vectors still have the same line of action, passing through the CP. There can be no moment developed about the CP. We can conclude that symmetrical airfoils do not generate pitching moments at any AOA. While no proof is offered here, it is also true that the CP does not move with a change in AOA for a symmetric airfoil.

Now consider a cambered airfoil operating at an AOA where it is developing no net lift (Fig. 3.10a). Upper surface lift and lower surface lift are numerically equal, but their lines of action do not coincide. A nose-down pitching moment develops from this situation. When the cambered airfoil develops positive lift (Fig. 3.10b), the nose-down pitching moment still exists. By reversing the camber it is possible to create an airfoil that has a nose-up pitching moment. Delta-wing aircraft have a reversed camber trailing edge to control the pitching moments.

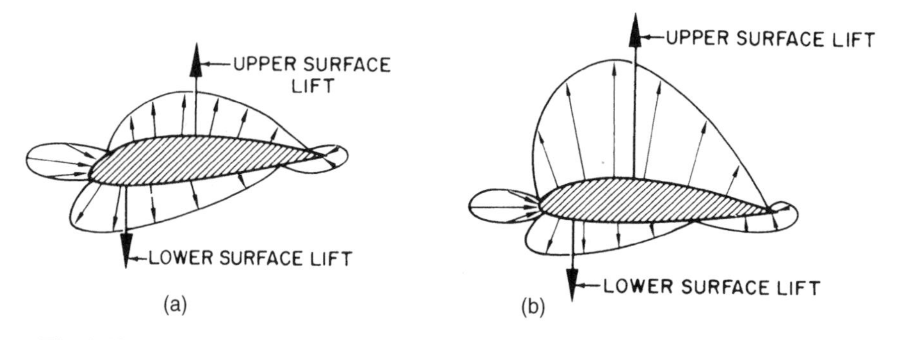

Fig. 3.10. Pitching moments on a cambered airfoil: (a) zero lift, (b) developing lift.

AERODYNAMIC CENTER

For cambered airfoils the CP moves along the chord line when the AOA changes. As the AOA increases, the CP moves forward and vice versa. This movement makes calculations involving stability and stress analysis very difficult. There is a point on a cambered airfoil where the pitching moment is a constant with changing AOA, if the velocity is constant. This point is called the *aerodynamic center* (AC).

The AC, unlike the CP, does not move with changes in AOA. If we consider the lift and drag forces as acting at the AC, the calculations will be greatly simplified. The location of the AC varies slightly, depending on airfoil shape. Subsonically, it is between 23 and 27% of the chord back from the leading edge. Supersonically, the AC shifts to the 50% chord.

Let's see how the lift forces act at the AC. If we replace the upper surface lift vector and the lower surface lift vector shown in Fig. 3.10b by a resultant "net lift" vector at the CP, we obtain Fig. 3.11a. Now we can place an upward vector equal to the lift at the AC and a downward vector of the same size also at the AC. We really haven't changed anything by this maneuver. This is shown in Fig. 3.11b. The downward vector at the AC and the upward vector at the CP create something that engineers call a *couple*.

A couple does not produce any net force but does produce a moment. The moment produced by the couple is the nose-down pitching moment, M, that we discussed above. Its value is equal to the lift, L, multiplied by the distance, x, between the CP and the AC:

$$M = Lx$$

Recall that this moment remains constant with change in AOA. So, if the AOA is reduced until the lift approaches zero, the CP must move backward until x approaches infinity. This, of course, is impossible, so the concept of CP is seldom used today.

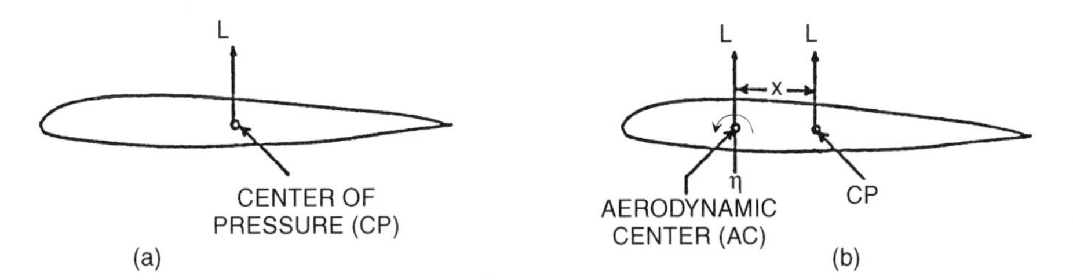

Fig. 3.11. Aerodynamic center concept.

Summary

1. The pitching moment at the AC does not change when the angle of attack changes (at constant velocity).

2. All changes in lift effectively occur at the AC.
3. AC is near 25% chord subsonically and at 50% supersonically.

SYMBOLS

AC Aerodynamic center
AF Aerodynamic force (lb)
AOA Angle of attack (degrees)
CP Center of pressure
D Drag (lb)
L Lift (lb)
RW Relative wind
α (alpha) Angle of attack (degrees)

PROBLEMS

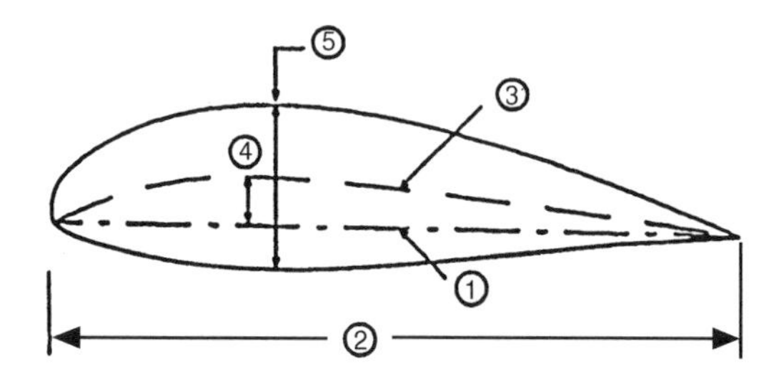

1. Number 3 on the drawing shows
   a. the chord line.
   b. the maximum camber.
   c. the thickness.
   d. the mean camber line.
2. Number 4 on the drawing shows
   a. the chord line.
   b. the thickness.
   c. the maximum camber.
   d. the mean camber line.
3. Number 5 on the drawing shows
   a. the maximum camber.
   b. the mean camber line.

   - **c.** the upper surface curvature.
   - **d.** the maximum thickness.

4. The Magnus effect explains why
   - **a.** a bowling ball curves.
   - **b.** a pitched baseball curves.
   - **c.** a golf ball slices.
   - **d.** Both (b) and (c)

5. Which of the below will develop positive lift?
   - **a.** A symmetrical airfoil at zero AOA
   - **b.** A nonrotating cylinder in a wind tunnel
   - **c.** A cambered airfoil at zero AOA

6. Which of the below will not produce a pitching moment?
   - **a.** A symmetrical airfoil at a positive AOA
   - **b.** A cambered airfoil that is developing zero lift
   - **c.** A cambered airfoil that is at a positive AOA
   - **d.** A symmetrical airfoil at zero AOA
   - **e.** Both (a) and (b)

7. The aerodynamic center (AC) is located at
   - **a.** 50% C subsonically and 25% C supersonically.
   - **b.** 25% C at all speeds.
   - **c.** 50% C at all speeds.
   - **d.** 25% C subsonically and 50% C supersonically.

8. For a cambered airfoil, the center of pressure (CP)
   - **a.** moves to the rear of the wing at low AOA.
   - **b.** moves backward as AOA increases.
   - **c.** moves forward as AOA increases.
   - **d.** Both (a) and (c)

9. Which of these statements is false? For a cambered airfoil;
   - **a.** the AC is where all changes in lift effectively take place.
   - **b.** there is no pitching moment at the AC.
   - **c.** the AC is located near 25% C subsonically
   - **d.** the pitching moment at the AC is constant with change of AOA (at constant airspeed).

10. For a symmetrical airfoil,
    - **a.** the center of pressure moves forward as AOA increases.
    - **b.** the center of pressure stays at the same place as AOA increases.
    - **c.** there is no pitching moment about the CP.
    - **d.** Both (b) and (c)

4 Lift

AERODYNAMIC FORCE EQUATIONS

The lift and drag equations depend on several factors, the most important of which are

1. Airstream velocity V (knots)
2. Airstream density ratio σ (dimensionless)
3. Planform area S (ft^2)
4. Profile shape of the airfoil
5. Viscosity of the air
6. Compressibility effects
7. Angle of attack α (degrees)

The first two factors determine the dynamic pressure, q, of the airstream (see Eq. 2.10):

$$q = \frac{\sigma V^2}{295} \quad \text{(psf)}$$

Factors 4, 5, and 6 influence the amount of drag that an airfoil will develop at a certain AOA. It is convenient to express the lift and drag forces in terms of nondimensional coefficients that are functions of AOA. These are called the *coefficient of lift* (C_L) and the *coefficient of drag* (C_D).

LIFT EQUATION

$$L = C_L qS = C_L \frac{\sigma V^2}{295} S \quad \text{(lb)} \tag{4.1}$$

Rewriting gives

$$C_L = \frac{L}{qS} = \frac{L/S}{q} = \frac{\text{lifting pressure}}{\text{dynamic pressure}}$$

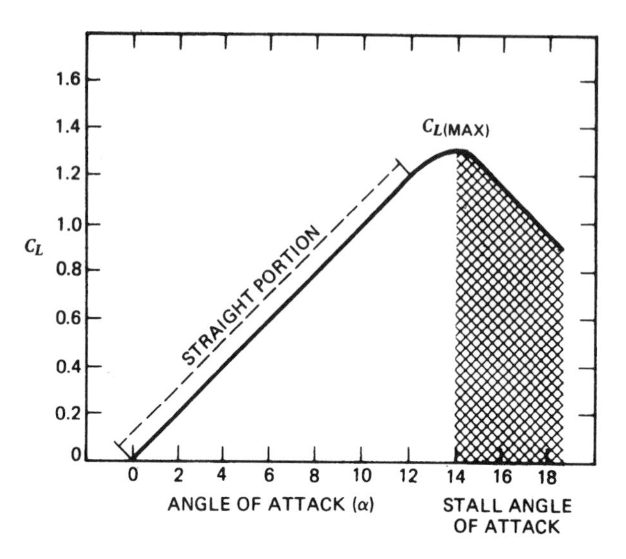

Fig. 4.1. C_L vs. AOA for a symmetrical airfoil.

C_L can be said to be the ratio between the lift pressure and the dynamic pressure." It is a measure of the effectiveness of the airfoil to produce lift. Values of C_L have been obtained experimentally for all airfoil sections and can be found in publications. such as Reference 1. Figure 4.1 shows a typical curve of C_L plotted against AOA for a symmetrical airfoil. Note that the curve passes through the (0, 0) point, indicating that no lift is developed at 0° AOA. The curve is essentially a straight line until it approaches its stall AOA. At the stall AOA the airfoil achieves its maximum lifting ability, $C_{L(max)}$. Stall is discussed later in this chapter.

An airfoil always stalls at the same AOA. For the airfoil shown in Fig. 4.1, this will be at 14° AOA. The stall AOA is independent of weight, attitude, and altitude if viscosity effects are ignored.

While the stall AOA is constant, the stall speed is not. Stall speed is affected by weight, wing loading, altitude, and other factors. Rewriting the lift equation 2.10 to solve for V gives

$$V = \sqrt{\frac{295L}{C_L \sigma S}}$$

Stall speed V_s occurs at $C_{L(max)}$, so

$$V_s = \sqrt{\frac{295L}{C_{L(max)} \sigma S}}$$

If the aircraft is in balanced, $1G$ flight, the lift is equal to the weight. During maneuvering flight, however, the G loading will not be equal to the weight, but will be the weight multiplied by the G value: $L = GW$.

Example Assume an aircraft with the C_L–α curve of Fig. 4.1 and the following data: gross weight = 15,000 lb, wing area $S = 340\,\text{ft}^2$, $C_{L(\max)} = 1.3$, and a sea level standard day. What is the stall speed?

$$V_s = \sqrt{\frac{(295)(15{,}000)}{(1.3)(1.0)(340)}} = 100 \text{ knots}$$

The stall speed varies as the square root of the lift (or weight under $1G$ flight), so the stall speed–weight relationship is

$$V_{s2} = V_{s1}\sqrt{\frac{W_2}{W_1}} \tag{4.2}$$

If the aircraft is flown at the same AOA, it will have the same lift coefficient, and not only will the stall speed vary with the square root of the weight, but all other speeds will vary by the same amount. Stall AOA does not vary with weight; stall speed does.

Example Assume that the aircraft discussed above has an approach speed of 114 knots at a gross weight of 15,000 lb. It develops a lift coefficient of

$$C_L = \frac{295W}{\sigma V^2 S} = \frac{(295)(15{,}000)}{(1.0)(114)^2(340)} = 1.0$$

Note from Fig. 4.1 that a 10° AOA must be held for this flight condition. If the weight of the aircraft is increased to 20,000 lb, the AOA remains the same but the approach speed of the aircraft must be increased, as indicated by Eq. 4.2:

$$V_2 = V_1\sqrt{\frac{W_2}{W_1}} = 114\sqrt{\frac{20{,}000}{15{,}000}}$$
$$= 131.6 \text{ knots}$$

The lift equation shows that for steady unaccelerated flight the lift is equal to the weight and is a constant. There are only two variables in the equation: C_L and V^2. The value of C_L is determined by the AOA and then the airspeed is determined. It follows that the control stick determines AOA and thus the airspeed. This is the fundamental concept in flying technique: AOA is the primary control of airspeed in steady flight. Rate of climb or descent must, therefore, be controlled by throttle position.

ANGLE OF ATTACK INDICATOR

The importance of AOA in determining aircraft performance cannot be overemphasized. We have discussed stall AOA, but these facts are of equal importance: (1) an aircraft has its maximum climb angle at a certain AOA, (2) it will achieve maximum rate of climb at another AOA, and (3) it will get maximum range at still another AOA. All aircraft performance depends on AOA, so modern, high-performance aircraft are fitted with AOA indicators. *These devices indicate the direction of the relative wind.* They consist of a probe, mounted on the side of the fuselage, which aligns itself with the relative wind, much like a weathervane, and cockpit indicators. The basic cockpit indicator merely has a pointer showing the AOA in uncalibrated units, rather than in degrees. This allows the same instrument to be used in different aircraft. The flight handbook will give the approach AOA, for instance, in units on the indicator, rather than in degrees.

The AOA indicating system often includes additional stall warning devices, such as stick or rudder shakers. Some fighter and attack aircraft are fitted with a stall indexer (Fig. 4.2). The indexer gives the pilot an easily understood visual command during landing approaches. If the angle of attack is too high, the low-speed symbol is illuminated. The pilot sees an arrow pointing down and pushes the nose down. The reverse happens if the AOA is too low.

AIRFOIL LIFT CHARACTERISTICS

The lift characteristics of airfoil sections are affected by thickness and location of maximum thickness, camber and its location, leading-edge radius, and other factors. Increasing the thickness, for example, results in higher local velocity airflow, which results in lower static pressure and more lift (Fig. 4.3).

Both of the airfoils in Fig. 4.3 have a maximum camber of 4% C, located at the 40% C position. The 4412 airfoil is 12% C thick, and the 4406 is 6% C

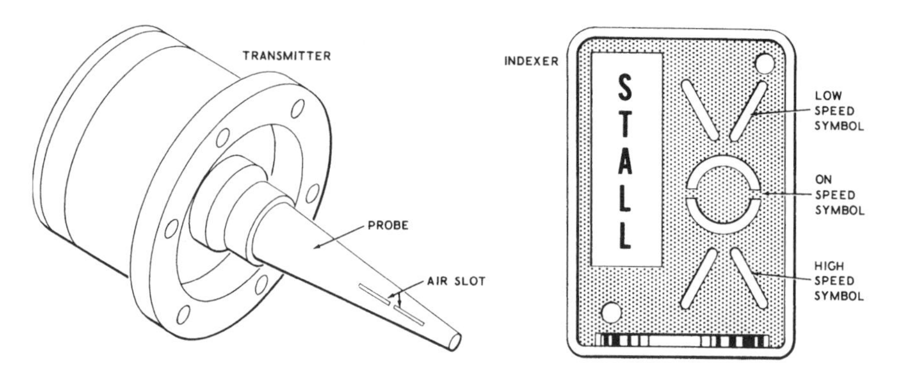

Fig. 4.2. AOA indicators.

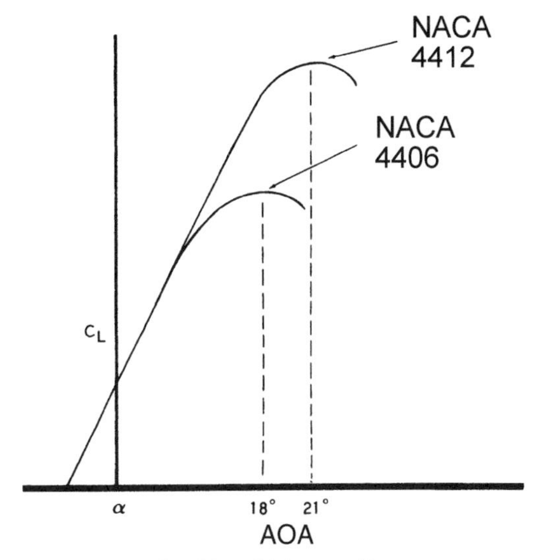

Fig. 4.3. Thickness effect.

thick. The thicker airfoil has a much higher value of $C_{L(max)}$, and the stall AOA is also higher. The 4412 airfoil is superior to the 4406 airfoil in producing lift, but it will also have more drag. The mission of a proposed aircraft influences the type of airfoil selected. Cambered airfoils produce lift at zero AOA, as can be seen from Fig. 4.3.

Nearly all airfoil sections have the same slope of their C_L–α curves. This is about a 0.1 increase in C_L for each degree of increase in AOA. Later, when we discuss wings, rather than airfoils, we will see the effects of aspect ratio and sweepback on the slope of these curves.

Figure 4.3 shows the curves for two cambered airfoils. To better compare cambered airfoils and symmetrical airfoils, we can show them together (Fig. 4.4). An increase in $C_{L(max)}$ and a decrease in stall AOA will be noted when camber is increased.

Example Using the curves in Fig. 4.4 calculate the stall speed at sea level standard day conditions for a 15,000-lb aircraft whose wing area is 340 ft^2. Calculate V_s for each airfoil section.

Symmetrical ($C_{L(max)} = 1.35$):

$$V_s = \sqrt{\frac{(295)(15{,}000)}{(1.35)(1)(340)}} = 98.2 \text{ knots}$$

Cambered ($C_{L(max)} = 1.55$):

$$V_s = \sqrt{\frac{(295)(15{,}000)}{(1.55)(1)(340)}} = 91.6 \text{ knots}$$

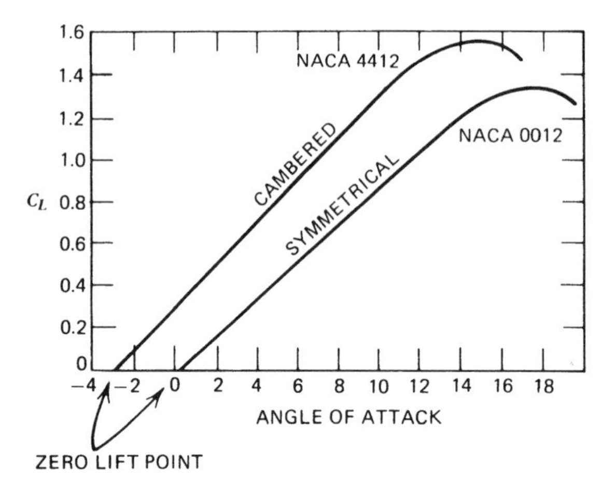

Fig. 4.4. Camber effect.

BOUNDARY LAYER THEORY

As we have seen, high AOA produces high coefficients of lift. Thus, lift to support an aircraft can be obtained at low airspeeds. To better understand what happens to the airflow when an aircraft is flown at or above stall AOA, we must understand the behavior of the air close to the wing (or airfoil) surface.

When an object, such as an airfoil, passes through an airstream, the viscosity of the air causes air particles next to the surface to be pulled along at approximately the speed of the airfoil. Particles of air slightly farther away will also be pulled along, but at a lesser velocity. The farther we move away from the surface, the lower the velocity of the particles. At some distance from the surface, a point is reached where the particles are not speeded up at all.

The layer of air from the surface of the airfoil to the point where there is no measurable air velocity is known as *the boundary layer*. The nature of the boundary layer determines the maximum lift coefficient and the stalling characteristics of the airfoil.

The beginning of airflow at the leading edge of a smooth airfoil surface produces a very thin layer of smooth airflow. This type of airflow is called *laminar flow* and is characterized by smooth regular streamlines. Fluid particles in this region do not intermingle. As the airflow moves back from the leading edge, the boundary layer thickens and becomes unstable. Small pressure disturbances cause the unstable airflow to tumble, and intermixing of the air particles takes place. This type of airflow is called *turbulent flow*.

Figure 4.5 illustrates the flow within the boundary layer on a flat plate and shows the increasing thickness of the boundary layer and transition from laminar to turbulent flow. The thickness is greatly exaggerated in the drawing.

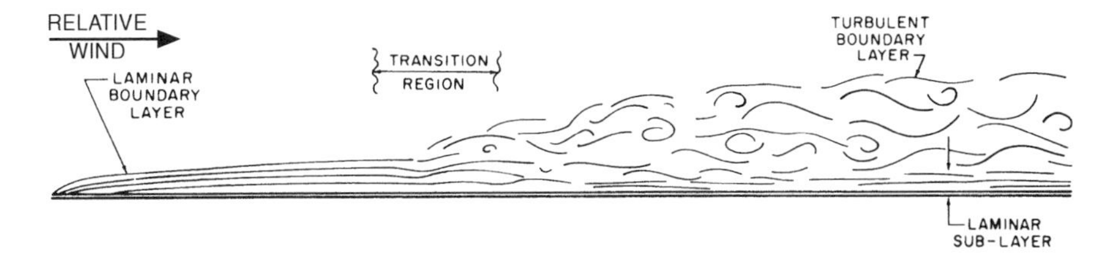

Fig. 4.5. Boundary layer composition.

You can observe laminar and turbulent airflow by lighting a cigarette in a draft-free room and noting the smoke as it rises, as shown in Fig. 4.6. Velocity profiles are drawn to help visualize the local velocity of an airstream in the boundary layer. Figure 4.7 shows one of these. The relative velocity at the surface is zero, and the relative velocities at various levels away from the surface increase until the outer edge of the boundary layer is reached.

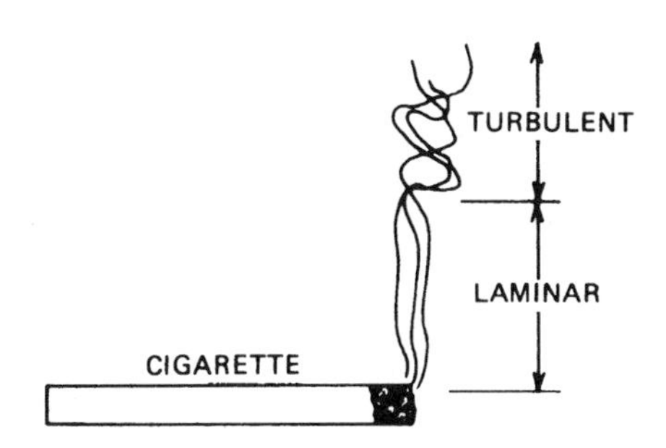

Fig. 4.6. Smoke pattern.

Velocity profiles are different in laminar and turbulent flow. Laminar profiles show a gradual decrease in the relative velocity from the outer edge of the boundary layer to the surface. Turbulent flow involves rapid intermixing of the air levels, and faster moving air speeds up the particles near the surface, resulting in profiles as shown in Fig. 4.8.

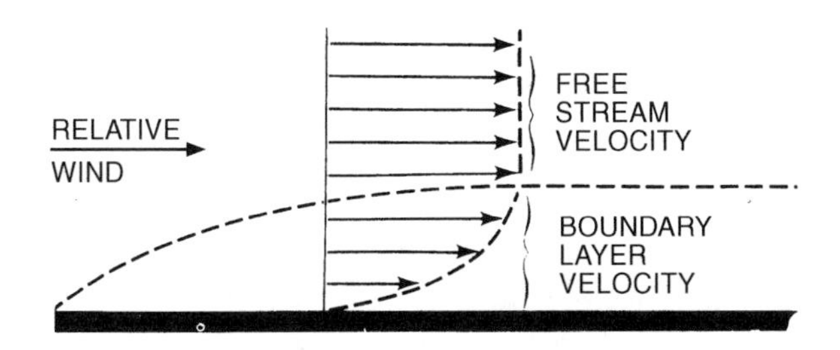

Fig. 4.7. Boundary layer velocity profile.

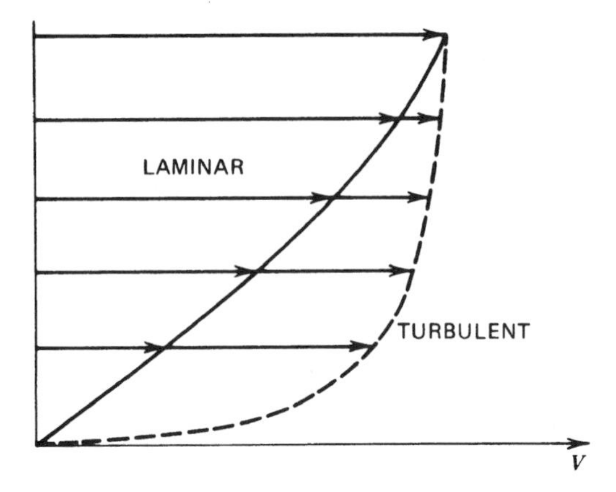

Fig. 4.8. Laminar and turbulent velocity profiles.

REYNOLDS NUMBER

In 1883 Osborn Reynolds carried out a series of experiments with the flow of water in pipes. By injecting dye into a stream of flowing water, he discovered that laminar and turbulent flow could be seen. He also discovered a relationship between the point where the fluid became turbulent and three properties of the fluid stream. The Reynolds number is defined by the following formula:

$$R_g = \frac{Vx}{\nu} \tag{4.3}$$

where

R_g = Reynolds number
V = stream velocity (fps)
x = distance downstream (ft)
ν = kinematic viscosity (ft^2/sec)

Reynolds' experiments showed that at low values of R_g the flow was laminar, and at higher values the flow was turbulent. When considering airflow about an airfoil, it is not possible to define the exact R_g when flow changes from laminar to turbulent. Both the contour of the airfoil and the smoothness of the surface influence the changeover point.

Studies on smooth, flat plates have been made, and a prediction of the type of flow is possible. Figure 4.9 shows the results of one study. Laminar airflow exists below R_g of 0.5 million, and turbulent flow exists for values greater than 10 million. Transition occurs between these two values. The consideration of Reynolds number is very important in correlating wind tunnel data of scale models with full-sized aircraft.

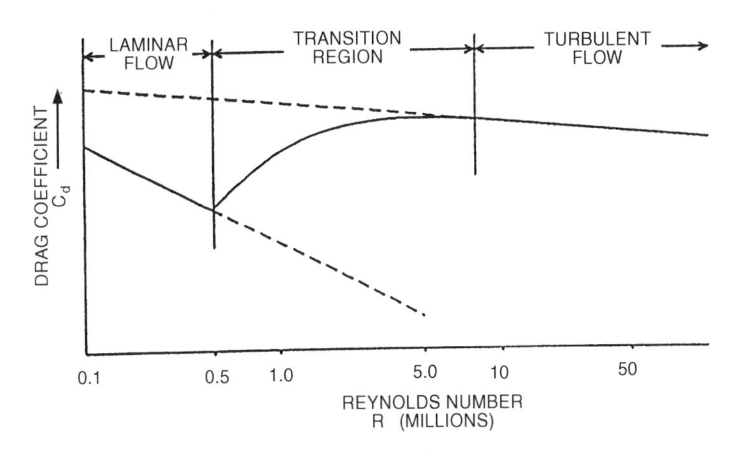

Fig. 4.9. Reynolds number effect on airflow on a smooth flat plate.

ADVERSE PRESSURE GRADIENT

Pressure distribution about a cambered airfoil that is producing lift is shown in Fig. 4.10. The minimum pressure point is shown by the heavy arrow. Airflow approaching the leading edge of the airfoil enters the high-pressure area near the stagnation point. As the air divides to pass over the top and under the bottom surfaces of the airfoil, it moves from a high-pressure area to a lower static pressure area.

The air moving over the top of the airfoil is important to this discussion. Until the air reaches the point of minimum pressure, it is in a favorable pressure gradient. The forces of nature are acting on the air favorably and it is

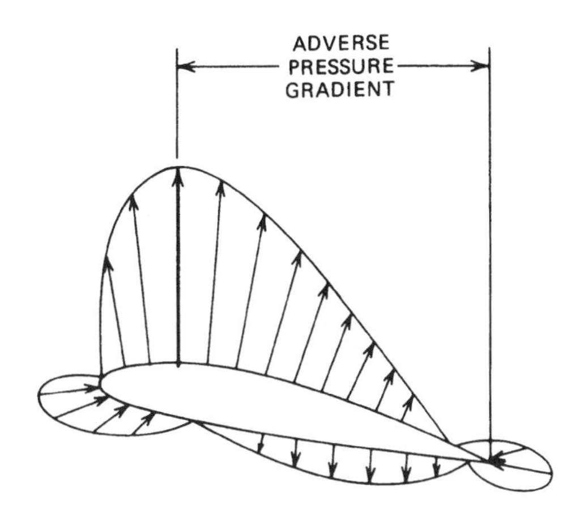

Fig. 4.10. Adverse pressure gradient.

accelerated and sped on its way. Once the air on the top of the airfoil passes the minimum static pressure point it is entering a region of higher static pressure. This is contrary to the laws of nature, and this region is called an *adverse pressure gradient*. The kinetic energy of the airflow is now being transformed into higher static pressure, and the velocity is reduced.

AIRFLOW SEPARATION

The airflow in the boundary layer is acted on by two forces:

1. Friction forces
2. Adverse pressure gradient forces

First, there are friction forces between the surface of the airfoil and the air particles, and there is friction between the particles themselves. Both of these tend to reduce the relative velocity to zero. Second, the air is slowed by the adverse pressure gradient. The slowing of the air creates a stagnated region close to the surface.

Airflow from outside the boundary layer will overrun the point of stagnation and a flow reversal results: The airflow moves forward, the boundary layer separates from the surface, lift is destroyed, and drag becomes excessively high. Airflow can be seen on an airfoil operating at high AOA if tufts of wool are glued to the upper surface. The tufts will reverse direction and flap violently when stall occurs. Figure 4.11 shows the velocity profiles in the boundary layer during airflow separation.

Consider a smooth sphere in an airflow (Fig. 4.12a). If the diameter is small, the Reynolds number will be small and laminar airflow will exist on the surface. This produces only a small amount of drag against the surface, but laminar flow allows the air to separate easily. When this happens, a wide wake forms behind the sphere. This wake produces high drag forces.

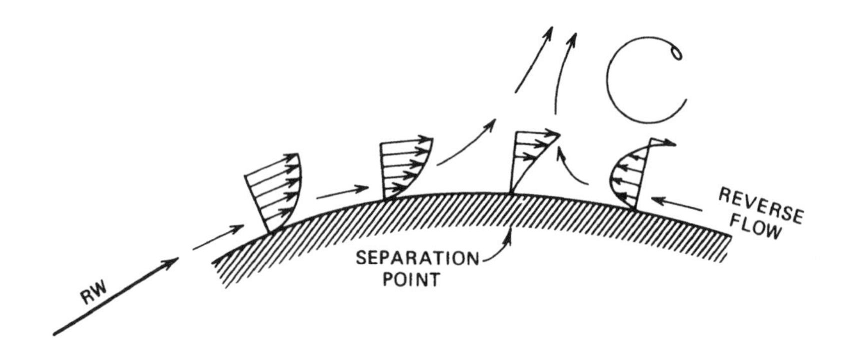

Fig. 4.11. Airflow separation velocity profiles.

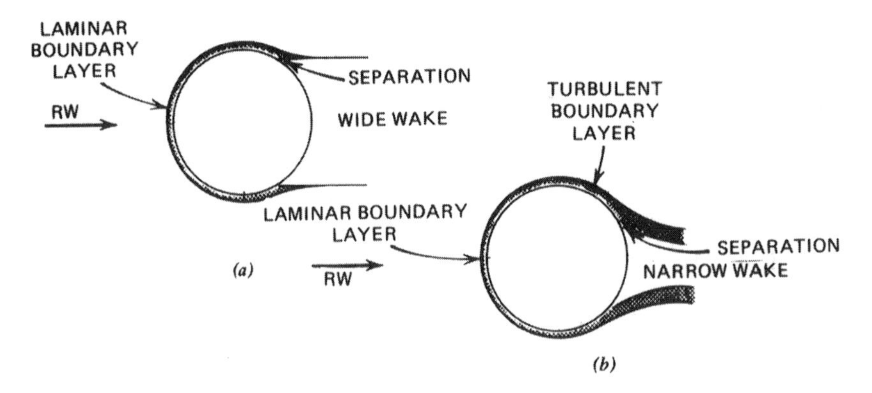

Fig. 4.12. Sphere wake drag: (a) smooth sphere, (b) rough sphere.

Now, assume the sphere's surface is rough (Fig. 4.12b). The airflow becomes turbulent very soon as it passes around the sphere. The surface drag is now increased, but turbulent air has more energy than laminar flow and will resist separation. Thus, the width of the wake is reduced and total drag is reduced. The dimples on a golf ball and the fuzz on a tennis ball are examples of how creating turbulent flow delays separation and reduces total drag.

STALL

Stall is airflow separation of the boundary layer from a lifting surface. It is characterized by reduction in lift and a rapid increase in drag. Two types of stall are of interest to the nonjet pilot. The most common stall is the slow-speed, high-AOA separation. It most often occurs during the takeoff and landing phases of flight and is dangerous because of the lack of altitude for recovery. This type of stall starts at the trailing edge of a wing and progresses forward. The part of the wing that stalls first is of great interest and depends on the planform shape of the wing. This is discussed in detail later.

The second kind of stall is the accelerated stall, which occurs with a sudden rapid increase in AOA. Sharp leading edges contribute to this type of stall. The rapid increase in AOA does not allow time for the airflow to turn the corner of the leading edge, and separation takes place there. To avoid this type of stall, don't make sudden nose-up movements of the control stick when recovering from unusual attitudes.

HIGH COEFFICIENT OF LIFT DEVICES

Landing and takeoff flaps, slots, slats, and other devices that enable an aircraft to take off and land at reduced speeds are often erroneously called high-lift devices. The lift of the aircraft during these maneuvers does not exceed its

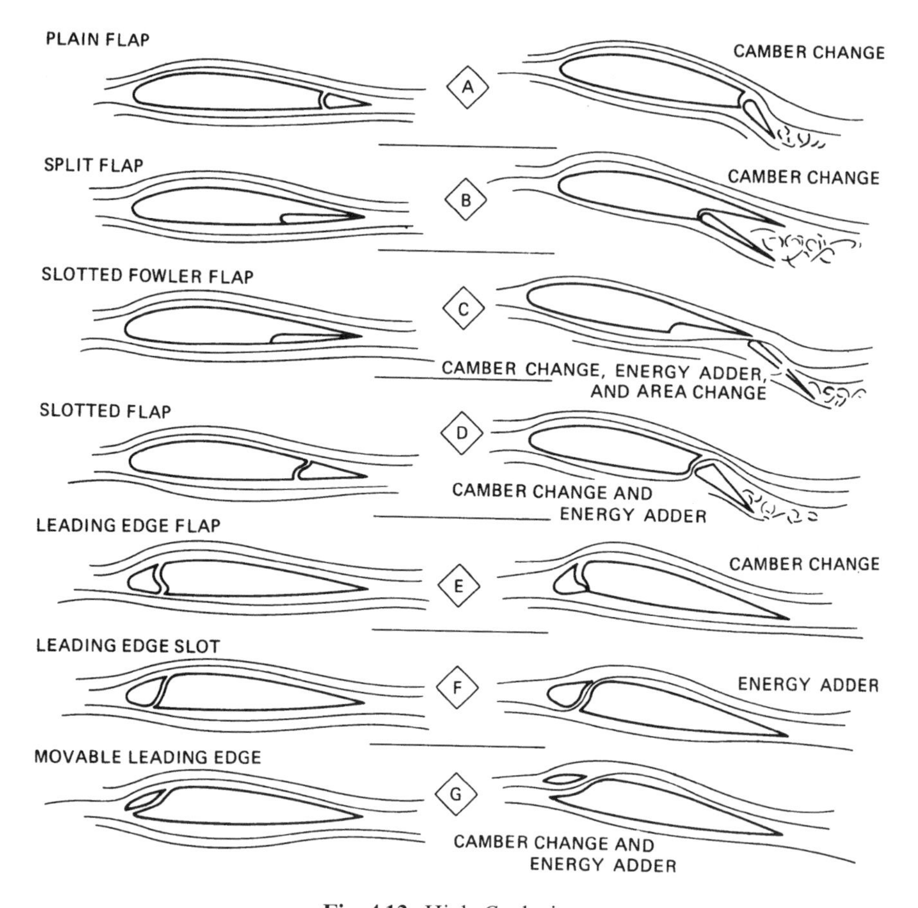

Fig. 4.13. High-C_L devices.

weight, so high lift is not required. But landing and takeoff speeds can be reduced by increasing the coefficient of lift. So, the proper term is *high coefficient of lift devices.*

In our discussion of airfoil lift characteristics we saw two ways that $C_{L(max)}$ of an airfoil could be increased: (1) by increasing its thickness and (2) by increasing its camber. No practical way of increasing its thickness in flight has been developed to date, so this possibility is not discussed here.

The first attempts to provide high C_L devices concentrated on increasing the camber of the wing. Devices that increase C_L by this method are classified as *camber changers*. Some examples of these are shown in Fig. 4.13. Figure 4.14 shows the effects of pure camber changers on the coefficient of lift curves.

Our discussion of boundary layer theory, separation, and stalls showed that airflow separation resulted when the lower levels of the boundary layer did not have sufficient kinetic energy. The air was slowed by friction and an adverse pressure gradient. If we can find a way to increase the kinetic energy in the

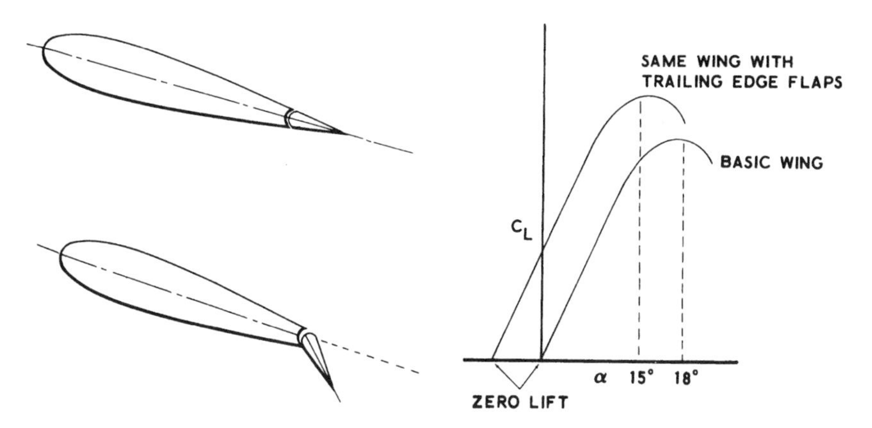

Fig. 4.14. Effect of a camber changer on the C_L–α curve.

boundary layer, flow separation can be suppressed. Then higher angles of attack can be used and higher values of C_L can be realized. Devices that use this principle are called *energy adders.*

There are several ways of adding energy to the boundary layer. The simplest way is by stirring up the boundary layer. This replaces the slow-moving laminar airflow next to the wing skin with faster-moving turbulent air. One method of producing turbulence in the boundary layer is the use of vortex generators. These are small airfoil-shaped vanes protruding upward from the wing as shown in Fig. 4.15. Vortex generators mix the turbulent outer layers of the boundary layer with slow-moving laminar lower layers thus reenergizing them. The advantages of reduced landing and takeoff speeds due to the increase in lift coefficient more than compensate for the increased parasite drag of the vortex generators.

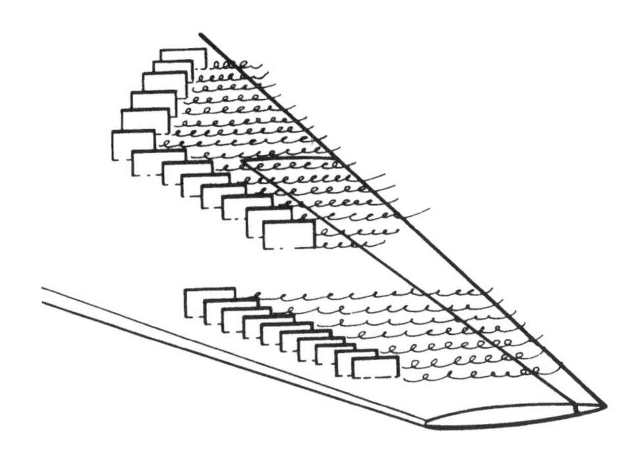

Fig. 4.15. Vortex generators.

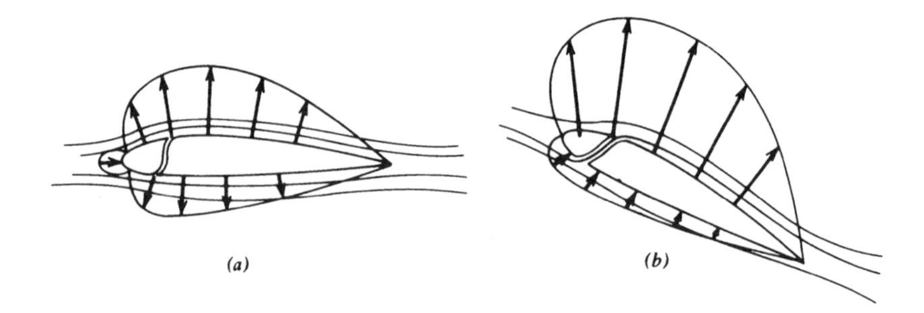

Fig. 4.16. Fixed slot at (a) low AOA and (b) high AOA.

A second method of adding kinetic energy to the boundary layer is to route high-pressure air from the stagnation region, near the leading edge of the wing, to the low-pressure area on top of the wing. This will invigorate the slow-moving lower layers of the boundary layer. Figure 4.16 shows this method, using a fixed slot at the leading edge of the wing. At low angles of attack the pressure on each end of the slot is about the same and little or no air flows through the slot (Fig. 4.16a). At high angles of attack the high-pressure stagnation point has moved down below the leading edge. The high-pressure differential causes a large airflow up through the slot and over the top of the wing (Fig. 4.16b).

The effect of pure energy adders on the C_L curve is shown in Fig. 4.17. The pure energy adders extend the curve to higher values of AOA and $C_{L(max)}$. The values of C_L at lower angles of attack are not affected. Most of the more complicated high-C_L devices use a combination of camber changer and energy adder techniques. Slotted leading- and trailing-edge flaps are examples of combination devices.

A third method of increasing the kinetic energy in the boundary layer is by supplying energy from some outside source. This type device is called boundary layer control, (BLC). There are two methods of doing this. The first is by using suction. A vacuum pump is attached to ducts, which lead to spanwise slots cut

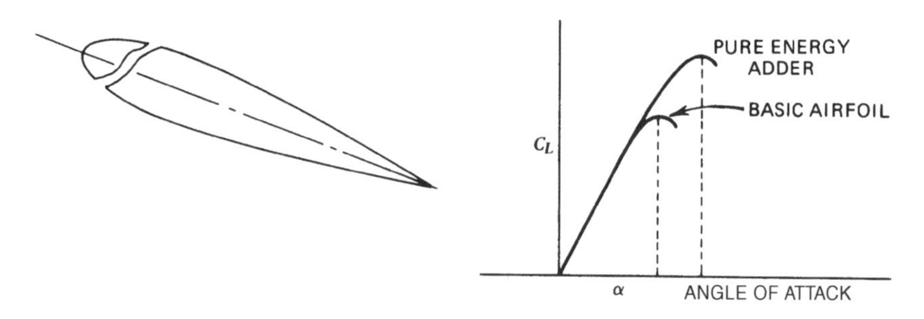

Fig. 4.17. Effect of an energy adder on the C_L–α curve.

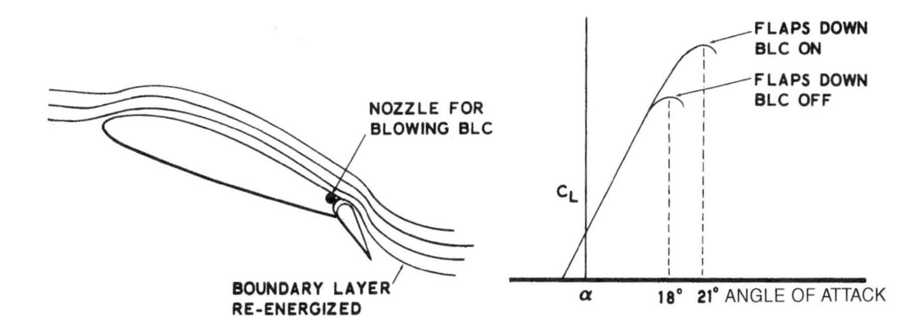

Fig. 4.18. Effect of blowing BLC over a trailing-edge flap on the C_L–α curve.

in the upper surface of the wing. When suction is applied, the lower layers of air in the boundary layer are removed and are replaced by faster moving air from the higher layers. This delays separation.

The more common type of BLC is the "blowing air" device (Fig. 4.18). High-pressure air is supplied from the compressor of the turbine engine and piped into rearward-facing slots located in front of the trailing-edge flaps. The blowing air re-energizes the boundary layer and delays separation.

SYMBOLS

C_D	Coefficient of drag (dimensionless)
C_L	Coefficient of lift
$C_{L(max)}$	Maximum value of C_L
G	Load factor (dimensionless)
R_g	Reynolds number (dimensionless)
S	Wing area (ft^2)
ν	(nu) Kinematic viscosity (ft^2/sec)

EQUATIONS

4.1 $$L = \frac{C_L \sigma V^2 S}{295}$$

4.2 $$V_2 = V_1 \sqrt{\frac{W_2}{W_1}}$$

4.3 $$R_g = \frac{Vx}{\nu}$$

PROBLEMS

1. As thickness of an airfoil is increased, the stall AOA
   a. is greater.
   b. is less.
   c. remains the same.
2. As camber of an airfoil is increased, its C_L at any AOA
   a. is less.
   b. remains the same.
   c. is greater.
3. Air in the boundary layer
   a. has zero velocity at the wing surface.
   b. is turbulent near the leading edge.
   c. is more apt to stall if it is turbulent.
   d. None of the above
4. Air in the boundary layer
   a. changes from laminar to turbulent at low Reynolds numbers.
   b. separates from the wing when its velocity is maximum.
   c. reverses flow direction when stall occurs.
   d. None of the above
5. Two things an airfoil designer can change to increase $C_{L(max)}$ are
   a. thickness and wing area.
   b. chord length and aspect ratio.
   c. camber and wing span.
   d. thickness and camber.
6. High values of the Reynolds number will more likely
   a. occur near the leading edge of an airfoil.
   b. indicate laminar flow.
   c. occur at lower airspeeds.
   d. show turbulent airflow.
7. Adverse pressure gradient on an airfoil is found
   a. from the point of maximum thickness to the trailing edge.
   b. near the stagnation point at the leading edge.
   c. from the point of minimum pressure to the trailing edge.
   d. Both (a) and (c).
8. At low-velocity stall, the airflow
   a. stops.

b. reverses direction.
c. speeds up.
d. moves toward the wingtips.

9. Airflow separation can be delayed by
a. making the wing surface rough.
b. using vortex generators.
c. directing high-pressure air to the top of the wing or flap through slots.
d. All of the above

10. Show which type of high-C_L device (camber changer, energy adder, or combination) each of the below is:
a. Blowing air (BLC) over a flap ____________________
b. Fixed slot ____________________
c. Slotted leading-edge flap ____________________
d. Plain flap ____________________

11. An aircraft has the C_L–α curve shown in Fig. 4.1. The following data applies:
Weight = 20,000 lb
Wing area, S = 340 ft^2
Density altitude = 10,000 ft
TAS = 242.4 knots
Find the aircraft's AOA for steady flight: ____________________

12. Calculate the stall speed for the aircraft in Problem 11: ____________________

13. If the airplane in Problem 11 burns 5000 lb of fuel during its flight, calculate its new stall speed: ____________________

14. An airplane has a rectangular wing with a chord length of 8 ft. If the airplane is flying at 200 knots TAS and at 20,000 ft density altitude, find the Reynolds number at the trailing edge of the wing.

5 Drag

Drag is the component of the aerodynamic force that is parallel to the relative wind and retards the forward motion of the aircraft. At subsonic speeds, there are two kinds of drag: *induced drag* and *parasite drag*. Compressibility drag is a third kind of drag, occurring at high speeds.

DRAG EQUATION

The basic drag equation is similar to the basic lift (Eq. 4.1):

$$D = C_D q S = \frac{C_D \sigma V^2 S}{295}$$

Rewriting gives

$$C_D = \frac{D/S}{q} = \frac{295D}{\sigma V^2 S} \tag{5.1}$$

The *coefficient of drag*, C_D, is the ratio of the drag pressure to the dynamic pressure. Values of C_D have been obtained experimentally for all airfoil sections and can be found in other publications (see Reference 1). Figure 5.1 shows typical curve for C_D vs. AOA for the same airfoil.

LIFT TO DRAG RATIO

An aircraft's lift/drag ratio, L/D, is a measure of its efficiency. An aircraft with a high L/D is more efficient than one with a lower L/D. The L/D can be found by dividing the values of C_L by those of C_D at the same AOA. This can be seen if we divide Eq. 4.1 by Eq. 5.1:

$$\frac{L}{D} = \frac{C_L}{C_D} \tag{5.2}$$

All the factors cancel except L, D, and their coefficients. Typical L/D curves are shown in Fig. 5.2.

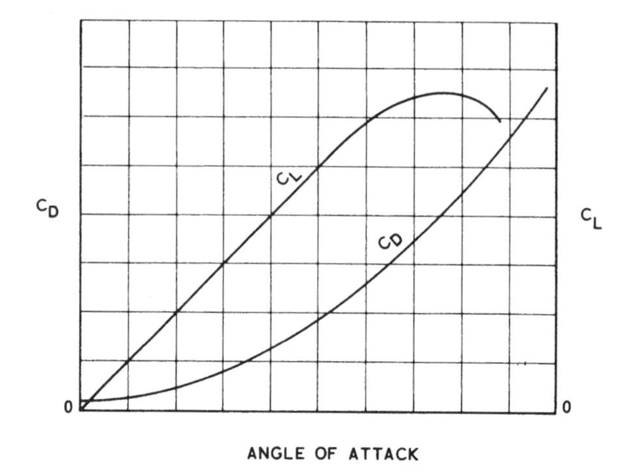

Fig. 5.1. C_L vs. AOA and C_D vs. AOA.

The highest point on the curve is important. It is called $(L/D)_{max}$ and occurs at the most aerodynamic efficient AOA. Minimum drag occurs at $(L/D)_{max}$. Under $1G$ flight, the lift of the aircraft equals the weight, so $L/D = W/D$. If the value of W/D is a maximum, then drag must be a minimum. Minimum drag is then equal to the weight of the aircraft divided by the value of $(L/D)_{max}$:

$$D_{min} = \frac{W}{(L/D)_{max}} \tag{5.3}$$

Several performance characteristics are present at $(L/D)_{max}$. For example,

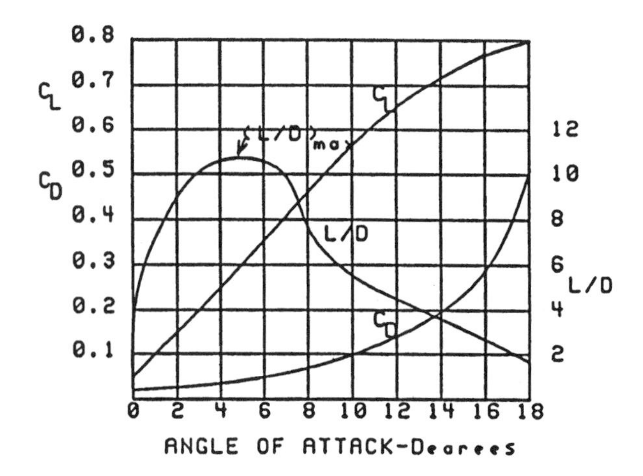

Fig. 5.2. Typical lift-to-drag ratios.

the best engine out glide ratio occurs here. Because lift is proportional to distance and drag is proportional to altitude, then L/D is proportional to glide ratio (distance/altitude). This means the numerical value of L/D equals the numerical value of the glide ratio.

Note that the value of $(L/D)_{max}$ and the angle of attack at which it occurs do not vary with aircraft weight or altitude. The airspeed for $(L/D)_{max}$ does vary with weight, as was shown in Eq. 4.2.

INDUCED DRAG

Induced drag, D_i, is the least understood type of drag, but it is the most important, especially in the critical low-speed region of flight. It is called the drag due to lift because it occurs only when lift is developed.

Up to this point we have been discussing airfoils, not wings. We must now consider the shape of the wing as viewed from above. This is called the planform of the wing. In describing the planform of a wing, several terms must be explained. These are shown in Fig. 5.3.

1. The *wing area*, S, is the plan surface area, including the area covered by the fuselage.
2. The *wing span*, b, is the distance from tip to tip.
3. The *average chord*, c_{av}, is the geometric average chord:

$$\text{Span} \times \text{average chord} = \text{wing area } (bc_{av} = S)$$

4. The *aspect ratio*, AR, is the span divided by the average chord:

$$\text{AR} = b/c_{av} = b^2/S$$

5. The *root chord*, c_r, is the chord measured at the aircraft centerline. The tip chord c_t, is the chord measured at the wingtip.
6. The *taper ratio*, λ (lambda), is the ratio of c_t to c_r:

$$\lambda = c_t/c_r$$

7. The *sweep angle*, Λ (uppercase lambda), is the angle between the line of 25% chord points and a perpendicular to the root chord.
8. The *mean aerodynamic chord*, MAC, is the chord drawn through the centroid (center of area) of the half-span area. The MAC and the c_{av} are not the same. If the actual wing was replaced by a rectangular wing having the same span and a chord equal to the MAC, the pitching moments of each would be identical.

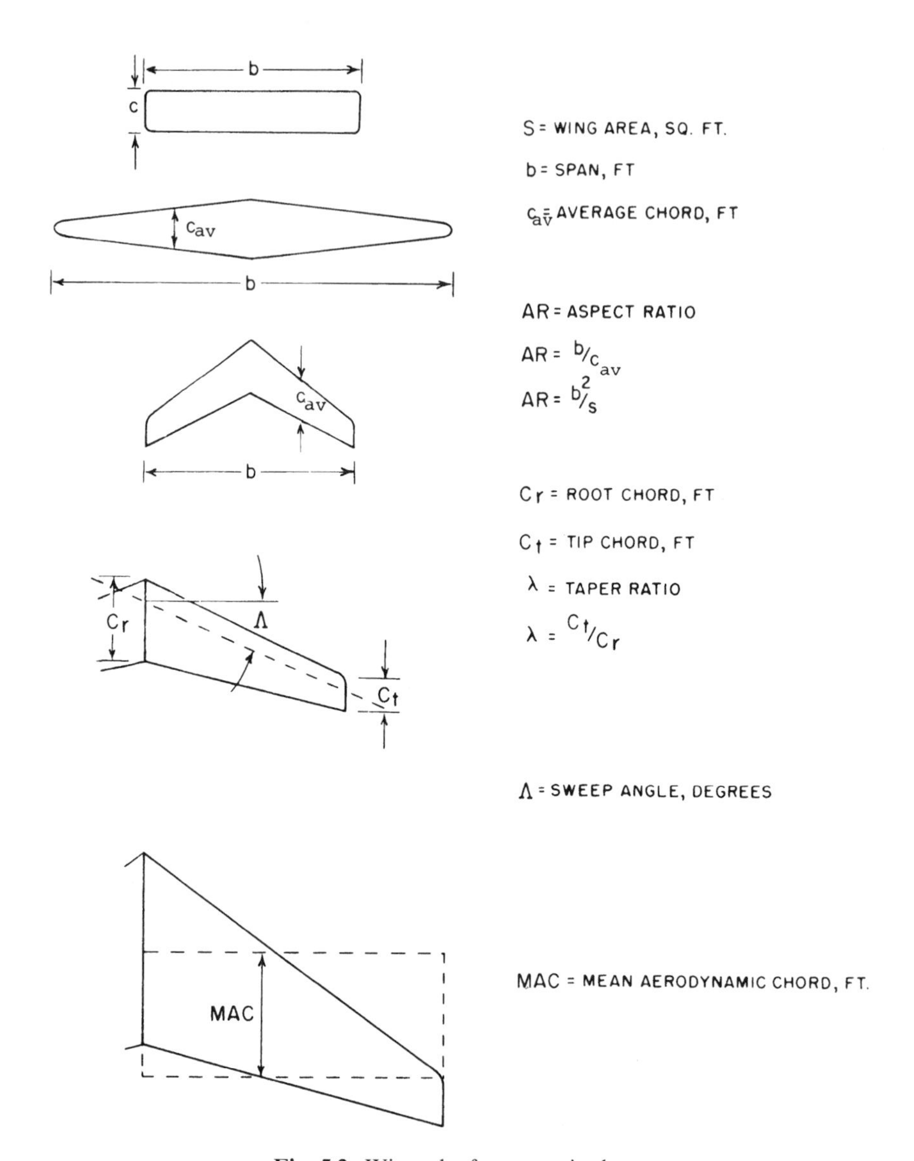

Fig. 5.3. Wing planform terminology.

Wingtip Vortices

A *wingtip vortex* is formed when higher pressure air below a wing flows around the wingtips into the low-pressure region on top of the wing that is developing lift. The vortices are strongest at the tips and become weaker toward the centerline of the aircraft (Fig. 5.4).

Consider an aircraft with an infinitely long wing. This *infinite wing* has no wingtips and so it has no wingtip vortices. The absence of wingtip vortices

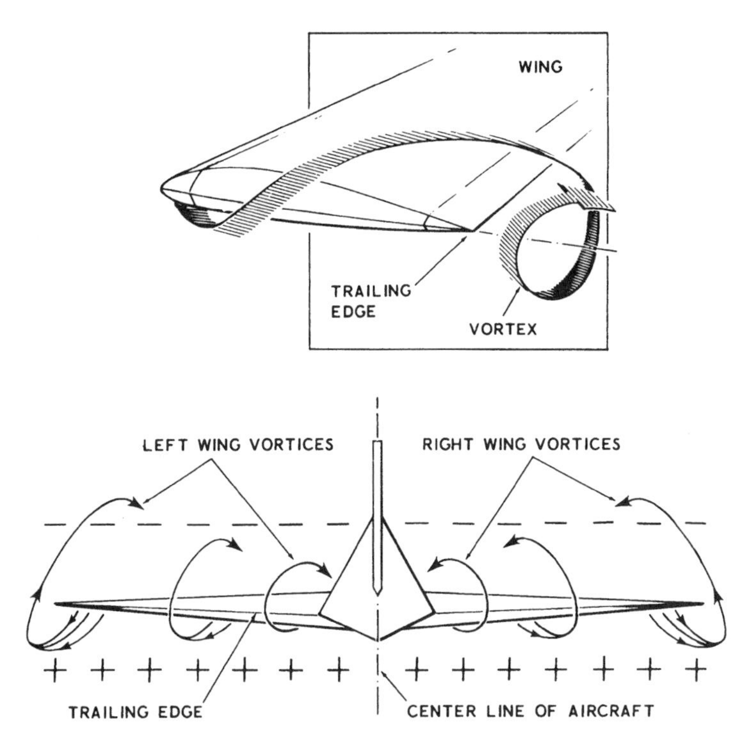

Fig. 5.4. Wingtip vortices.

means that the upwash in front of the wing and the downwash behind the wing cancel each other, so there is no net downwash behind the wing (Fig. 5.5). The relative wind (RW) ahead of this wing is horizontal, and the RW behind the wing is also horizontal. The RW at the aerodynamic center (AC) is the average of these two and also must be horizontal. Lift is 90° to the RW, so it acts

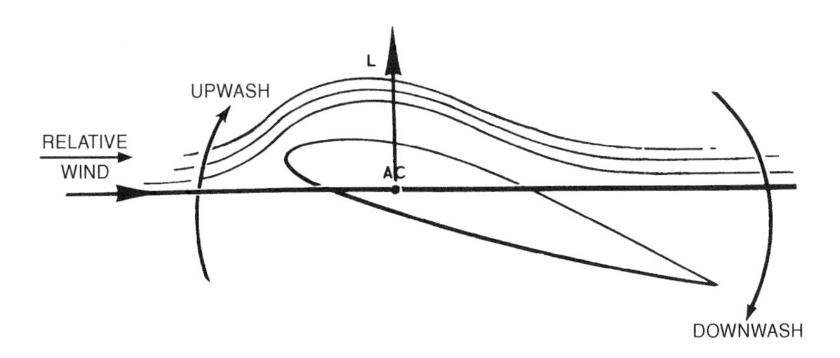

Fig. 5.5. Airflow about an infinite wing.

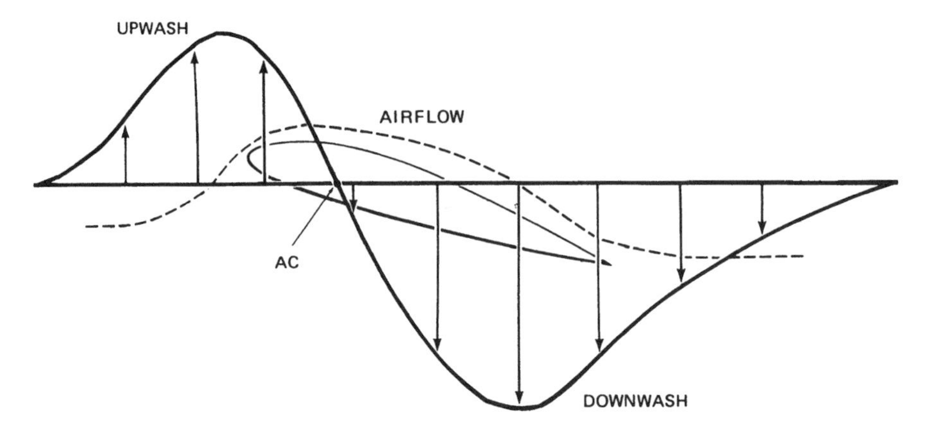

Fig. 5.6. Vertical velocity vectors of an infinite wing.

vertically. The vertical components of the local RW for an infinite wing are plotted in Fig. 5.6. Note that the RW at the AC has no upward or downward velocity and is horizontal.

A wing with wingtips is called a *finite wing*. A finite wing that is developing lift will have tip vortices, as described previously. These vortices force air down behind the wing. The vertical components of the local RW for a finite wing that is producing lift are shown in Fig. 5.7.

The local RW is the vector sum of the free stream RW and the vertical velocity vectors (Fig. 5.7). The RW far ahead of the aircraft is not disturbed by upwash, so for an aircraft flying horizontally it is also horizontal. The local RW behind the wing is depressed by the final downwash velocity, $2w$. The angle that the RW leaving the wing makes with the free stream RW is called the downwash angle. It can be seen in Fig. 5.8.

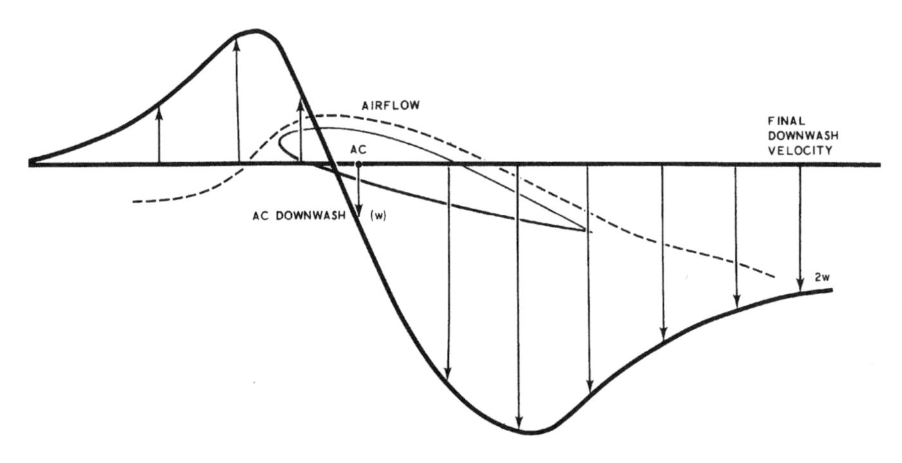

Fig. 5.7. Vertical velocity vectors of a finite wing.

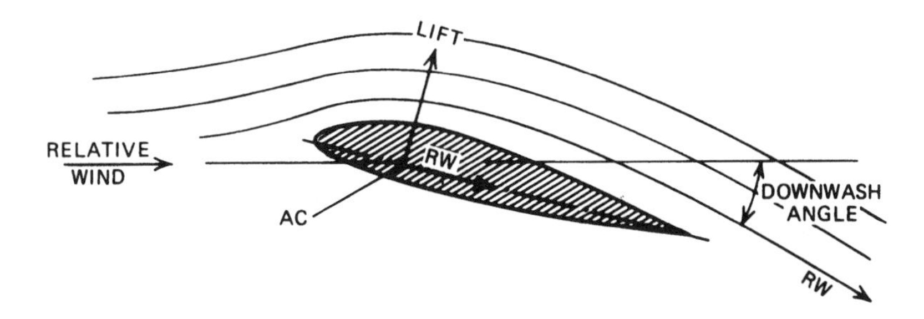

Fig. 5.8. Airflow about a finite wing.

The local RW at the AC is the average of the approaching and departing relative winds.The downwash angle of the local RW at the AC is, therefore, one-half of the trailing-edge downwash angle. The lift vector at the AC, which is perpendicular to the local RW, is tilted back by the same angle. The angle that the lift vector is tilted back is called the *induced angle of attack*, α_i.

The finite wing is operating at a lower AOA than the infinite wing due to the depressed RW at the AC. Thus, if the finite wing is to produce the same lift coefficient as the infinite wing, the AOA, with respect to the free stream RW, must be increased by α_i. The AOA of the infinite wing is α_0; the AOA of the finite wing is α:

$$\alpha = \alpha_0 + \alpha_i \tag{5.4}$$

This is shown vectorially in Fig. 5.9. The vertical vector, effective lift, is the lift vector of the infinite wing, and the tilted vector, total lift, is the lift vector of

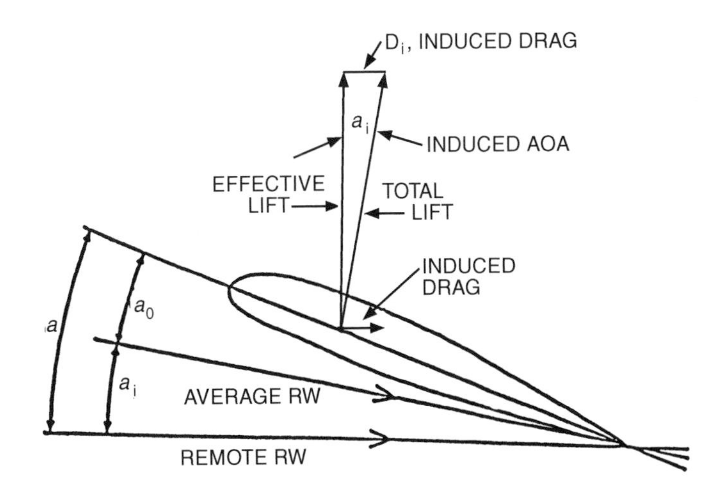

Fig. 5.9. Relative wind and force vectors on a finite wing.

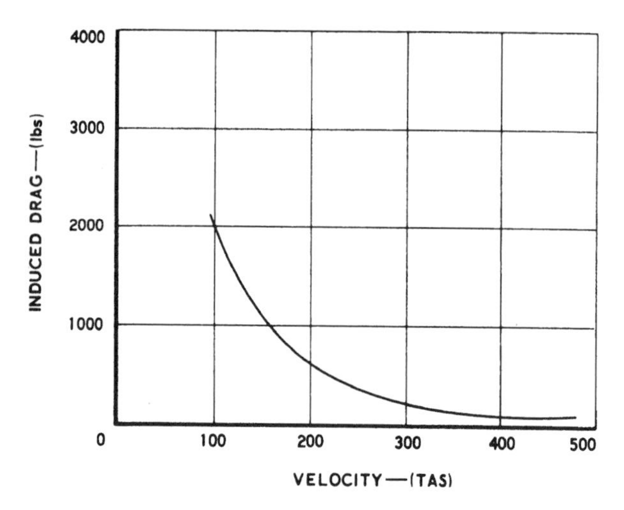

Fig. 5.10. Induced drag versus velocity.

the finite wing. The component of the tilted lift vector that is parallel to the remote RW is the induced drag D_i. Induced drag is influenced by lift coefficient and aspect ratio. It increases directly as the square of C_L and inversely as the aspect ratio.

The most critical condition, where induced drag is greatest, is during low-speed (high-C_L) flight with a low aspect ratio (short-wingspan) aircraft. Equation 5.5 shows that induced drag varies inversely with the velocity squared (Fig. 5.10):

$$\frac{D_{i_2}}{D_{i_1}} = \left(\frac{V_1}{V_2}\right)^2 \tag{5.5}$$

Induced Drag Summary

1. Wingtip vortices are formed by higher pressure air beneath a wing moving into the lower pressure air above the wing.
2. Wingtip vortices cause the airflow behind the wing to be pushed downward; this is called downwash.
3. Downwash causes the RW behind the wing to be deflected downward at a downwash angle.
4. The RW at the AC is influenced by the downwash, and it is deflected downward by one-half of the downwash angle.
5. The lift vector is tilted backward by the induced angle of attack, α_i, which is numerically equal to one-half of the downwash angle.
6. The rearward component of the tilted lift vector is induced drag.

LAMINAR FLOW AIRFOILS

In the late 1930s a radically new design of the airfoil was discovered, the *laminar flow airfoil.* In Fig. 5.11 the C_D–C_L curves are compared for two symmetrical airfoils, an NACA (National Advisory Committee for Aeronautics) "4-digit series" and an NACA " 6-series" (laminar flow). The most obvious difference in these curves is the "drag bucket" that exists for the laminar flow airfoil. As long as this airfoil operates at low C_L (low AOA), the airflow remains laminar much farther back from the leading edge than on the conventional airfoil. Laminar flow has much less skin friction drag than turbulent flow, and thus less C_D. At higher C_L values, the advantage of the laminar airfoil disappears. The sharper leading edge of the laminar airfoil causes early separation and a resulting increase in drag coefficient.

The laminar flow airfoil was used on the B-24 Liberator bombers in World War II. It was not an overwhelming success for several reasons:

1. The airplanes were operating from muddy dirt airfields. Mud thrown on the leading edges of the wings by the prop wash of other planes disrupted the laminar flow.
2. Shrapnel damage had a similar effect.
3. Often enemy gunfire caused the loss of one or two engines. This required that the airplane fly at low airspeeds and high angles of attack. Thus, flight in the "drag bucket" was impossible.

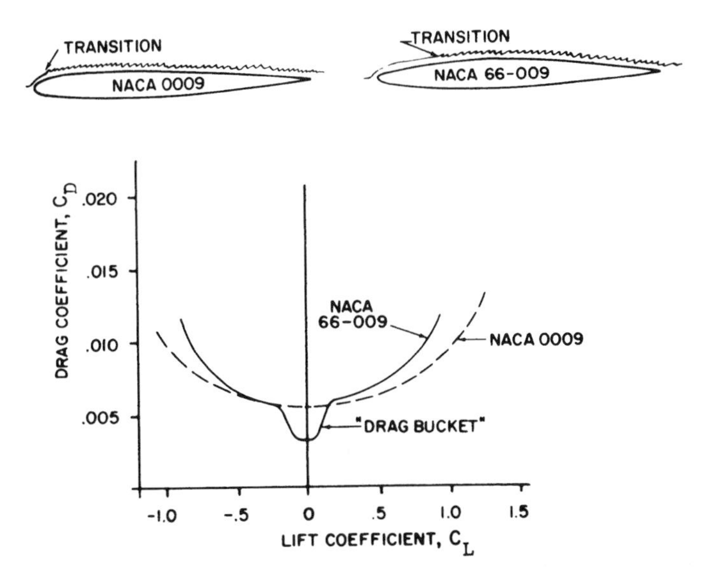

Fig. 5.11. Comparison of drag characteristics of conventional and laminar flow airfoils.

GROUND EFFECT

Three effects can occur when an airplane is flown close to the runway or to a water surface. Collectively they are called *ground effect* and they are noticeable when the aircraft is about one wingspan or less above the surface. The closer the aircraft is to the surface, the greater the effects will be.

The first and most noticeable effect is caused by the surface physically reducing the downwash, which reduces both the upwash and the wingtip vortices. Without so much downwash, the local relative wind at the AC is closer to the remote RW. The lift vector is closer to the infinite wing lift and the induced drag is reduced. The reduction in induced drag coefficient as a function of height, *h*, above the runway to the wingspan, *b*.

Low-winged aircraft are affected more by ground effect than high-winged aircraft. This reduction of drag is clearly seen on light aircraft in their tendency to "float" during landings. To reduce this floating tendency, the aircraft approach should be made with some throttle on, instead of at idle, then the throttle should be closed as the aircraft is flared. The reduction of drag is compensated for by a reduction in thrust, and there is little or no floating action.

Ground effect changes the C_L vs. AOA curves and the thrust-required curves (Fig. 5.12). The reduction of thrust required results in a reduction in fuel flow. This effect can be used in emergency flying over water or flat ground if the fuel available is critical. A word of caution to pilots of turbine engine aircraft: Fuel consumption is high for these aircraft at low altitude, so the benefits of ground effect may be nullified by poor fuel consumption.

The other effects from flying in ground effect are less well known but influence many aircraft. The second effect concerns pitching moments that develop when entering or departing from ground effect. A nose-up pitching moment is required to rotate an aircraft for takeoff or to flare the aircraft during landing. This requires a download on the tail. To obtain this pitching moment the pilot applies back stick so that the horizontal tail is effectively operating at a negative AOA. The downwash from the wing accentuates this negative AOA.

When the aircraft enters ground effect, on landing, the downwash is decreased, the negative AOA is reduced, the downward lift on the tail is reduced, and the aircraft experiences a nose-down pitching moment. High-tailed aircraft will not experience this pitching moment if the tail is above the wing downwash and is not influenced by it. During takeoff, as the aircraft leaves ground effect, the opposite pitching moment occurs. Here, the increased downwash hits the horizontal tail and causes a download on it and a nose-up pitching moment results. Unless corrective action is taken, this will cause a sharp increase in drag that, when coupled with the increase in induced drag that results when an aircraft leaves ground effect, may prevent a successful takeoff.

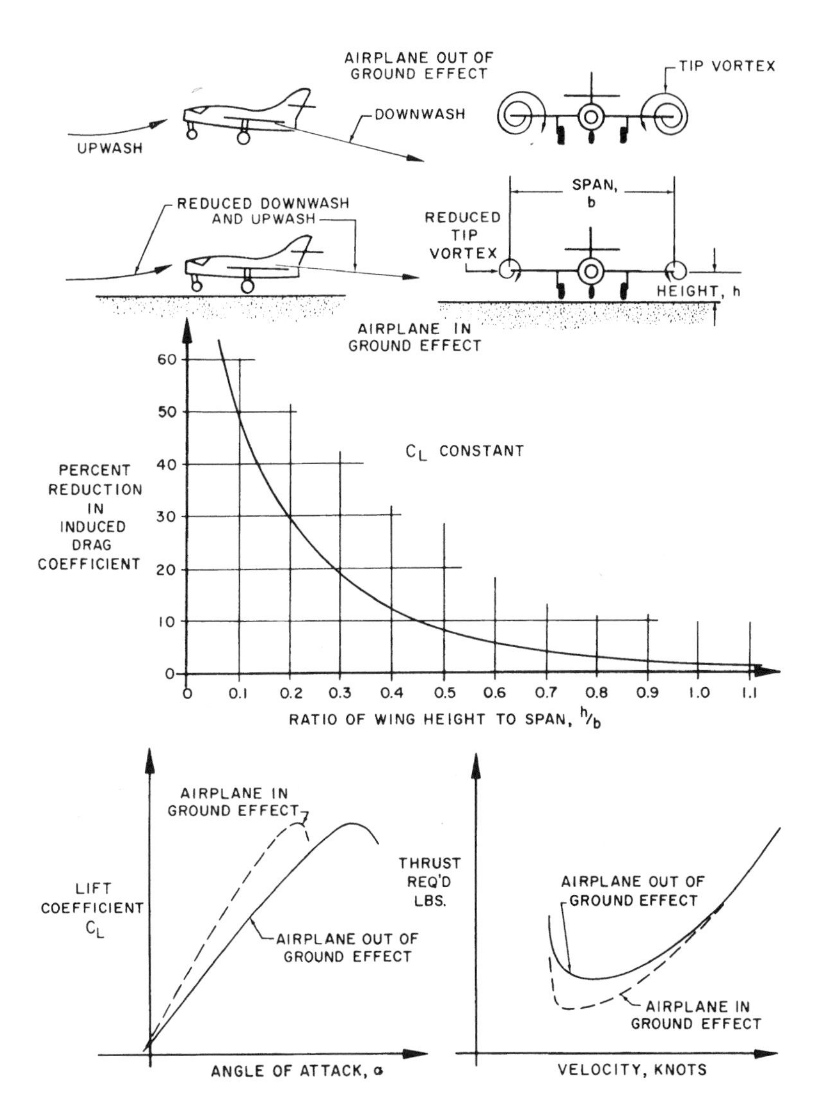

Fig. 5.12. Ground effect.

The third effect of flying in ground effect occurs to aircraft that have static airspeed ports located below the wings. There is a small, but measurable, increase in static pressure below the wings as an aircraft enters ground effect. This creates a false reading from the airspeed indicator. It will read lower than the actual airspeed on entering ground effect. This goes unnoticed during takeoffs and landings, but on long flights over water, pilots trying to take advantage of ground effect may be upset to see the airspeed decrease as the aircraft approaches the water.

Ground Effect Summary

On entering ground effect;

1. Induced drag is decreased.
2. Nose-down pitching moments occur.
3. The airspeed indicator reads low.

On leaving ground effect;

1. Induced drag is increased.
2. Nose-up pitching moments occur.
3. The airspeed indicator will read higher (correctly).

PARASITE DRAG

Parasite drag, D_p, is easily understood, but difficult to measure. There are several types of parasite drag and the total parasite drag is not a simple addition of each component. Five types of parasite drag are discussed here:

1. *Skin friction drag* is caused by the viscous friction within the boundary layer. The total area of the aircraft skin that is exposed to the airstream will be affected by this type of drag. Skin smoothness also greatly affects this drag. Flush-head rivets, waxed and polished surfaces, and removal of aluminum oxide help reduce skin friction drag.
2. *Form drag* is influenced by the shape or form of the aircraft. Streamlining the fuselage, engine nacelles, pods, and external stores helps reduce form drag.
3. *Interference drag* is caused by the interference of boundary layers from different parts of the airplane. If the drag of two component parts of an airplane is measured individually and then the parts are assembled, the drag of the assembly will be greater than the drag of the parts. The boundary layer interference is the reason for this. Smooth fairings at surface junctions reduce this type of drag.
4. *Leakage drag* is caused by differential pressure inside and outside the aircraft. Air flowing from a higher pressure inside the fuselage through a crack or door seal will create an airstream that impinges on the airflow around the aircraft and creates drag. Door and window sills are sealed with masking tape before an air race to lessen this drag.
5. *Profile drag* is of particular interest to helicopter pilots. It is the drag of the moving rotors, and it develops anytime the rotors are in motion. This drag can exist even if the aircraft is not in motion or developing lift.

Parasite Drag Calculations

Parasite drag is hard to measure, but there is a way of simplifying this measurement. Consider the basic drag formula (5.1) with subscript p added to indicate parasite.

$$D_{\mathrm{p}} = C_{D_{\mathrm{p}}} q S$$

At any AOA the value of $C_{D_{\mathrm{p}}}$ will be a constant and, of course, S is a constant. A new constant, called the *equivalent parasite area*, f, is now introduced.

$$f = C_{D_{\mathrm{p}}} S$$

Because $C_{D_{\mathrm{p}}}$, is dimensionless, the dimensions of f are the same as S (ft^2). The equivalent parasite area is often called the "barn door" area of the aircraft. This is not technically correct, but it may help you visualize the area as that seen by an observer of an approaching aircraft.

The value of f for an aircraft can be determined by placing a model in a wind tunnel and adjusting the AOA until zero lift is developed, then measuring the drag. With no lift, there is no induced drag, so all the drag will be parasite. If the density ratio and tunnel velocity are known, then the equivalent parasite area can be found by

$$f = \frac{D_{\mathrm{p}}}{q}$$

For low angles of attack, the value of f can be considered a constant, so

$$D_{\mathrm{p}} = f q = \frac{f \sigma V^2}{295}$$

Parasite drag varies directly as the velocity squared:

$$\frac{D_{\mathrm{p}2}}{D_{\mathrm{p}1}} = \left(\frac{V_2}{V_1}\right)^2 \tag{5.6}$$

Figure 5.13 shows a plot of parasite drag variation with airspeed.

TOTAL DRAG

Total drag, D_{T}, is the sum of the induced drag and the parasite drag:

$$D_{\mathrm{T}} = D_{\mathrm{i}} + D_{\mathrm{p}}$$

As we mentioned earlier, the drag is the component of the aerodynamic force

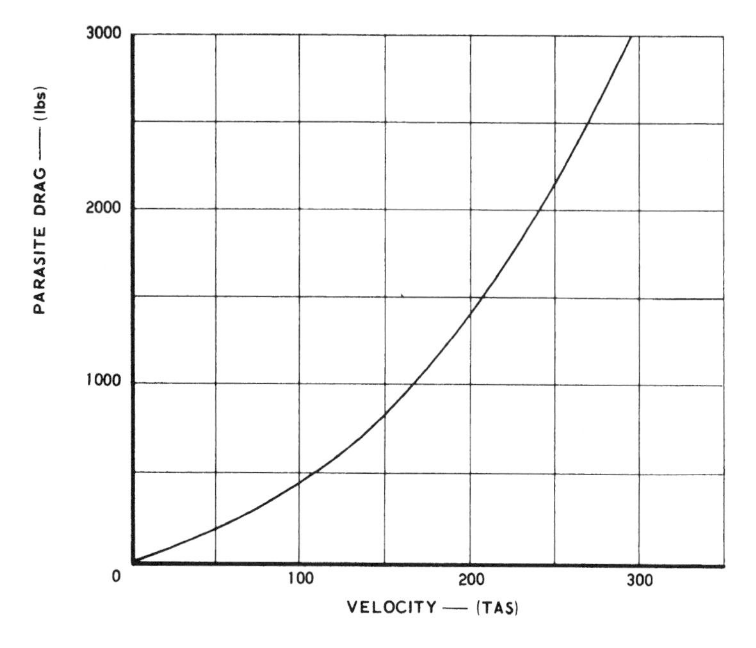

Fig. 5.13. Parasite drag–airspeed curve.

that is parallel to the free stream RW. It is made up of both the induced drag vector and the parasite drag vector (Fig. 5.14).

The total drag curve for an airplane with the induced drag curve in Fig. 5.11 and the parasite drag curve in Fig. 5.13 is shown in Fig. 5.15. Note that the minimum drag or $(L/D)_{\max}$ point occurs where the induced drag and parasite drag curves intersect. Therefore, at this velocity, $D_i = D_p = \frac{1}{2}D_T$. An equation for calculating total drag can be derived by combining Eqs. 5.5 and 5.6.

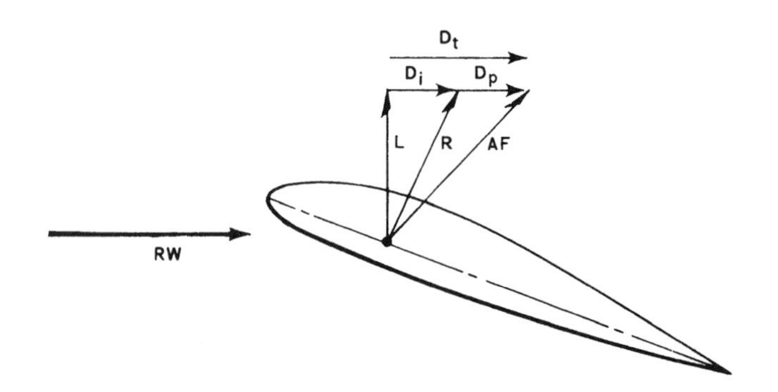

Fig. 5.14. Drag vector diagram.

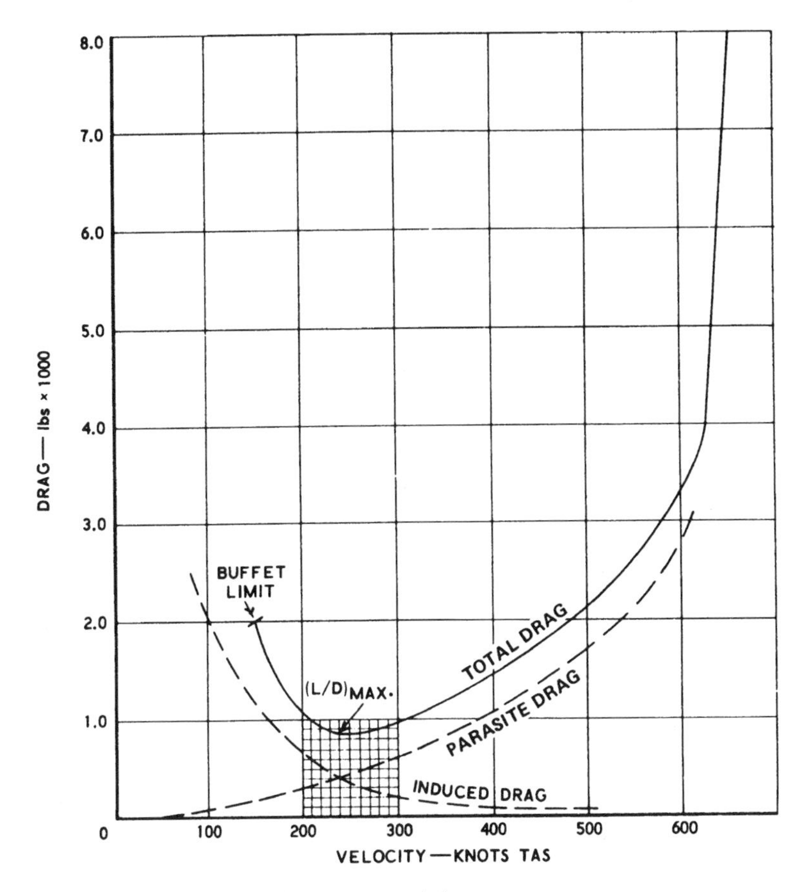

Fig. 5.15. Total drag curve.

Let condition 1 be at $(L/D)_{\mathrm{max}}$ and condition 2 be any other point on the curve. Also remember that $D_{\mathrm{i}_1} = D_{\mathrm{p}_1} = \frac{1}{2}D_{\mathrm{min}}$.

$$D_{\mathrm{T}} = \frac{1}{2} D_{\mathrm{min}} \left[\left(\frac{V_1}{V_2} \right)^2 + \left(\frac{V_2}{V_1} \right)^2 \right] \tag{5.7}$$

The limitations of this curve must be emphasized. In the low-speed region, the high AOA required results in more "equivalent parasite area" than was calculated. The value of f is not constant. The value of parasite and total drag will be greater than the equation shows. In the high-speed region, faster airplanes may encounter compressibility effects and resultant "wave drag." This was not considered in Eq. 5.7. The total drag curves are not accurate at airspeeds below about $1.3V_{\mathrm{s}}$ or above about 0.7 Mach.

DRAG REDUCTION

Most drag on modern turbojet airliners and other large jet aircraft is caused by skin friction drag and induced drag. These forms of drag are roughly 75% of the total drag and reducing them is a major factor in decreasing fuel consumption. In our earlier discussion of induced drag we saw that wingtip vortices cause much of the downwash and resultant backward tilting of the lift vector. Reducing the wing vortices has been a problem for some time. Earlier attempts to do this are shown in Fig. 5.16. More modern devices are called winglets.

Winglets

Winglets were developed to solve the problem of the parasite drag caused by other types of vortex suppression. Without any such device the airflow around the tip is nearly circular. It flows outward toward the tip from beneath the wing, goes upward around the wingtip, and then back inward on the top surface. Although this rotation is reduced by placing the winglet in the path of the airflow, the winglets are usually placed some distance back from the leading edge. This is shown in the side view of Fig. 5.17.

The remaining circular flow creates an inward velocity, V_i, which can be vectorially added to the aircraft's relative velocity, V_{ac}, to obtain the relative wind, RW, shown in the top view of Fig. 5.17. The aerodynamic force, drag, and lift vectors are also shown. The lift vector is tilted forward and by dropping a line from the arrow of the lift vector perpendicular to the winglet's chord line, we get a vector, $-D$. This is larger and opposite to the drag vector, showing that the winglet actually is not causing drag on the airplane.

Reduction of Skin Friction Drag

The major way that skin friction drag can be reduced is by maintaining the laminar airflow over the skin surfaces. Skin smoothness is one obvious method. Laminar flow airfoils is another method of delaying boundary layer transition

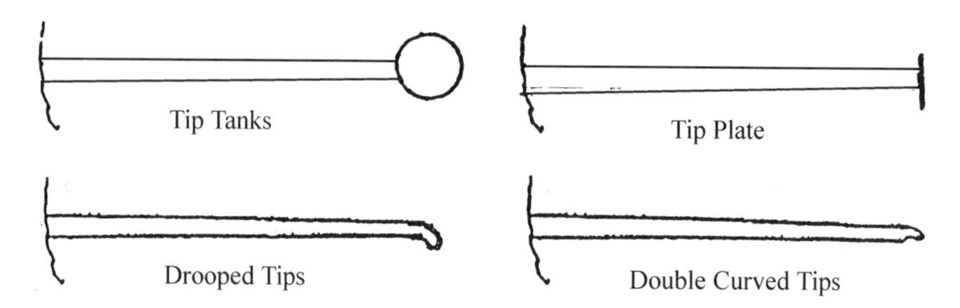

Fig. 5.16. Wingtip vortex reduction methods.

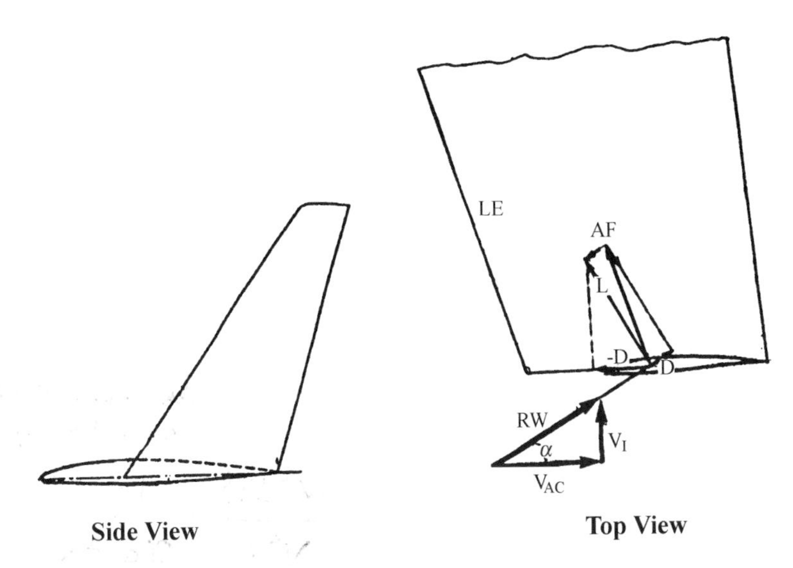

Fig. 5.17. Winglets.

from laminar to turbulent flow. A third method is the use of *laminar flow control*, LFC.

LFC requires removal of the thickening boundary layer by suction. Thin slots or holes are cut in the upper wing surfaces and suction is applied by mechanical means to remove the lower layers of the boundary layer and laminar flow is maintained. Full-scale tests have been run and full wing laminar flow was achieved, but the results were not completely successful. The cost and complexity of cutting the tiny slots, controlling the suction amounts, the need for double skin surfaces, the added weight and complexity of the suction engines and ducting, and other problems have prevented commercial development to date.

SYMBOLS

AR	Aspect ratio
b	Wing span (ft)
C	Chord (ft)
C_{av}	Average chord
C_r	Root chord
C_t	Tip chord
D_i	Induced drag (1b)
D_p	Parasite drag
D_T	Total drag

f Equivalent parasite area (ft^2)
$(L/D)_{\text{max}}$ Maximum lift/drag ratio
MAC Mean aerodynamic chord (ft)
$2w$ Downwash velocity (knots)
α_0 AOA of infinite wing
α_i Induced AOA of finite wing
λ (lambda) Taper ratio
Λ (uppercase lambda) Sweep angle (degrees)

EQUATIONS

AR= aspect ratio
e= efficiency

$$\frac{W^2 295}{V^2 S} \cdot \frac{1}{\pi e \, AR}$$

5.1 $D = C_\text{D} q S = \dfrac{C_\text{D} \sigma V^2 S}{295}$

5.2 $\dfrac{L}{D} = \dfrac{C_\text{L}}{C_\text{D}}$

5.3 $D_{\text{min}} = \dfrac{W}{(L/D)_{\text{max}}}$

5.4 $\alpha = \alpha_0 + \alpha_\text{i}$

5.5 $\dfrac{D_{\text{i}_2}}{D_{\text{i}_1}} = \left(\dfrac{V_1}{V_2}\right)^2$

5.6 $\dfrac{D_{\text{p}2}}{D_{\text{p}1}} = \left(\dfrac{V_2}{V_1}\right)^2$

5.7 $D_\text{T} = \dfrac{1}{2} D_{\text{min}} \left[\left(\dfrac{V_1}{V_2}\right)^2 + \left(\dfrac{V_2}{V_1}\right)^2\right]$

PROBLEMS

1. The coefficient of drag can be defined as
- **a.** the ratio of lift forces to drag forces.
- **b.** a measure of the efficiency of the airplane.
- **c.** the ratio of drag forces to lift forces.
- **d.** the ratio of drag pressure to dynamic pressure.

2. Laminar flow airfoils have less drag than conventional airfoils
- **a.** in the high-C_L region.
- **b.** in the high-AOA region.
- **c.** in the high-speed region.
- **d.** in the landing phase of flight.

3. Laminar flow airfoils have less drag than conventional airfoils because
   a. the adverse pressure gradient starts farther back on the airfoil.
   b. the airfoil is thinner.
   c. more of the airflow is laminar.
   d. Both (a) and (c)

4. Lift/drag ratio is
   a. a measure of the aircraft's efficiency.
   b. a maximum when the drag is a minimum.
   c. numerically equal to the glide ratio.
   d. All of the above

5. An airplane flying at $C_{L(max)}$ will have
   a. more parasite drag than induced drag.
   b. more induced drag than parasite drag.
   c. equal amounts of parasite and induced drag.
   d. Not enough information is given to evaluate the drag.

6. Induced drag is
   a. more important to low aspect ratio airplanes than to high aspect ratio airplanes.
   b. reduced when the airplane enters ground effect.
   c. reduced if the airplane has winglets.
   d. All of the above

7. Induced drag results from the lift vector being tilted to the rear, because
   a. the tip vortices cause downwash behind the wing.
   b. the relative wind behind the wing is pushed downward.
   c. the local relative wind at the AC is depressed.
   d. All of the above

8. An airplane with a heavy load __________ when lightly loaded.
   a. can glide farther than
   b. cannot glide as far as
   c. can glide the same distance as

9. If an airplane has a symmetrical wing that has an angle of incidence of 0° during takeoff, all the drag is
   a. profile drag.
   b. parasite drag.
   c. wave drag.
   d. induced drag.

10. A low-tailed airplane with static ports beneath the wing leaves ground effect after takeoff. It will have:

a. increased drag, nose-up pitch, lowered IAS.

b. increased drag, nose-up pitch, higher IAS.

c. decreased drag, nose-down pitch, lowered IAS.

d. decreased drag, nose-up pitch, higher IAS.

11. A 10,000-lb aircraft has a $(L/D)_{max}$ value of 9.8; find its minimum drag.

12. If the aircraft in Problem 11 has a wing area $S = 200\,\text{ft}^2$ and the value of $C_L = 0.5$ at $(L/D)_{max}$, find the sea level airspeed where D_{min} occurs.

13. For the aircraft in Problem 11 find D_i and D_p at $(L/D)_{max}$.

14. For the aircraft in Problem 11 complete the following table and plot D_i, D_p, and D_T on a sheet of graph paper.

V_2	$(V_2/V_1)^2$	$D_p = \frac{1}{2}D_{min}(V_2/V_1)^2$	$(V_1/V_2)^2$	$D_1 = \frac{1}{2}D_{min}(V_1/V_2)^2$	$D_T = D_p + D_i$
125					
150					
172					
200					
300					
400					

6 Jet Aircraft Basic Performance

The performance capabilities of an aircraft depend on the relationship of the forces acting on it. The principal forces are lift, weight, thrust, and drag. If these forces are in equilibrium, as shown in Fig. 6.1, the aircraft will maintain steady-velocity, constant-altitude flight. If any of the forces acting on the aircraft change, the performance of the aircraft will also change. To better understand the relationship between these forces and performance, we will study aircraft performance curves in this chapter. Some of the items of performance that can be obtained from these curves are

1. Maximum level flight velocity
2. Maximum climb angle
3. Velocity for maximum climb angle
4. Maximum rate of climb
5. Velocity for maximum rate of climb
6. Velocity for maximum endurance
7. Velocity for maximum range

This chapter and the next one show the construction and use of the performance curves for *thrust-producing aircraft*. Power-producing aircraft (propeller-driven aircraft) are discussed in Chapters 8 and 9.

THRUST-PRODUCING AIRCRAFT

Some aircraft produce thrust directly from the engines. Turbojet, ramjet, and rocket-driven aircraft are examples of thrust producers. Fuel consumption is roughly proportional to the thrust output of these aircraft and, because range and endurance performance are functions of fuel consumption, the thrust required is of prime interest.

Thrust-Required Curve

Each pound of drag requires a pound of thrust to offset it. In Chapter 5 we discussed how the drag curve for an aircraft is drawn. We can now call this same curve a *thrust-required curve.* The curves used to illustrate the drag or thrust-required (T_r) curves in this chapter are for the T-38 supersonic turbojet trainer. Because this aircraft encounters a large amount of compressibility drag

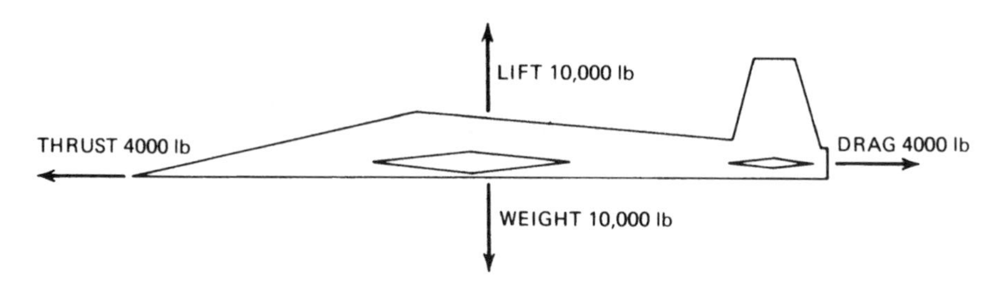

Fig. 6.1. Aircraft in equilibrium flight.

at high speeds, a sharp increase in the thrust required will be noted in this region.

Figure 6.2 shows the drag curve for the T-38. The drag curve shows the amount of drag of the aircraft at various airspeeds. Figure 6.2 shows that the aircraft has a drag of 2000 lb at the buffet limit (stall speed). At about 485 knots

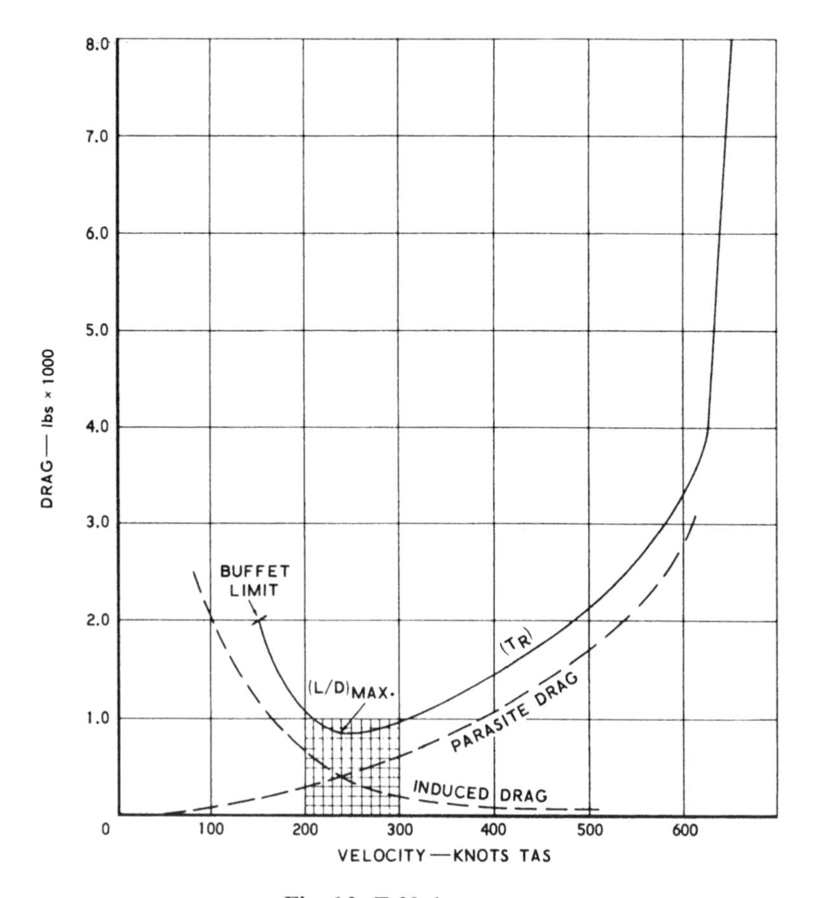

Fig. 6.2. T-38 drag curve.

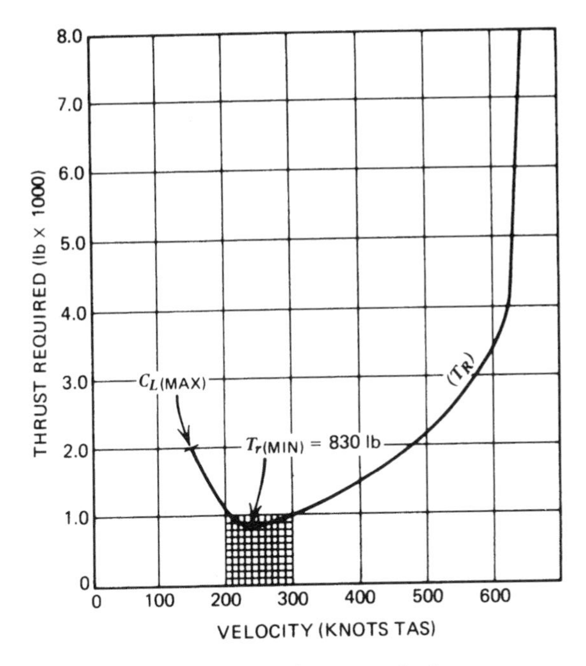

Fig. 6.3. T-38 thrust required.

the drag is also 2000 lb. Minimum drag occurs at $(L/D)_{max}$, which is at 240 knots for this weight aircraft, and has a value of 830 lb. As was mentioned above, the thrust required is equal to the drag at this speed. The curve shown in Fig. 6.3 is identical to the one shown in Fig. 6.2, except the induced and parasite drag curves are omitted and the vertical scale is relabeled as " thrust required." The curves in Figs. 6.2 and 6.3 are drawn for a 10,000-lb airplane in the clean configuration at sea level on a standard day.

PRINCIPLES OF PROPULSION

All aircraft powerplants have certain common principles, based on Newton's second and third laws. You will recall that the second law can be written as $F = ma$. Simply stated, an unbalanced force F acting on a mass m will accelerate a the mass in the direction of the force. The force is provided by the expansion of the burning gases in the engine. The mass is the mass of the air passing through the jet engine or through the propeller/rotor system. The acceleration is the change in the velocity of the intake air per unit of time.

Newton's third law states that for every action force, there is an equal and opposite reaction force. This reaction force provides the *thrust*, T, to propel the aircraft:

$$F = T = Q(V_2 - V_1) \tag{6.1}$$

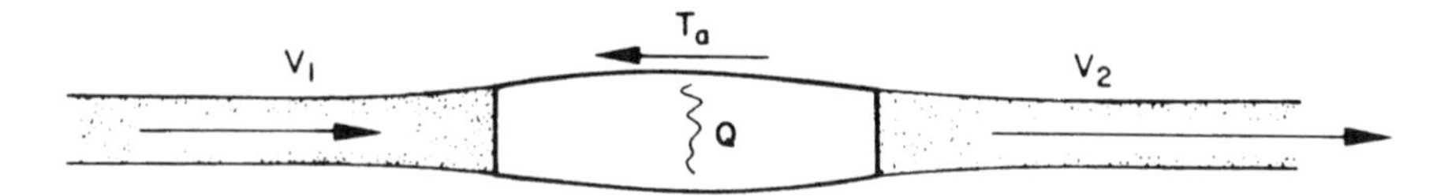

Fig. 6.4. Engine thrust schematic.

where

T = thrust (lb)

Q = mass airflow = p AV (Eq. 2.10) (slugs/sec)

V_1 = inlet (flight) velocity (fps)

V_2 = exit velocity (fps)

Figure 6.4 shows a schematic of the process of producing thrust in a jet-type engine. T_a represents thrust available.

Equation 6.1 means that the thrust output of an engine can be increased by either (1) increasing the mass airflow or (2) increasing the exit velocity/flight velocity (V_2/V_1) ratio. The first alternative requires that the cross-sectional area of the engine be increased to allow more air to be processed. The second alternative requires increasing the relative velocity, $V_2 - V_1$. This imparts higher kinetic energy to the gases. The kinetic energy is wasted in the exhaust airstream, resulting in decreased efficiency of the engine.

Propulsion efficiency, η_p, can be expressed by

$$\eta_p = \frac{2V_1}{V_2 + V_1} \tag{6.2}$$

As V_2 increases, propulsion efficiency goes down. This is shown in Fig. 6.5.

Early turbojet engines were used on jet fighters. To keep the drag of the engines within reason, the engines could not have a large frontal area. This limited the amount of air that could be processed, and high exhaust velocities were required to obtain the needed thrust. As larger aircraft were developed, it was possible to use larger engines. Fanjet engines and large Q values improved the efficiency of jet engines.

THRUST-AVAILABLE TURBOJET AIRCRAFT

Turbojet aircraft engines are rated in terms of static thrust. The aircraft is restrained from moving, and the thrust is measured and converted to standard sea level conditions. It can be seen from the thrust equation 6.1, $T_a = Q(V_2 - V_1) = p\,AV_1(V_2 - V_1)$, that as the aircraft gains velocity, the mass flow increases, but the acceleration through the engine decreases (V_2 is nearly constant).

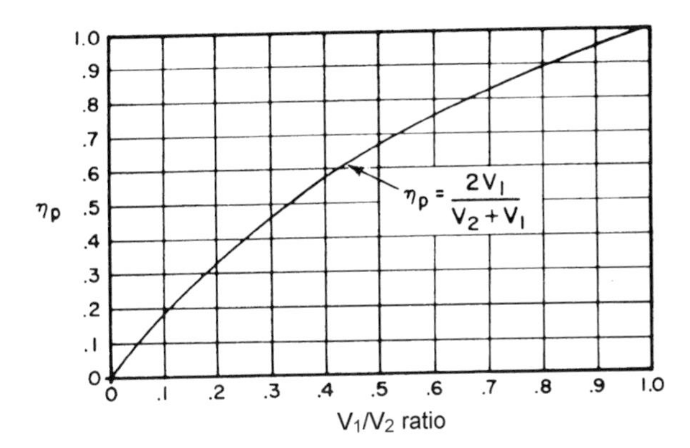

Fig. 6.5. Propulsion efficiency.

The thrust available is nearly a constant with airspeed and is considered to be a constant in this discussion. Thrust varies, however, if the rpm of the engine is changed from the 100% value. The reduction of thrust is not linear with the reduction of rpm (Fig. 6.6). Thrust is also reduced with an increase in altitude. Equation 6.1 indicates that the mass flow decreases as the density of the air decreases. Temperature is also involved. Lower temperatures at altitude improve efficiency, so the loss in thrust is not as great as the decrease in density (Fig. 6.7).

SPECIFIC FUEL CONSUMPTION

Specific fuel consumption c_t is the fuel flow per pound of thrust developed by a

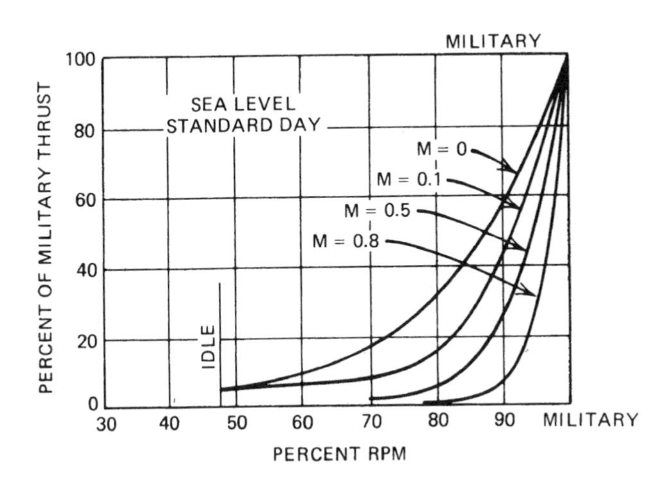

Fig. 6.6. T-38 installed thrust.

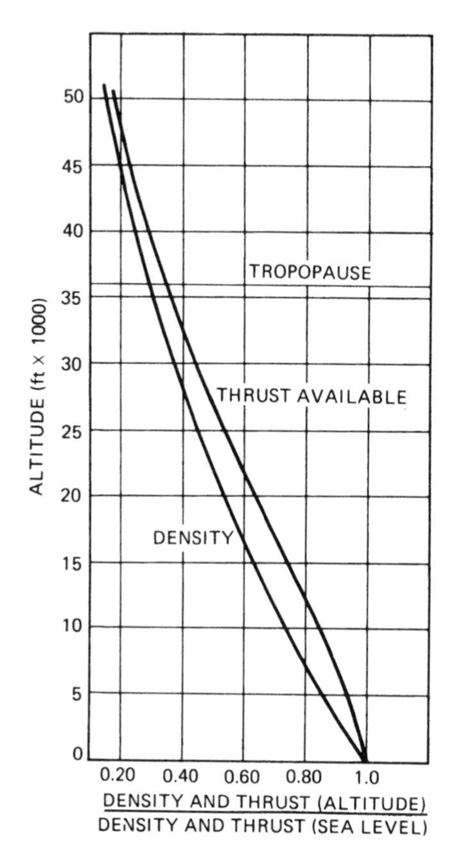

Fig. 6.7. T-38 thrust variation with altitude.

jet engine:

$$c_t = \frac{\text{Fuel flow (lb/hr)}}{\text{Thrust (lb)}} \tag{6.3}$$

Specific fuel consumption is an indication of the efficiency of an engine. Low values of c_t , are desirable.

Turbojet engines are designed to operate at high rpm and will produce low specific fuel consumption values in this region. Figure 6.8 shows the variation of c_t with rpm. The lowest values of c_t occur between 95 and 100% rpm, and the increase in c_t is very rapid as rpm is decreased.

The beneficial effect of lower temperature at altitude results in decreased specific fuel consumption (Fig. 6.9). Above the tropopause the temperature is constant, so no reduction in c_t will occur. In fact, the compressor will not operate efficiently because the air density is reduced, causing an increase in c_t.

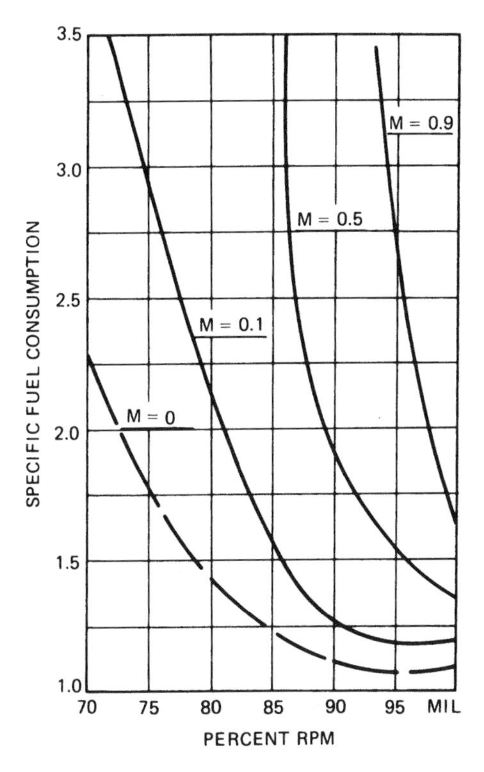

Fig. 6.8. T-38 c_t–rpm.

FUEL FLOW

The total fuel flow, FF, equals the thrust of the engine, T, multiplied by the specific fuel consumption, c_t:

$$\mathrm{FF} = Tc_t \tag{6.4}$$

Thrust varies with rpm and altitude (Figs. 6.6 and 6.7). Specific fuel consumption varies with these same factors (Figs. 6.8 and 6.9). Fuel flow at altitude as a ratio of that at sea level, at a fixed rpm, is shown in Fig. 6.10.

THRUST-AVAILABLE–THRUST-REQUIRED CURVES

Performance depends on the relationship of the thrust-available and thrust-required curves. If the thrust available is equal to the thrust required, the aircraft can fly straight and level, but cannot climb or accelerate (without losing altitude). An excess of T_a over T_r is required for these maneuvers. The thrust-available and thrust-required curves are plotted for the T-38 at 10,000 lb

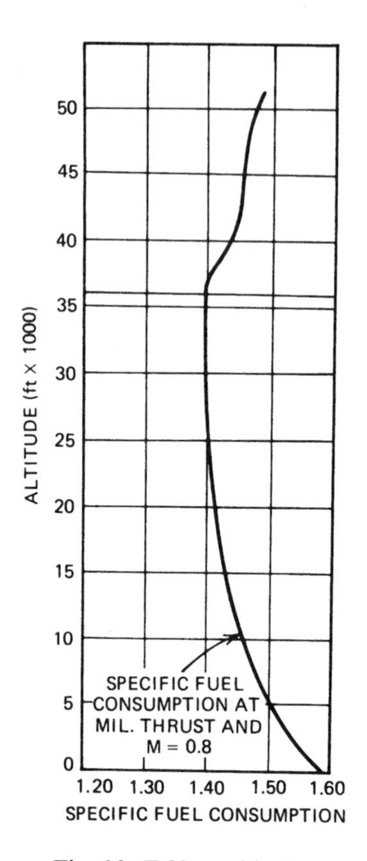

Fig. 6.9. T-38 c_t–altitude.

gross weight, clean configuration at sea level, in Fig. 6.11. The thrust available is plotted for two different conditions (100% rpm and 95% rpm) at sea level. The reduction of T_a with rpm was explained earlier (Fig. 6.6).

These curves show that a 5% reduction in rpm results in a reduction of T_a from 4100 to 2100 lb. This aircraft could not maintain altitude at any speed if the thrust is reduced below 830 lb. This is a ratio of 830/4100 = 0.20 of the T_a at 100% rpm, sea level conditions. Figure 6.6 shows that this occurs at about 88% rpm at sea level and Mach (M) = 0.5. Absolute ceiling occurs when T_a falls to 830 lb. Figure 6.7 shows that this occurs at about 47,500 ft.

ITEMS OF AIRCRAFT PERFORMANCE

Straight and Level Flight

When an aircraft is in steady (unaccelerated) level flight, it must be in equilibrium. Lift must be exactly equal to the weight of the aircraft, and the

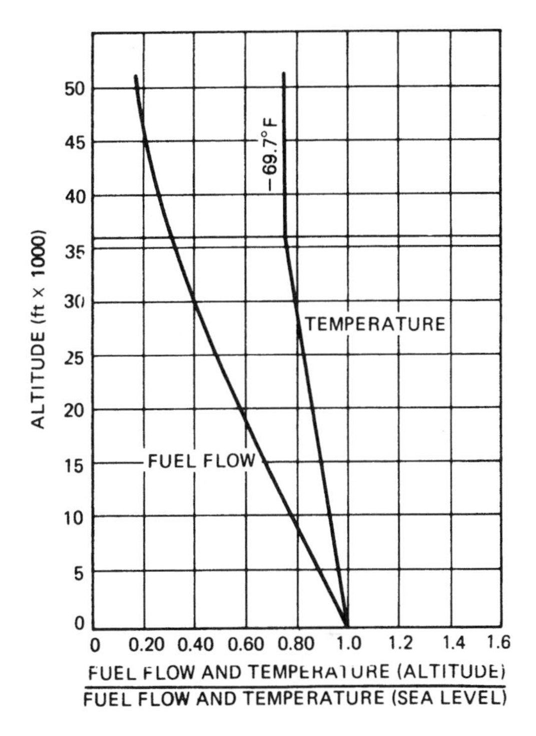

Fig. 6.10. T-38 fuel flow–altitude.

thrust on the aircraft must be exactly equal to the drag. The thrust is equal to the drag when the thrust available is equal to the thrust required. This happens when the T_a and T_r curves intersect. There are two possibilities for such an intersection: One is at the high-speed region of flight and the other is at the low-speed area. The problems in the low-speed area are discussed in later chapters. The conditions for flight at maximum level velocity, V_{max}, are discussed here.

For nonafterburner flight the thrust available for the T-38 is 4100 lb. From Fig. 6.11 it can be seen that the high-speed intersection point of T_a and T_r is at 605 knots true airspeed. With lesser thrust settings the V_{max} will be less. For 95% rpm the value of V_{max} is 500 knots TAS.

Climb Performance

Every flight requires that the pilot climb the aircraft to some altitude. Mission requirements dictate the climb schedule. For instance, an intercept mission would require that altitude be gained as rapidly as possible, while a mission of long range would require that much distance be covered in the climb.

There are two basic types of climb: delayed climb and steady-velocity climb. In a delayed climb the pilot builds up airspeed (kinetic energy) before starting

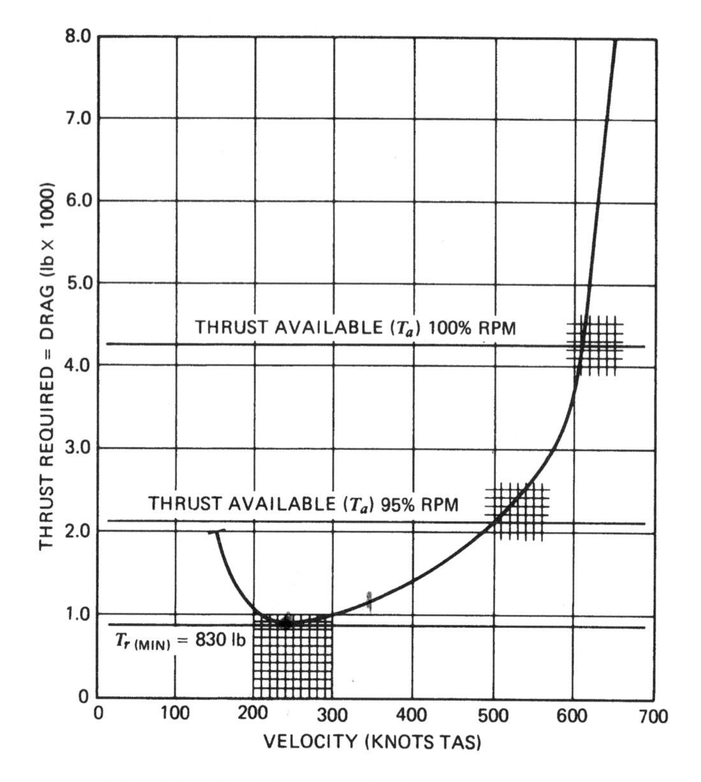

Fig. 6.11. T-38 thrust available–thrust required.

to climb. The pilot then "zooms" the aircraft to altitude, which converts some of the kinetic energy to potential energy (height). This maneuver is used mostly by fighter aircraft pilots in intercept tactics, setting climb to altitude records, and other maneuvers.

Steady-velocity climb, is used much more often. This type of climb is an equilibrium condition, with all the forces along the flight path being balanced (Fig. 6.12). The forces acting along the flight path are thrust, T, and drag, D. Lift acts 90° to the flight path and the weight, W, acts toward the center of the earth. The climb angle is designated by the Greek letter gamma, γ. The weight vector is replaced by two other vectors. One is perpendicular to the flight path ($W \cos y$) and the other is parallel to the flight path ($W \sin \gamma$). Note that W sin y actually acts at the aircraft's CG, not as shown in the figure. For steady velocity the forces along the flight path must be balanced. Thus,

$$T - D - W \sin \gamma = 0 \tag{6.5a}$$

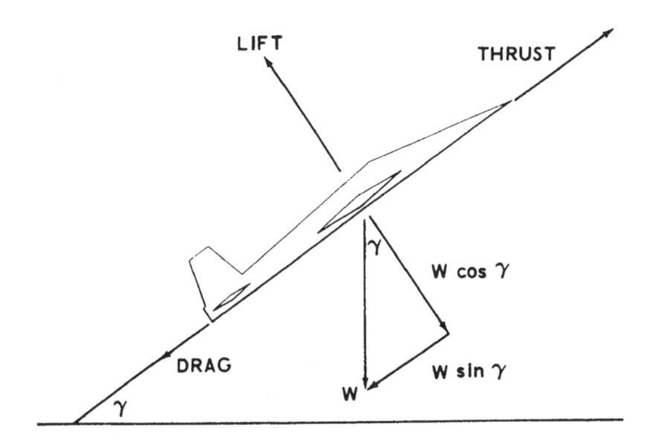

Fig. 6.12. Forces acting on a climbing aircraft.

Angle of Climb

For the angle of climb, Eq. 6.5a is rewritten as

$$\sin \gamma = \frac{T - D}{W} = \frac{T_a - T_r}{W} \tag{6.5b}$$

As an angle increases in size, the sine of the angle increases in value. The maximum climb angle will occur at $(T_a - T_r)_{max}$ where maximum excess thrust exists. For a turbojet, with the thrust available assumed to be constant with velocity, this will be where T_r (drag) is a minimum. This, of course, is at $(L/D)_{max}$ (Fig. 6.13).

When an aircraft is in steady flight, horizontally, the lift is equal to the weight of the aircraft. Figure 6.12 shows that the lift is less than the weight if the aircraft is in a climb. This is possible because some of the thrust supports some of the weight of the aircraft. The steeper the climb is, the less the lift supports the weight. The extreme case occurs when the aircraft climbs straight

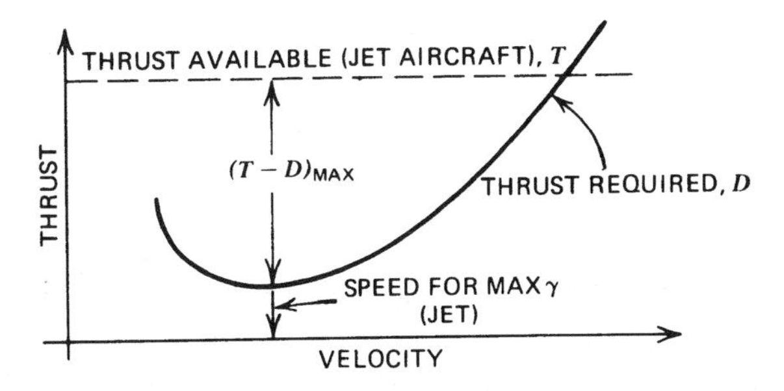

Fig. 6.13. Velocity for maximum climb angle.

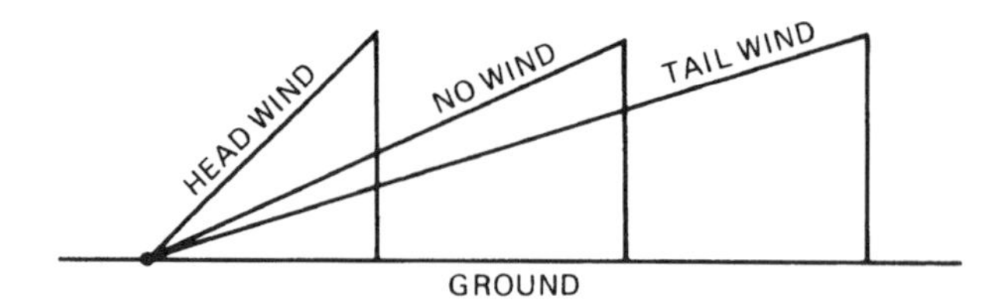

Fig. 6.14. Wind effect on climb angle to the ground.

up ($\gamma = 90^\circ$). Then the lift would be zero and the thrust would support the entire weight of the aircraft and overcome the drag. A thrust/weight ratio of at least 1 is required for this maneuver. Obstacle clearance is affected by headwind and tailwind as shown in Fig. 6.14. Once the airplane is airborne, the groundspeed is reduced by a headwind and increased by a tailwind. With reduced groundspeed, the time to reach the obstacle is increased, and more altitude is gained.

One other factor in obstacle clearance is the time required to reach the $(L/D)_{max}$ velocity. After the jet takes off, it must accelerate to this velocity and, in doing so, precious ground distance is used up. Even though the jet is finally climbing at its highest climb angle, it may not clear the obstacle. The aircraft may clear the obstacle if the climb is started sooner, at a lower speed, and at a lesser climb angle (Fig. 6.15).

The aircraft on the left in Fig. 6.16 is flying at maximum climb angle. The airspeed of this aircraft is relatively slow compared to the aircraft on the right. The aircraft on the right is flying at a smaller climb angle, but the much higher airspeed more than compensates for the lower angle, and it is at a higher altitude. The aircraft on the right has a higher rate of climb, ROC. This is shown in Fig. 6.17.

We can solve the right triangle in Fig. 6.17 for the rate of climb:

$$\text{ROC} = V_k \sin\gamma \qquad \text{(knots)}$$

$$= V_k \frac{T_a - T_r}{W} \qquad \text{(knots)}$$

$$= 101.3 V_k \frac{T_a - T_r}{W} \qquad \text{(fpm)} \tag{6.6}$$

The velocity for maximum rate of climb cannot be determined by a simple examination of the $T_a - T_r$ curves. Equation 6.7 shows that ROC depends on velocity *and* excess thrust, both of which are variables. However, by selecting various airspeeds and finding the corresponding values of T_a and T_r from Fig. 6.11, then solving Eq. 6.7, a plot of ROC versus velocity can be made similar to Fig. 6.18 and the velocity for maximum ROC can be found.

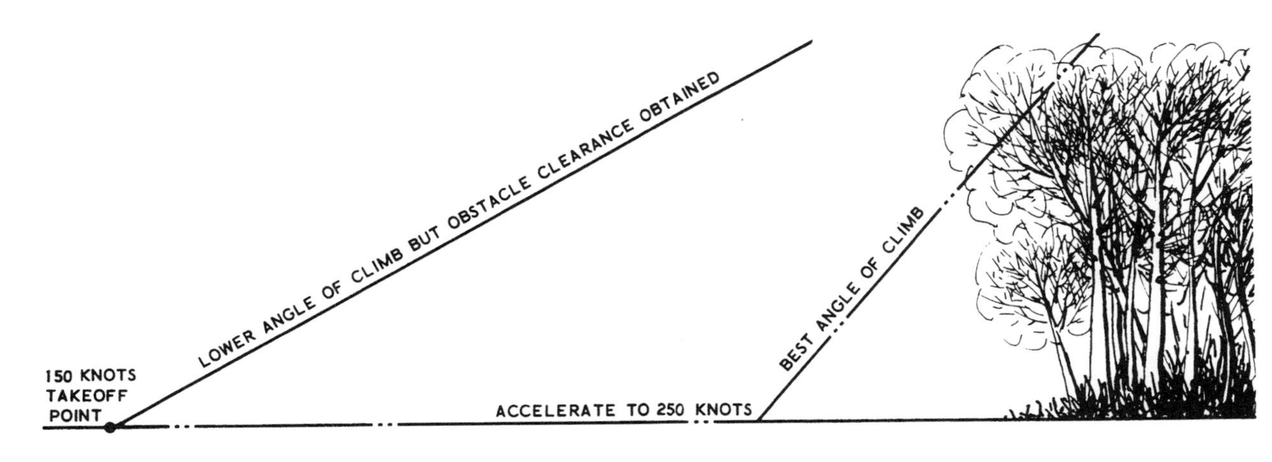

Fig. 6.15. Obstacle clearance for jet takeoff.

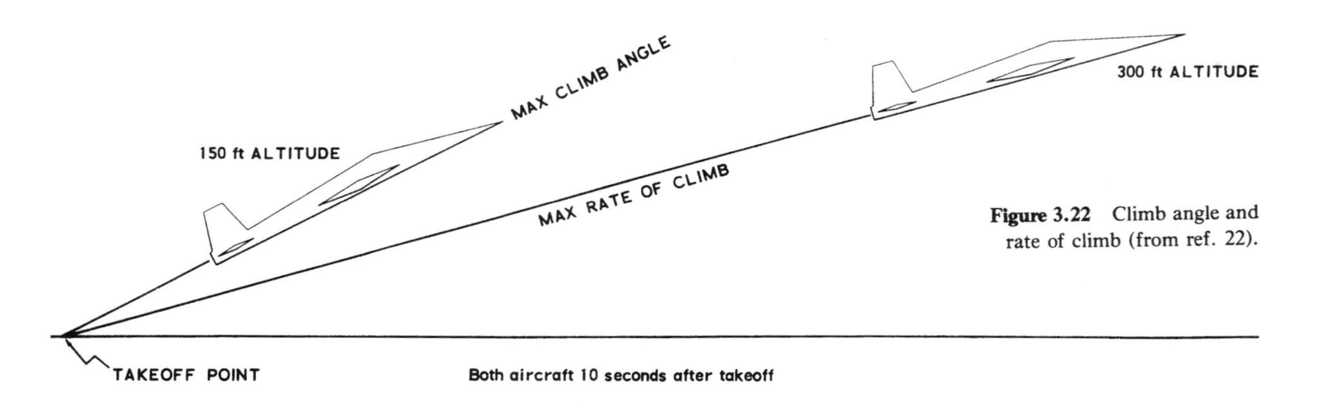

Fig. 6.16. Climb angle and rate of climb.

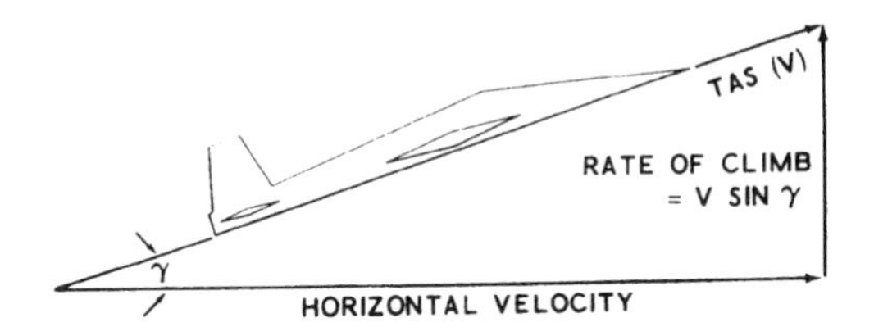

Fig. 6.17. Rate of climb velocity vector.

Endurance

The endurance of an aircraft is the amount of time that it can remain airborne. Maximum endurance is independent of distance covered. It is useful when holding due to weather, air traffic, etc. Endurance is inversely proportional to fuel flow. Long endurance results from low fuel flow.

Figure 6.19 shows the thrust-required curve for the T-38 with the fuel flow also plotted on the vertical scale. The minimum fuel flow occurs at the minimum thrust required. This, again, is $(L/D)_{max}$. Maximum endurance for a jet aircraft also occurs at $(L/D)_{max}$. For jet aircraft, we have seen that the following items of performance occur at $(L/D)_{max}$:

1. Minimum drag
2. Maximum engine-out glide range
3. Maximum climb angle
4. Maximum endurance

Obviously it is important to know the airspeed or AOA for $(L/D)_{max}$. Airspeed will vary with aircraft weight, but AOA will not. Indicated airspeed for $(L/D)_{max}$ will not vary with altitude.

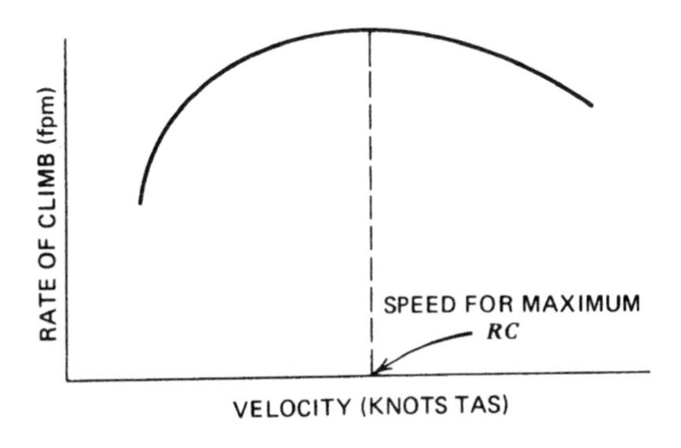

Fig. 6.18. Velocity for maximum rate of climb.

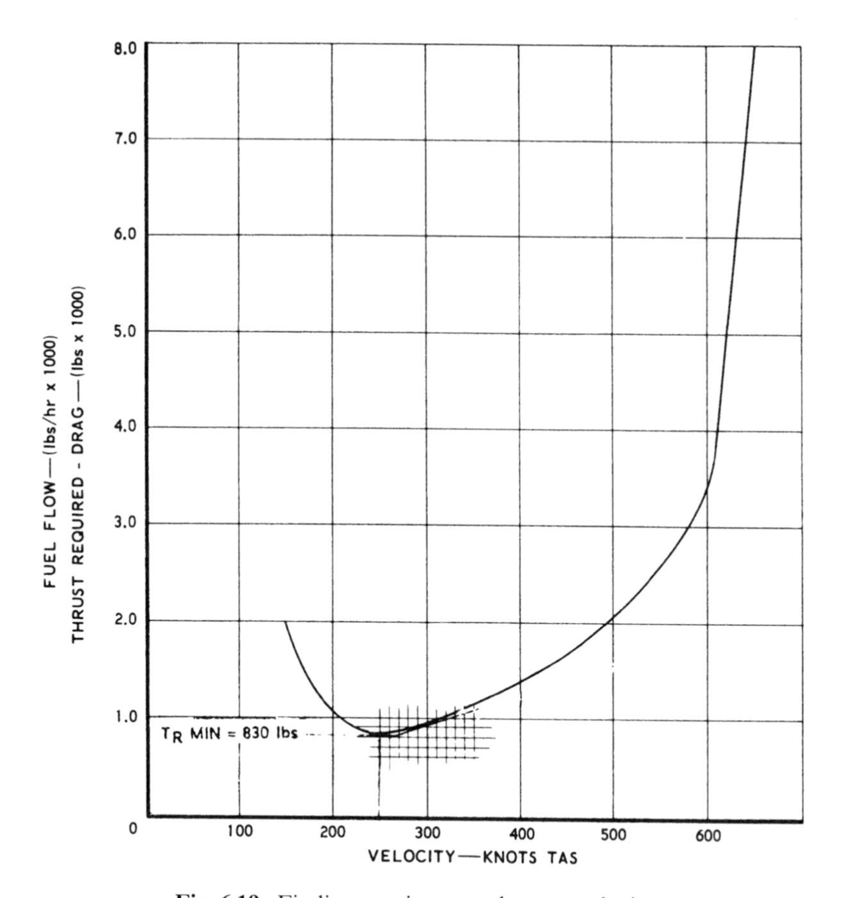

Fig. 6.19. Finding maximum endurance velocity.

Specific Range

To get maximum range, an aircraft must fly the maximum distance with the fuel available; thus, the specific range, SR, must be a maximum. Specific range can be defined by the following relationship:

$$\text{Specific range} = \frac{\text{nmi}}{\text{lb fuel}}$$

$$= \frac{\text{nmi/hr}}{\text{lb fuel/hr}}$$

$$= \frac{\text{Speed (knots)}}{\text{Fuel flow (lb/hr)}}$$

which means that

$$(\text{SR})_{\text{max}} \text{ occurs at } \left(\frac{\text{FF}}{V}\right)_{\text{min}}$$

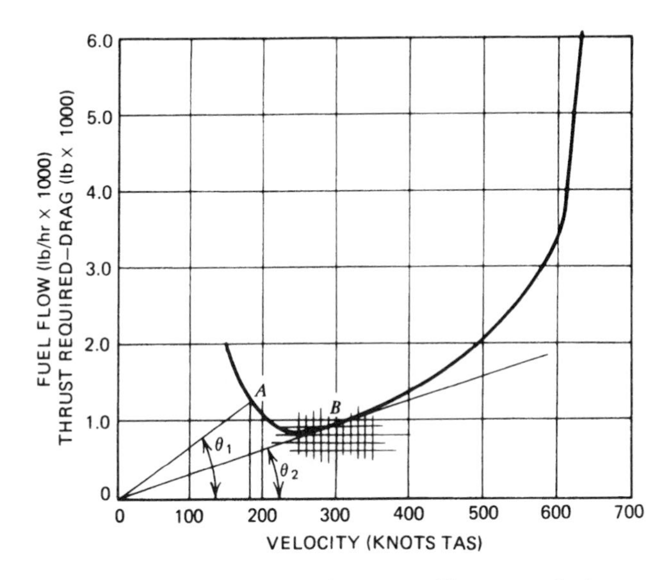

Fig. 6.20. Finding the maximum specific range velocity.

Graphically, the maximum specific range point can be shown on a fuel flow versus velocity curve, such as that in Fig. 6.20. Select any point on the T_r curve, such as A. Draw a straight line from A to the origin and drop a vertical line from A to the velocity scale. Label the angle at the origin of the resulting right triangle, angle θ_1. The tangent of θ_1 equals the opposite side of the triangle divided by the adjacent side. The opposite side is a measure of the fuel flow, and the adjacent side of the triangle is the aircraft's velocity:

$$\tan \theta_1 = \frac{\text{FF}}{V}$$

Hence, if FF/V is a minimum, $\tan \theta$ must be a minimum, which means that θ must be a minimum. Only one point on the curve meets this requirement—the point where the line from the origin is tangent to the T_r curve at point B. θ_2 is a minimum at B, and maximum specific range will occur if the aircraft is flown at this velocity. Maximum specific range occurs at a velocity corresponding to the tangent point of a straight line drawn from the origin to the curve. This is not $(L/D)_{\max}$, but is the point where $\sqrt{C_L/C_D}$ is a maximum.

Wind Effect on Specific Range

An aircraft flying into a headwind is in an unfavorable environment and thus should fly at higher true airspeed to reduce the effect of the headwind. Likewise, an aircraft flying with a tailwind is being helped on its way and thus should slow down to take advantage of the wind.

How much the airspeed should be altered is shown in Fig. 6.21. This illustration shows the correct airspeed for maximum range with a 100-knot

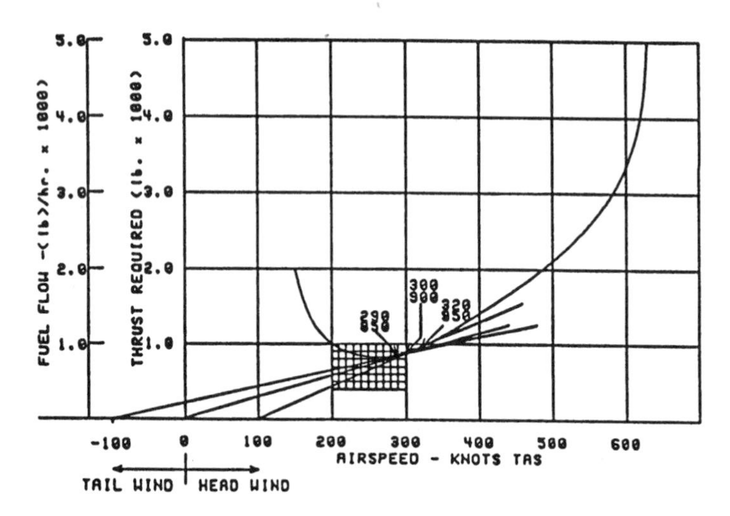

Fig. 6.21. Wind effect on specific range.

headwind or tailwind. To find the corrected airspeed for a 100-knot headwind, plot the headwind on the velocity scale to the right of the origin. Use this point as a new origin and draw a tangent line to the curve. This will give the new airspeed for best range. In the example the no-wind airspeed for maximum range was 300 knots and the corrected airspeed is 320 knots.

Correcting for a tailwind is done similarly. The tailwind is plotted on the velocity scale, but to the left of the origin. A new tangent line will show the correct airspeed of 290 knots.

Total Range

The total range of an aircraft depends on the fuel available and the specific range. Specific range is a variable. As the weight of an aircraft decreases with fuel consumption, the specific range improves. This variation is shown in Fig. 6.22. The total range is measured by the area under the curve, between the beginning and ending cruise weights. A way of approximating the total range is to find the average specific range and multiply it by the fuel burned.

SYMBOLS

c_t Specific fuel consumption (lb fuel/hr per lb thrust)

FF Fuel flow (lb/hr)

Q Mass airflow (slugs/sec)

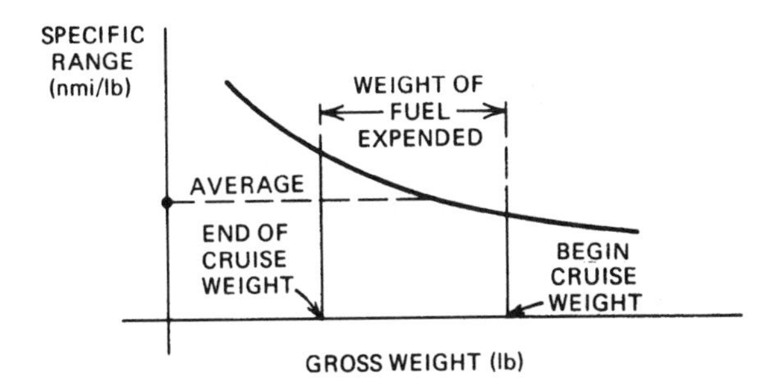

Fig. 6.22. Total range calculation.

ROC	Rate of climb (fpm)
SR	Specific range (nmi/lb fuel)
T_a	Thrust available (lb)
T_r	Thrust required
V_1	Intake (flight) velocity (fps)
V_2	Exit velocity
γ (gamma)	Climb angle (degrees)
η_p (eta)	Propulsion efficiency (%/100)

EQUATIONS

6.1 $$T = Q(V_2 - V_1)$$

6.2 $$\eta_p = \frac{2V_1}{V_2 + V_1}$$

6.3 $$c_t = \frac{\text{Fuel flow (lb/hr)}}{\text{Thrust (lb)}}$$

6.4 $$\text{FF} = Tc_t$$

6.5a $$T - D - W \sin \gamma = 0$$

6.5b $$\sin \gamma = \frac{T - D}{W} = \frac{T_a - T_r}{W}$$

6.6 $$\text{ROC} = 101.3V_k \frac{T_a - T_r}{W} \quad \text{(fpm)}$$

PROBLEMS

1. The equation $T_a = Q(V_2 - V_1)$ shows that
   a. you can get more thrust from a jet engine by using water injection.
   b. you can get more thrust from a jet engine by decreasing the exhaust pipe area.
   c. Neither (a) nor (b)
   d. Both (a) and (b)
2. Q in $T_a = Q(V_2 - V_1)$ is the mass flow (slugs/sec) The mass flow depends on
   a. the cross-sectional area of the turbojet engine inlet.
   b. the engine inlet velocity.
   c. the density of the air at the same point.
   d. All of the above
3. The equation $\eta_p = 2V_1/(V_2 + V_1)$ shows that decreasing the exhaust pipe area will
   a. decrease the propulsive efficiency of a jet engine.
   b. increase the propulsive efficiency.
   c. increase the thrust of the engine.
   d. decrease the thrust of the engine.
4. Specific fuel consumption of a turbine engine at 35,000 ft altitude compared to that at sea level is
   a. less.
   b. more
   c. the same.
5. The thrust available from a jet engine at 35,000 ft altitude compared to that at sea level is
   a. less.
   b. more.
   c. the same.
6. Fuel flow for a jet at 100% rpm at altitude compared to that at sea level is
   a. less.
   b. more.
   c. the same.
7. A pilot is flying an airplane at the speed for best range under no wind conditions. A tailwind is encountered. To get best range now the pilot must
   a. speed up by an amount less than the wind speed.
   b. slow down by an amount equal to the wind speed.
   c. slow down by an amount more than the wind speed.
   d. slow down by an amount less than the wind speed.

8. To obtain maximum range a jet airplane must be flown at

a. a speed less than that for $(L/D)_{max}$.

b. a speed equal to that for $(L/D)_{max}$.

c. a speed greater than that for $(L/D)_{max}$.

9. Maximum rate of climb for a jet airplane occurs at

a. a speed less than that for $(L/D)_{max}$.

b. a speed equal to that for $(L/D)_{max}$.

c. a speed greater than that for $(L/D)_{max}$.

10. Maximum climb angle for a jet aircraft occurs at

a. a speed less than that for $(L/D)_{max}$.

b. a speed equal to that for $(L/D)_{max}$.

c. a speed greater than that for $(L/D)_{max}$.

11. A single engine turbojet aircraft is flying at 296 knots TAS at sea level. The mass flow rate through the engine is 10 slugs/sec. The exit velocity from the engine is 800 fps. Find

a. The thrust of the engine

b. The propulsive efficiency

12. A jet airplane has thrust available as shown in Fig. 6.11 and specific fuel consumption as shown in Fig. 6.8. Find the fuel flow at sea level (military rpm and $M = 0.70$).

13. Using Fig. 6.10 find the fuel flow for the airplane in Problem 12 at the tropopause.

14. A jet airplane has the $T_a - T_r$ curves shown in Fig. 6.11. The airplane data are gross weight = 10,000 lb, standard sea level day, and clean configuration. Find

a. V_{max} at 95% rpm

b. V for maximum climb angle

c. Sine of maximum climb angle

d. Rate of climb at 400 knots

e. V for best endurance

f. V for maximum range

g. V for maximum range with 100 knots of headwind

Boeing 717-200 (Courtesy of the Boeing Company).

7 Jet Aircraft Applied Performance

The aircraft performance items that we discussed in Chapter 6 were for one weight, clean configuration, and sea level standard day conditions. In this chapter we alter these conditions and examine the resulting performance. To simplify the explanation, only one condition will be altered at a time.

VARIATIONS IN THE THRUST-REQUIRED CURVE

The basic thrust-required curve for the T-38 (Fig. 6.2) was drawn for a 10,000-lb airplane in the clean configuration, at sea level on a standard day. We now show how the curve changes for variations in weight, configuration, and altitude.

Weight Changes

Changing the weight of an aircraft changes the induced drag much more than the parasite drag. The only change in the parasite drag is caused by small changes in equivalent parasite area, f, at varying angles of attack. Only changes in induced drag are considered in Fig. 7.1.

The total drag is also increased by the addition of weight. The intersection of the induced drag and the parasite drag curves still marks the point of $(L/D)_{max}$, but it is moved up and to the right. In fact, all points on the thrust-required curve are moved up and to the right. Because induced drag is greater than parasite drag in the low-speed region of flight, the curve is moved by a greater amount in the low-speed region than in the high-speed region.

The total drag curves are shown as the thrust required curves in Fig. 7.2. Both of the curves were obtained by altering the basic T_r curve for the 10,000-lb T-38 (Fig. 6.2). First the airspeed for each point on the curve was corrected using Eq. 4.2. Second, the drag was corrected by using the fact that the value of L/D is not changed by weight. This was discussed on page 63 where it was stated: "Note that the value of $(L/D)_{max}$ and the angle of attack at which it occurs do not vary with aircraft weight or altitude."

Not only does $(L/D)_{max}$ not vary but other L/D values do not vary either. They are functions of AOA only. So, if L/D does not vary with weight, then

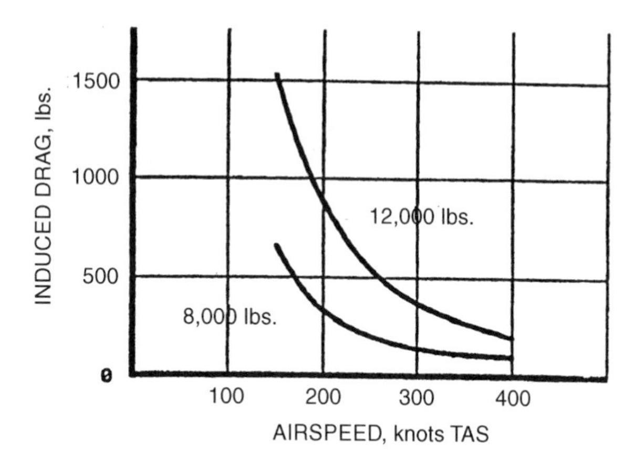

Fig. 7.1. Effect of weight change on induced drag.

changing the weight changes the drag proportionately and

$$\frac{L_2}{D_2} = \frac{L_1}{D_1}$$

Under $1G$ flight $L = W$, so

$$\frac{D_2}{D_1} = \frac{W_2}{W_1} \tag{7.1}$$

Figure 7.2 was drawn by altering Fig. 6.2 for different weights. The airspeeds were changed using Eq. 4.2 and the corresponding drags, T_r, were changed using Eq. 7.1. Later in this chapter we discuss how the pilot must use this curve to obtain maximum performance from the aircraft as the weight changes.

Configuration Changes

When a pilot lowers the landing gear (and/or flaps), the equivalent parasite area is greatly increased. This increases the parasite drag of the aircraft, as shown in Fig. 7.3, but it has little effect on the induced drag. The effect on the thrust-required curve is to move it upward and to the left. The amount that the curve moves is not easily calculated. The equivalent parasite area must be known for the aircraft in the dirty configuration, and the new parasite drag values must be calculated. The T_r curve can then be found by adding the new curve values to the old D_i curve values and plotting.

The thrust-required curves for several configurations of the T-38 are shown in Fig. 7.4. In discussing the effect of weight change we saw that the values of L/D remained constant at the same AOA. The value of $(L/D)_{max}$ does not

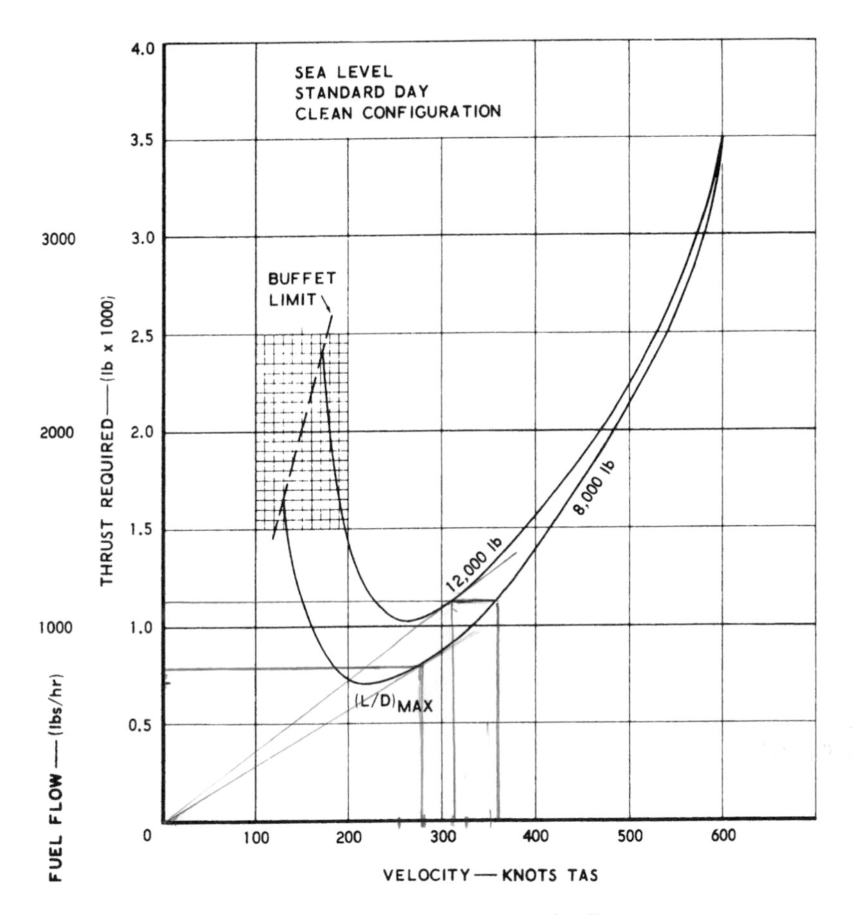

Fig. 7.2. Effect of weight change on the T_r curve.

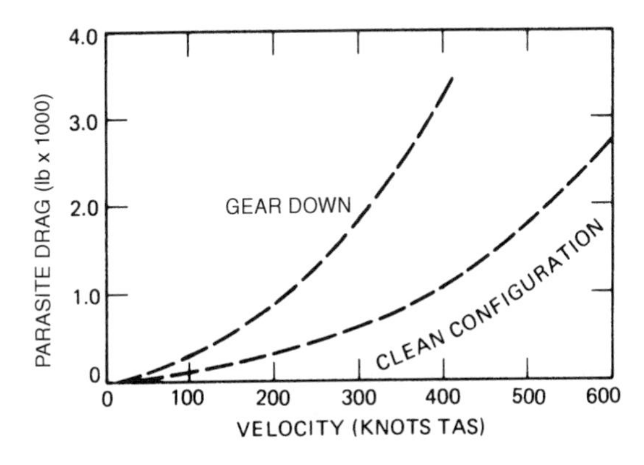

Fig. 7.3. Effect of configuration on parasite drag.

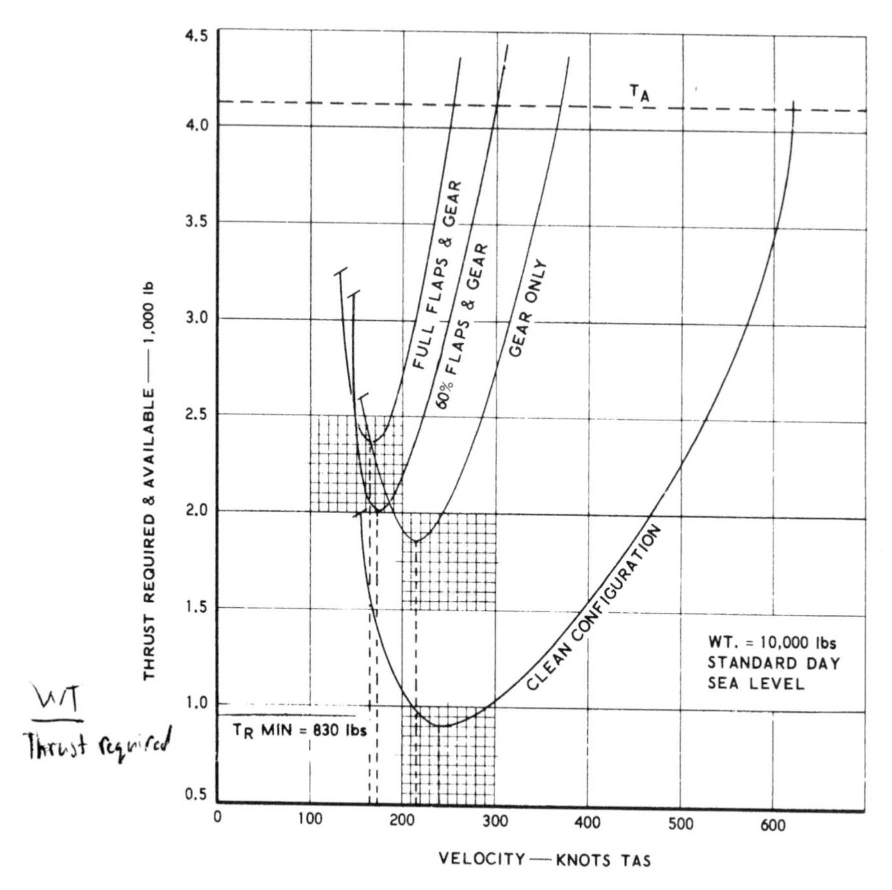

Fig. 7.4. Effect of configuration on the T_r curve.

change as weight varies, but it does change as the configuration changes. For the clean configuration, the value of $(L/D)_{max}$ in Fig. 7.4 is 12.05. In the full flaps and gear configuration it is only 4.26.

Altitude Changes

Drag on an airplane depends on the dynamic pressure:

$$D = C_D q S$$

At any given AOA, the value of the drag coefficient, C_D, is a constant and the wing area, S, is also a constant. Therefore, the drag will also be a constant if the dynamic pressure is a constant. If we plotted drag versus equivalent airspeed, there would be only one drag curve for all altitudes. We could say that "drag at altitude equals drag at sea level," but we would have to qualify

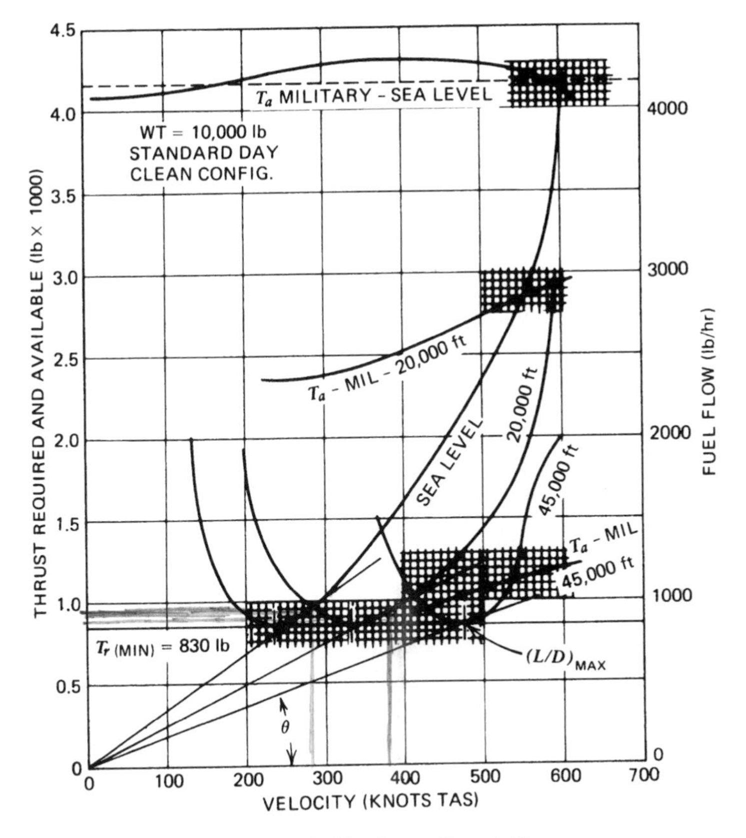

Fig. 7.5. Effect of altitude on T_r and T_a curves.

that statement by adding "at the same equivalent airspeed." Because the curves are drawn for true airspeed, different curves must be drawn for each altitude. If the drag does not change but the true airspeed at which it occurs changes, the only correction we must make is one for airspeed:

$$\frac{V_2}{V_1} = \sqrt{\frac{\sigma_1}{\sigma_2}} \tag{7.2}$$

The curves in Fig. 7.5 show that the minimum thrust required is 830 lb for both altitudes. Of course, this means that the value of $(L/D)_{max}$ is not changed by altitude. The true air speed for minimum drag does change as shown in Eq. 7.2. If V_1 is at sea level, where $\sigma_1 = 1.0$, and Eq. 7.2 is rewritten, then

$$V_2 = \frac{V_1}{\sqrt{\sigma_2}}$$

In addition to the T_r changes shown in Fig. 7.5, we must also change the T_a at 20,000 ft. This is done in accordance with Fig. 6.7.

VARIATIONS OF AIRCRAFT PERFORMANCE

Straight and Level Flight

Weight Change Increase in weight means an increase in drag, but T_a is not changed, so V_{max} is decreased. A weight decrease will reduce the drag and the T_a–T_r intersection (V_{max}) will be at a higher airspeed.

Configuration Change V_{max} is usually limited by structural strength of landing gear and flaps; therefore, V_{max} is reduced by a dirty configuration.

Altitude Change V_{max} true airspeed is only slightly changed by altitude. The T_r curve moves to the right with an increase in altitude and the T_a decreases. V_{max} EAS will be less at altitude as T_a decreases.

Climb Performance

Angle of Climb

Weight Change Increased weight means increased drag, while T_a is not changed. The $(T - D)/W$ term in Eq 6.3 is reduced and a lower angle of climb results. When weight is reduced the opposite occurs and climb angle is increased.

Configuration Change Drag increases when gear and flaps lowered, so a smaller angle of climb results.

Altitude Increase Thrust available decreases with increase in altitude and drag remains the same, so $T - D$ is smaller and angle of climb is smaller at altitude.

Rate of Climb

Weight Change Rate of climb is the velocity times the sine of the climb angle. As we saw above the $T - D$ term reduces with a weight increase and the extra drag also slows the aircraft, so ROC is reduced with increased weight and increased with decreased weight.

Configuration Change A dirty aircraft reduces climb performance. The aircraft should be cleaned up, as soon as practical.

Altitude Increase T_r (drag) does not change with an increase in altitude but T_a is decreased, thus $T - D$ is reduced. The velocity increases with altitude. The product of the two terms, however, decreases. When the T_a reduces to the point where it is tangent to the T_r curve, the absolute ceiling is reached and the ROC is zero.

Endurance

Weight Change More drag is developed with increased weight. Fuel flow is proportional to T_r (drag), so endurance is reduced, and vice versa.

Configuration Change Pull up the gear and flaps to improve endurance. If gear or flaps cannot be retracted, slow down to reduce parasite drag.

Altitude Increase Is the same, but fuel flow is less with altitude so endurance is improved. A word of caution: You may burn more fuel climbing to a higher altitude than if you had remained at the lower altitude.

Specific Range

Weight Change As fuel is burned, weight is reduced, and the thrust-required curve moves downward and to the left (Fig. 7.6). The tangent lines indicate

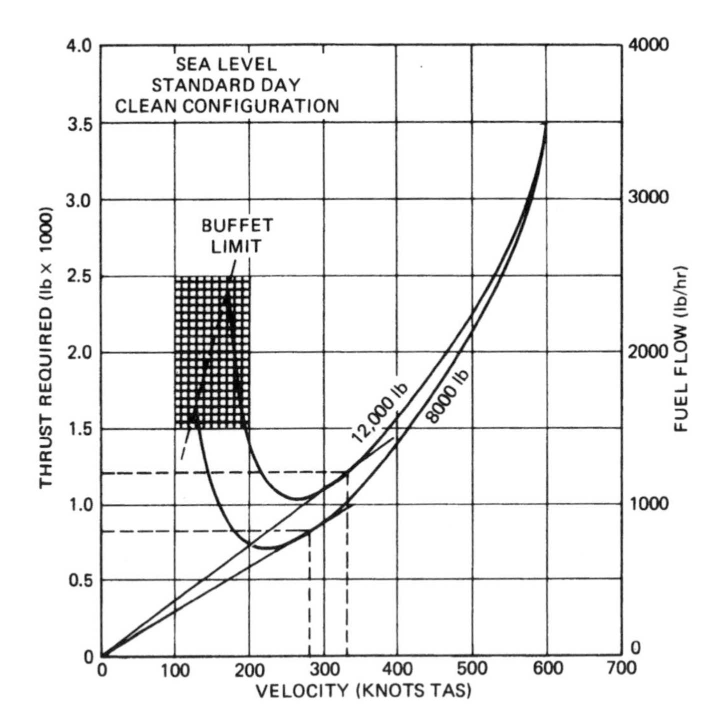

Fig. 7.6. Effect of weight change on specific range.

velocity for SR_{max}. Two important changes in the flight schedule must be made by the pilot as weight decreases, if maximum range is to be obtained: The thrust and the airspeed of the aircraft must be reduced. However, a reduction in throttle will result in a reduction of thrust available, which will automatically lower the airspeed. The natural tendency of a pilot to keep the throttle setting the same as weight is reduced, and to allow the airspeed to increase must be suppressed. For maximum range, as fuel is burned, *reduce airspeed.*

Example From Fig. 7.6, the SR at 12,000 lb is

$$SR_{12} = \frac{V}{FF} = \frac{325}{1200} = 0.2708 \text{ nm/lb}$$

If the pilot did not reduce thrust but allowed airspeed to increase as fuel was burned, the airspeed would be 375 knots when the weight was 8000 lb. Fuel flow would still be 1200 lb/hr, and the SR would be

$$SR_8 = \frac{375}{1200} = 0.3125 \text{ nm/lb}$$

By reducing thrust to 825 lb and airspeed to 285 knots, the resulting SR would be

$$SR_8 = \frac{285}{825} = 0.3455 \text{ nm/lb}$$

Not reducing thrust results in a 15.4% increase in SR, but reducing thrust results in a 27.6% increase.

Configuration Change If it is necessary to fly in the dirty configuration, slow the airplane to the new tangent point to reduce parasite drag.

Altitude Increase

There are several advantages in flying jet aircraft at altitude. Up to the tropopause, the lowering of temperature increases engine efficiency. This was discussed earlier in our study of specific fuel consumption and was illustrated in Fig. 6.9. A second advantage in reducing specific fuel consumption results from operating the engine at high rpm. As thrust is reduced at altitude, higher rpm is required to produce the thrust required and lower c_t results. This is shown in Fig. 6.8.

The third advantage can be seen in Fig. 7.7. The fuel flow remains about the same for the three altitudes shown, but the velocity is substantially higher as altitude increases. The angle θ decreases as altitude increases, thus indicating better SR. An 88% increase in SR is shown at 45,000 ft versus sea level.

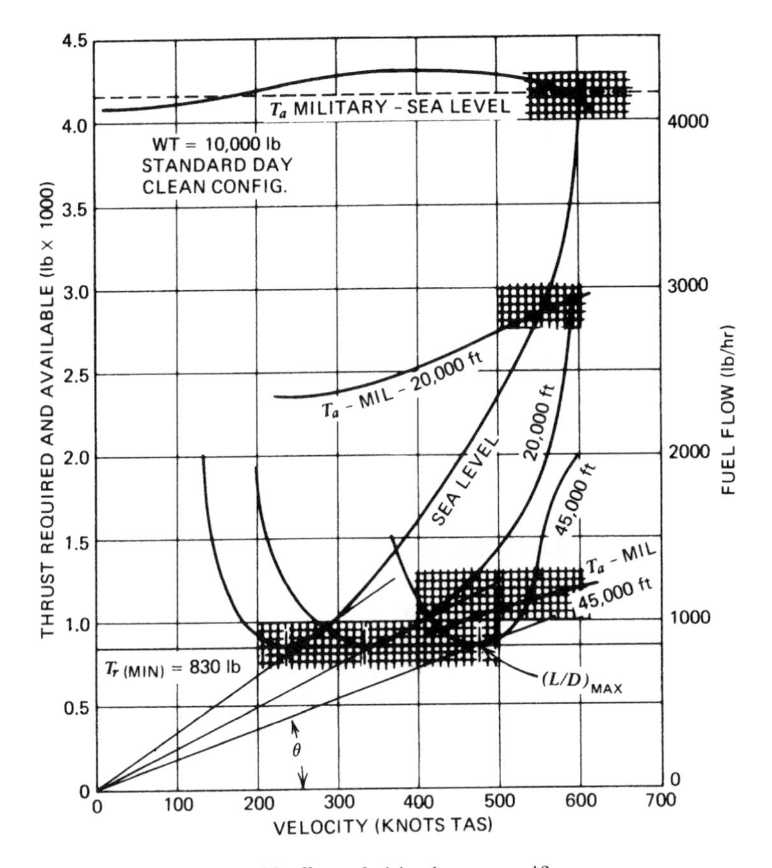

Fig. 7.7. T-38 effect of altitude on specific range.

Cruise–Climb Flight

When air traffic control will allow flight that is not restricted to a set altitude, a further increase in aircraft range (for jet aircraft) can be obtained by using a cruise–climb technique. Cruise flight for turbojet aircraft occurs at or above the tropopause. For maximum range an aircraft is flown at $(\sqrt{C_L}/C_D)_{max}$. With values of C_L, C_D, and TAS held constant, both lift and drag are directly proportional to the density ratio, σ.

Above the tropopause thrust is also proportional to the density ratio. As fuel is consumed, weight is reduced and, if the airplane was allowed to climb, it would remain in equilibrium because lift, drag, and thrust all vary in the same manner. Constant velocity above the tropopause means constant Mach number. Fuel flow decreases with altitude, as was seen in Fig. 6.10, and thus specific range is increased.

The procedure is then to establish proper maximum range airspeed at the tropopause. Maintain constant Mach number and allow the altitude to

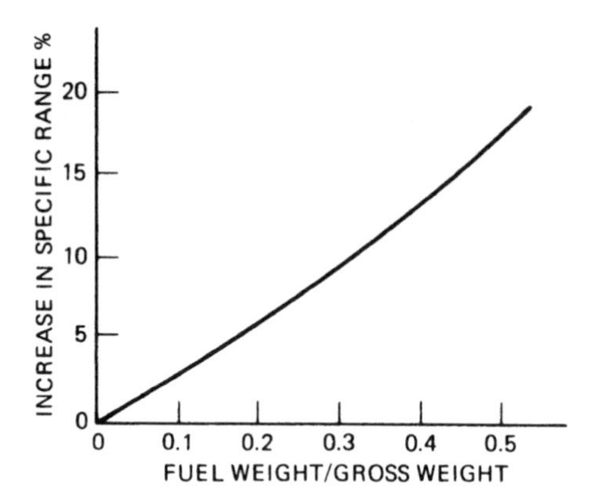

Fig. 7.8. Range improvement using cruise–climb.

increase as weight is reduced. The improvement in range by using cruise–climb flight over constant altitude flight is a function of the weight of fuel used during cruise divided by the aircraft's gross weight at the beginning of cruise (Fig. 7.8).

Note: In our discussion of performance in thrust-producing aircraft the word "power" was never used. Many pilots use the terms "power" and "thrust" interchangeably. This is a gross error. *Thrust-producing aircraft do not produce power.* They do not use *power curves.* Pilots will often make statements such as "add power" or "backside of the power curve." All such statements are correct only for power-producing aircraft (discussed in the next two chapters), not for thrust-producing aircraft.

EQUATIONS

7.1 $\dfrac{D_2}{D_1} = \dfrac{W_2}{W_1}$

7.2 $V_2 = \dfrac{V_1}{\sqrt{\sigma_2}}$ (V_1 is at sea level on standard day)

PROBLEMS

1. If the weight of a jet airplane is increased then

a. parasite drag increases more than induced drag.

b. induced drag decreases more than parasite drag.

c. both parasite and induced drag increase by the same amount.

d. induced drag increases more than parasite drag.

2. If the weight of a jet airplane is reduced as fuel is burned, the T_r curve

a. moves down and to the right.

b. moves up and to the right.

c. moves down and to the left.

d. moves up and to the left.

3. If a jet airplane is in the gear down configuration, the increase in

a. parasite drag is more than that of the induced drag.

b. induced drag is more than that of the parasite drag.

c. both types of drag is the same.

4. If it is impossible to raise the landing gear of a jet airplane, to obtain best range, the airspeed must be ______ from that for the clean configuration.

a. increased

b. decreased

c. not changed

5. From Fig. 7.4, the glide ratio $(L/D)_{max}$ for the airplane in the full flaps and gear down configuration is

a. 12.05.

b. 5.4.

c. 5.0

d. 4.2.

6. The minimum drag for a jet airplane does not vary with altitude.

a. True

b. False

7. Figure 7.5 shows an increase in specific range with altitude because

a. T_r decreases while fuel flow decreases.

b. T_r remains the same while fuel flow decreases.

c. T_r remains the same while T_a decreases.

d. fuel flow remains about the same while airspeed increases.

8. A jet airplane is flying to obtain maximum specific range. As fuel is burned the pilot must

a. reduce throttle but maintain the same airspeed.

b. maintain throttle setting and let the plane accelerate.

c. reduce throttle and airspeed.

d. maintain throttle and reduce airspeed.

9. A lightly loaded airplane will be able to glide farther but at a lower airspeed than when it is heavily loaded.

a. True.

b. False.

10. Which of the following words should never be used in the discussion of jet aircraft?

a. Power

b. Horsepower

c. Power curve

d. All of the above

11. Using Fig. 7.2, find the velocity for best range for the airplane at 12,000 and 8000 lb.

12. Using Fig. 7.2, calculate the specific range for the airplane at both weights if it is flying at the best-range airspeed.

13. Using Fig. 7.2, calculate the specific range of the airplane if the throttle is not retarded from the 12,000-lb best-range position, the airspeed is allowed to increase as fuel is burned, and weight is reduced to 8000 lb. Compare your answer to that for Problem 12 and state your conclusions.

14. Using Fig. 7.5, calculate the specific range for this airplane if it is flying at best-range airspeed at sea level and at 20,000 ft altitude.

8 Propeller Aircraft—Basic Performance

All aircraft in flight must produce thrust to overcome the drag of the aircraft. In turbojets and other thrust-producing aircraft thrust is produced directly from the engine. In aircraft that have propellers (or rotors), the engine does not produce thrust directly. These aircraft are called *power producers* because the engine produces power, which turns the propeller. The propeller then develops an aerodynamic force as it turns through the air. This force is thrust.

Fuel consumption of power-producing aircraft is roughly proportional to the power produced, instead of the thrust produced. Range and endurance performance are functions of fuel consumption, and so the power required to fly the aircraft is of prime importance.

The relationship between thrust and power was discussed in Chapter 1:

$$\text{Thrust} = \text{Force} = \text{A push or pull} \quad \text{(lb)}$$

$$\text{Work} = \text{Force} \times \text{Distance} \quad \text{(ft-lb)}$$

$$\text{Power} = \frac{\text{Work}}{\text{Time}} = \frac{\text{Force} \times \text{Distance}}{\text{Time}} \quad \text{(ft-lb/sec)}$$

$$\frac{\text{Distance}}{\text{Time}} = \text{Velocity}$$

$$\text{Power} = \text{Force} \times \text{Velocity} = \text{FV}_{\text{fps}} = \text{TV}_{\text{fps}} \quad \text{(ft-lb/sec)}$$

$$\text{Horsepower} = \frac{\text{Power}}{550} = \frac{\text{TV}_{\text{fps}}}{550}$$

If V is in knots,

$$\text{HP} = \frac{\text{TV}_{\text{k}}}{325}$$

POWER-REQUIRED CURVES

Equation 1.13 allows us to convert the drag (or T_{r}) curve into the horsepower-required, P_{r}, curve. Equation 1.13 can be rewritten as

$$P_{\text{r}} = \frac{DV_{\text{k}}}{325}$$

Before we convert the total drag curve into a total power-required curve, let us see how the power required to overcome the induced drag, P_{ri}, and the power required to overcome the parasite drag, P_{rp} are affected. The P_{ri} curve varies with velocity as

$$\frac{P_{ri_2}}{P_{ri_1}} = \frac{V_1}{V_2} \tag{8.1}$$

The induced drag varies as the inverse of the velocity ratio squared. The induced power required varies as the inverse of the velocity ratio to the first power. To draw the P_{ri} curve, convert the induced drag values, at selected velocities, from the induced drag curve and plot them. The curve is labeled "induced power req'd" in Fig. 8.1b. It is much flatter than the induced drag curve seen in Fig. 8.1a. No large increase in induced power is required at low speeds as there was for induced drag. The P_{rp} curve varies with velocity as

$$\frac{P_{rp2}}{P_{rr1}} = \left(\frac{V_2}{V_1}\right)^3 \tag{8.2}$$

The parasite drag varied directly as the velocity ratio squared. The parasite power required varies directly as the velocity ratio cubed. To draw the P_{rp} curve, convert the parasite drag values at selected velocities to the corresponding power required values and plot them. The curve is steeper than the parasite drag curve and is labeled "parasite power req'd" in Fig. 8.1b.

The total power-required curve is simply the addition of the P_{ri} and the P_{rp} curves. The P_r curve is flatter in the low-speed region than the T_r curve, but it is steeper in the high-speed region. This explains why propeller aircraft perform better than turbojets in the low-speed region, but lack the high power required to fly at high speeds. The intersection of the P_{ri} and P_{rp} curves is at the velocity of the $(L/D)_{max}$ point, but this is no longer the lowest point on the curve. A tangent line from the origin to the curve locates the $(L/D)_{max}$ point.

Figure 8.1 shows both the total drag, T_r, and total power required, P_r, curves for the same aircraft. The conversion of T_r curves to P_r curves is made by using Eq. 1.13. For this discussion we have chosen a "typical" propeller airplane whose power-required curve is shown in Fig. 8.2.

PRINCIPLES OF PROPULSION

The principles of propulsion are based on Newton's second and third laws. The second law is written as

$$F = ma$$

Simply stated, an unbalanced force, F, acting on a mass, m, will accelerate, a, the mass in the direction of the force. In propeller aircraft the force is supplied

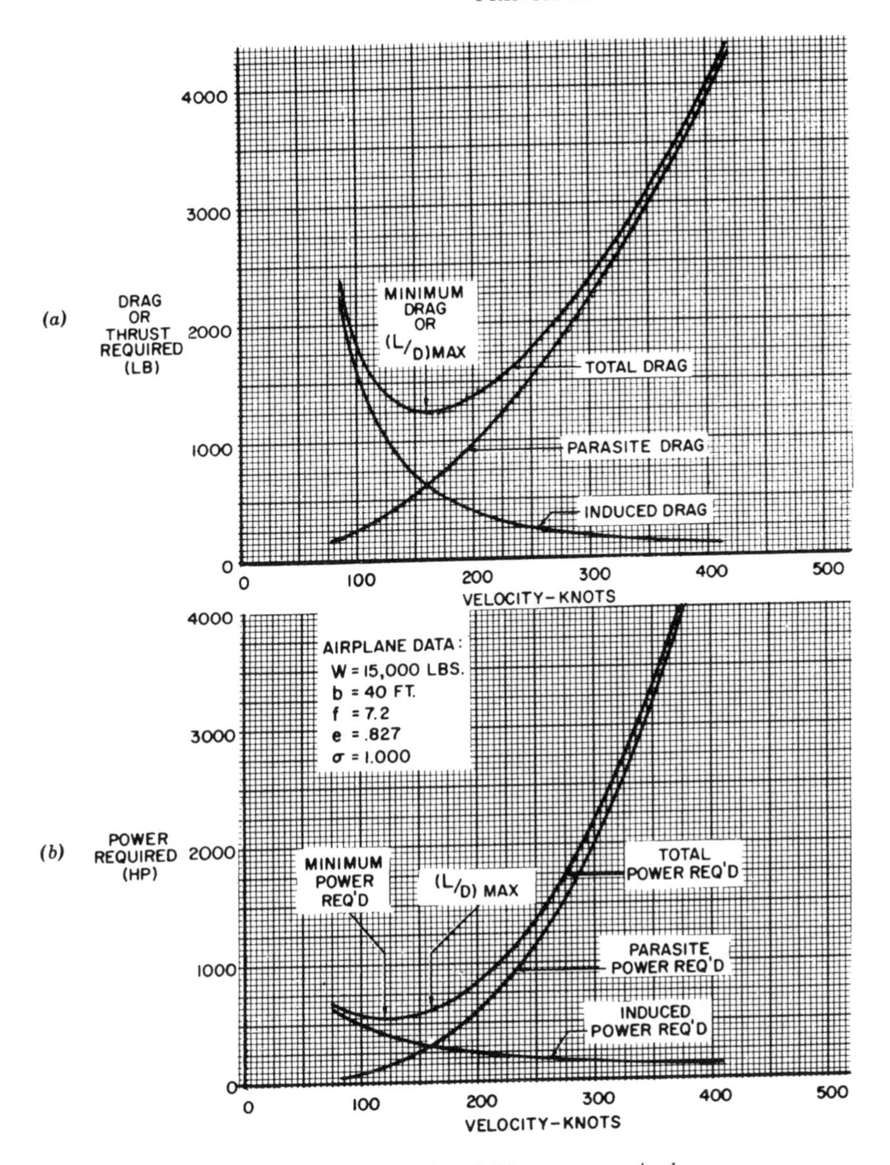

Fig. 8.1. (a) Thrust-required and (b) power-required curves.

by the engine shaft, which rotates the propeller. The propeller moves the air mass through the propeller disk and accelerates it.

Newton's third law states that for every action force, there is an equal and opposite reaction force. It is this reaction force that provides the thrust, T, to propel the aircraft. The thrust was shown in Eq. 6.1 to be

$$T_a = Q(V_2 - V_1)$$

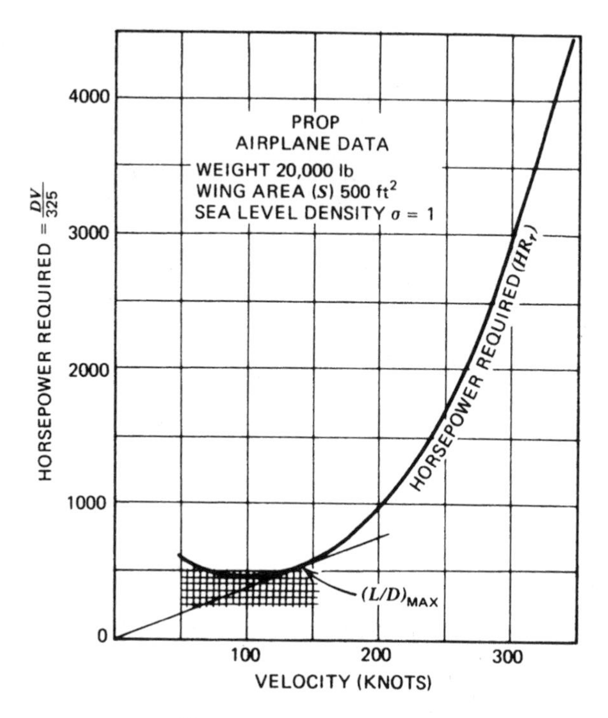

Fig. 8.2. Power required.

where

T_a = thrust available (lb)
Q = mass airflow = pAV (slugs/sec) (Eq. 2.6)
V_1 = inlet (flight) velocity (fps)
V_2 = final velocity (fps)

Figure 8.3 shows a schematic of the process of producing thrust from a propeller.

Propulsion efficiency, η_P, was expressed in Eq. 6.2 as

$$\eta_P = \frac{2V_1}{V_2 + V_1}$$

High exhaust velocity does produce high thrust, but it also produces low efficiency. High mass airflow also produces high thrust and does not reduce propulsion efficiency. Propeller aircraft process large quantities of air with only a small acceleration of the air compared with turbojets. Therefore, propeller aircraft are much more efficient than turbojets.

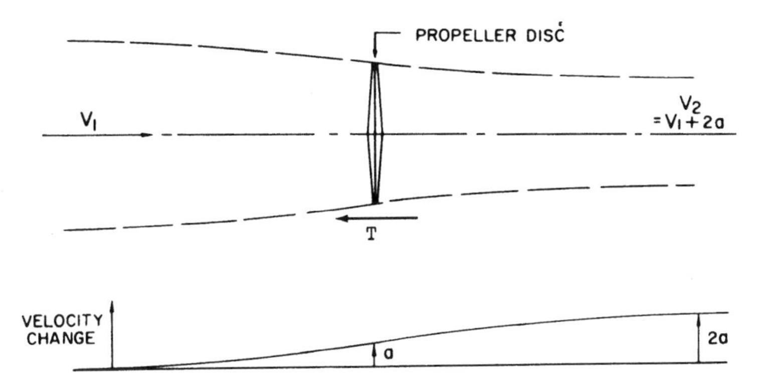

Fig. 8.3. Thrust from a propeller.

POWER AVAILABLE

Several types of horsepower are produced by a power-producing engine and propeller combination:

- *Brake horsepower* (BHP) got its name from an early device used to measure horsepower, called a *prony brake.* It is the horsepower measured at the crankshaft (piston engines) or at the turbine shaft (turbine engines).
- *Shaft horsepower* (SHP) is less than the BHP because of gearing losses in reducing engine rpm to propeller rpm. It is measured at the propeller shaft.
- *Thrust horsepower* (THP) is the usable horsepower. It is less than the SHP because of propeller efficiency loss. *Do not confuse THP with thrust.* Thrust horsepower is a type of horsepower and must be converted to thrust units by Eq. 1.13.

Turboprop aircraft produce both power and thrust, but the amount of thrust produced directly by the engine is only about 15% of the total. They are, therefore, classified as power producers. The amount of thrust that they produce is converted into horsepower units and added to the shaft horsepower. The result is called *equivalent shaft horsepower* (ESHP):

$$\text{ESHP} = \text{SHP} + \frac{TV}{325\eta} \tag{8.3}$$

Propeller efficiency η, is defined as the ratio of power output to power input:

$$\eta = \frac{\text{THP}}{\text{SHP}} \tag{8.4}$$

Propellers can be classified into four general groups:

1. *Fixed pitch* This is a one pitch setting blade propeller, usually of two-blade design. The blade achieves its maximum efficiency at only one airspeed.
2. *Adjustable pitch* The pitch setting can be adjusted on the ground and requires that the engine be stopped and, in some cases, that the propeller be removed from the aircraft.
3. *Controllable pitch* The pitch of this type can be changed in the air by the pilot. It usually is restricted to either the takeoff setting (low pitch) or to cruise setting (high pitch). The pilot control is usually mechanical, although some models are hydraulically or electrically controlled.
4. *Constant speed* This type is hydraulically or electrically controlled. The pilot selects the desired rpm with a lever, and a control governor automatically changes the blade angle to maintain a constant propeller rpm. This type is the most efficient, but a maximum of about 92% efficiency is the best that has been obtained to date. Turboprop engine rpm is electronically controlled and is set at a constant rpm setting. Power is controlled by changing fuel flow and blade angle.

Power Available Versus Velocity

In our discussion of thrust available from turbojet engines in Chapter 6, we saw that the thrust output did not vary, to any great extent, with the velocity of the aircraft. A similar relationship between brake horsepower (and shaft horsepower) and the aircraft velocity exists in power-producing aircraft. However, the thrust horsepower does vary significantly with velocity, because the propeller efficiency varies with velocity. Thrust horsepower is called power available, P_a. Figure 8.4 shows a plot of power available and power required. Fuel flow is also plotted on the vertical scale.

Variations With Power and Altitude

In discussing power available from power-producing engines, we must consider three types of powerplants: turboprops, supercharged reciprocating engines, and nonsupercharged reciprocating engines.

Turboprops In Chapter 6 we discussed the effects on thrust output of turbojet engines when rpm and altitude were changed. Generally, these effects apply to any gas turbine engine (including power producers), so they will not be repeated here. In summary, we can say that as altitude increases, the power available from a turboprop engine decreases and the fuel flow is also decreased.

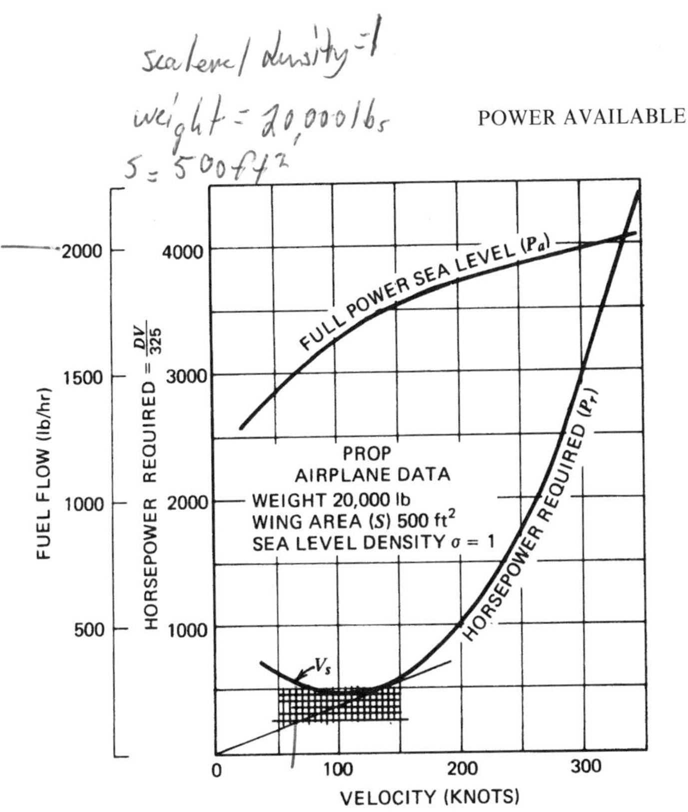

Fig. 8.4. Power required and power available.

Supercharged Reciprocating Engines Reciprocating engines do not operate at high rpm settings (except for takeoff power). The specific fuel consumption of these engines is called brake-specific fuel consumption, c_b, where c_b is the fuel flow per brake horsepower of the engine. It is lowest in the 40–60% power range:

$$c_b = \frac{\text{Fuel flow (lb/hr)}}{\text{Brake horsepower}} = \frac{\text{FF}}{\text{BHP}} \tag{8.5}$$

A very serious problem in reciprocating engines is called *detonation*. Detonation results from a sudden, unstable decomposition of fuel at high temperature and/or pressure. Under certain combinations of these, the mixture ahead of the advancing flame front can suddenly explode instead of smoothly burning. This explosion will send strong pressure waves throughout the combustion chamber that are many times higher than normal combustion pressures. Heavy detonation can cause immediate severe structural damage to the engine. Detonation can be detected by a rapid rise in cylinder head pressure, a drop in manifold pressure, loss of power, and loud explosive noises (similar to the "ping" in car engines). Reducing throttle may actually increase power output by stopping the detonation.

Nonsupercharged Reciprocating Engines Nonsupercharged reciprocating engines are called *normally aspirated engines.* They lose power as altitude is increased and the air density is reduced. At about 19,000 ft altitude, they can develop only about one-half of the sea level horsepower. This is because of the corresponding reduction in air density with altitude. Modern turbo superchargers use exhaust gases to rotate a centrifugal compressor. They can increase the power available so that sea level power can be obtained up to a critical altitude of about 19,000 ft.

ITEMS OF AIRCRAFT PERFORMANCE

Straight and Level Flight

Equilibrium conditions on an aircraft exist when the power developed by the engine is equal to the power required by the airframe for the same set of flying conditions. When these conditions are met, the aircraft will fly at a constant altitude and airspeed. Unaccelerated maximum velocity will occur at the intersection of the full power-available curve and the power-required curve. This is shown in Fig. 8.4 and is at about 335 knots TAS. Of course, the assumption is made that other aircraft restrictions as to maximum velocity do not restrict flight at this airspeed.

Climb Performance

Modern propeller aircraft are not used as fighters and do not use "zoom" techniques of jet fighters. Therefore, steady-velocity climb is assumed in this discussion. In steady-velocity climb, the aircraft is in equilibrium, with all the forces along the flight path in balance (Fig. 8.5). The forces acting along the

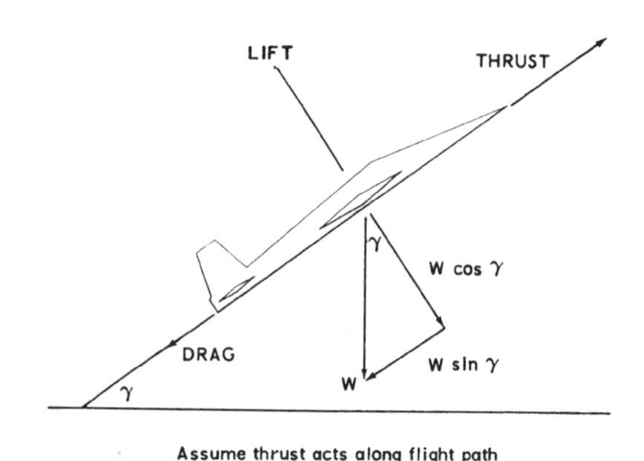

Fig. 8.5. Forces on a climbing aircraft.

flight path are thrust (acting forward), drag, and $W \sin\gamma$ (acting to the rear). $W \sin\gamma$ actually acts at the aircraft's CG, not as shown in the figure. For steady velocity to exist, these must be balanced. From Eq. 6.3 we saw that, for equilibrium

$$T - D - W \sin\gamma = 0$$

Rearranging we get

$$\sin\gamma = \frac{T - D}{W} = \frac{T_a - T_r}{W}$$

Angle of Climb

From Eq. 6.3 it can be seen that the maximum climb angle is obtained when the excess thrust is a maximum. However, to avoid the confusion of using thrust curves for power-producing aircraft, a different approach is presented here, ignoring propeller efficiency losses.

The basic equation 1.13 for converting thrust or drag units to horsepower units can be rewritten as

$$T = \frac{325\text{HP}}{V_k}$$

Adding subscripts yields

$$T_a = \frac{325P_a}{V_k} \quad \text{and} \quad T_r = \frac{325P_r}{V_k}$$

By substituting in Eq. 6.3 we have

$$\sin\gamma = \frac{325(P_a - P_r)}{V_k W} \tag{8.6}$$

If we select values of P_a and P_r from Fig. 8.4 at several arbitrary velocities, we can calculate the sine of the corresponding climb angle using Eq. 8.6. The climb angle itself then can be plotted against velocity as shown in Fig. 8.6. Maximum climb angle occurs at the stall speed for this aircraft. Contrast this to the thrust-producing aircraft, which has its maximum climb angle at $(L/D)_{max}$. The advantage of the propeller aircraft over the jet in short field obstacle clearance is seen.

The old saying "Hang it on the props" is really true. A discussion of wind effect on obstacle clearance is similar to that for thrust-producing aircraft (see Fig. 6.14) and so it is not repeated here.

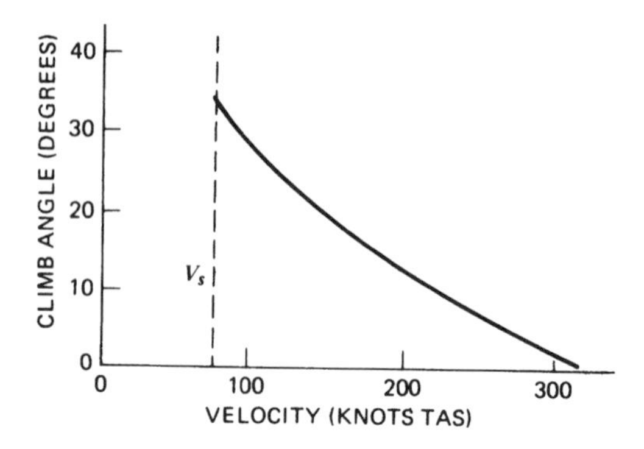

Fig. 8.6. Climb angle versus velocity.

Rate of Climb

Rate of climb, ROC, is the vertical component of the flight velocity shown in Fig. 8.7:

$$\mathrm{ROC} = V \sin \gamma$$

$$= \frac{V_\mathrm{k}(T - D)}{W} \quad \text{(knots)}$$

$$= \frac{101.3 V_\mathrm{k}(T - D)}{W} \quad \text{(fpm)}$$

Substituting gives

$$TV_\mathrm{k} = 325 P_\mathrm{a} \quad \text{and} \quad DV_\mathrm{k} = 325 P_\mathrm{r}$$

Therefore,

$$\mathrm{ROC} = \frac{33{,}000(P_\mathrm{a} - P_\mathrm{r})}{W} \tag{8.7}$$

Equation 8.7 shows that the maximum rate of climb, $\mathrm{ROC}_{\mathrm{max}}$ occurs at the velocity where maximum excess power occurs (Fig. 8.8).

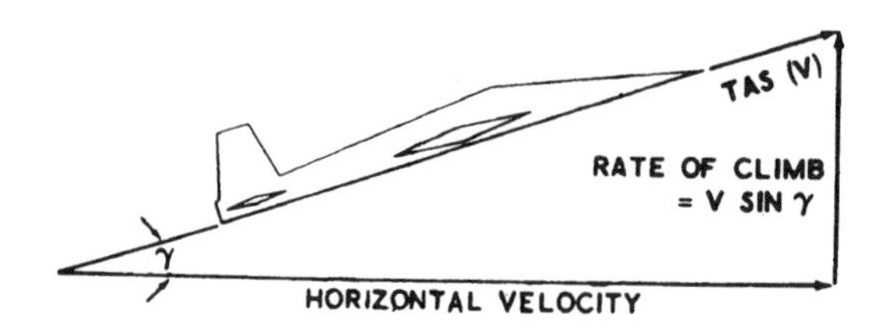

Fig. 8.7. Rate of climb velocity vector.

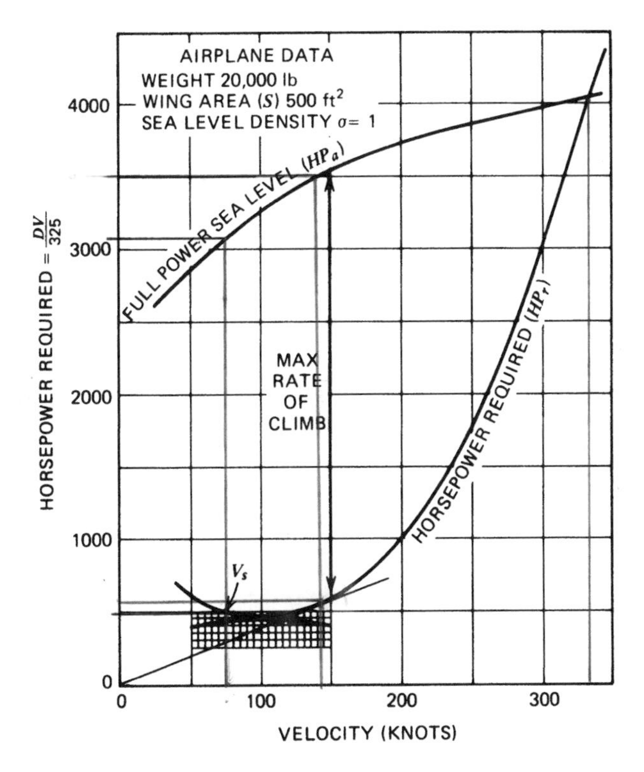

Fig. 8.8. Finding the maximum rate of climb.

Endurance

To obtain maximum endurance, minimum fuel flow is required. Time in flight, not distance covered, is the objective:

$$\text{Specific endurance} = \frac{\text{Fuel available (lb)}}{\text{Fuel flow (lb/hr)}} \quad \text{(hours)}$$

Figure 8.9 is used to show the velocity to fly for both maximum endurance and maximum specific range for a propeller aircraft. Because fuel flow is roughly proportional to the power required, minimum P_r will produce minimum fuel flow. As mentioned earlier, this is not $(L/D)_{max}$ for a propeller aircraft. It is merely called *minimum power required*. Mathematically it is the point where $C_L^{3/2}/C_D$ is maximum.

Specific Range

To get maximum range, the aircraft must fly the maximum distance with the fuel available. The specific range must be a maximum. Specific range, SR, can

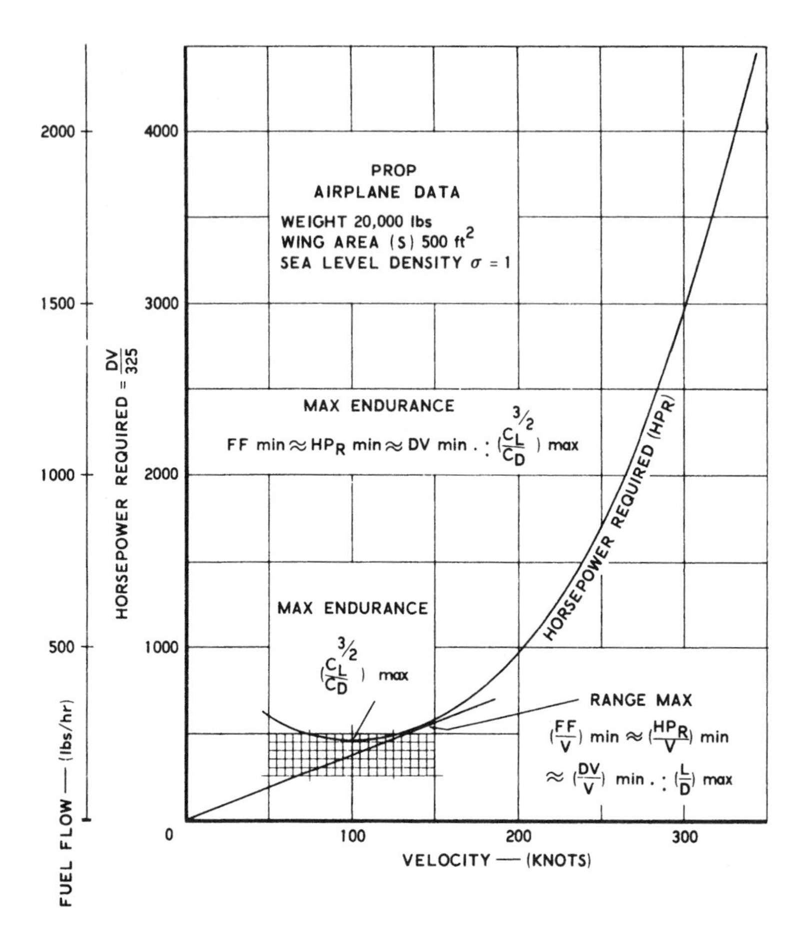

Fig. 8.9. Finding the maximum endurance and range.

be defined by the following relationship:

$$\mathrm{SR} = \frac{\text{Nautical miles flown}}{\text{Pounds of fuel used}} = \frac{\text{Airspeed (knots)}}{\text{Fuel flow (lb/hr)}}$$

For maximum specific range,

$$\mathrm{SR}_{\max} = \left[\frac{V}{\mathrm{FF}}\right]_{\max} \quad \text{or} \quad \mathrm{SR}_{\max} \text{ occurs at } \left[\frac{\mathrm{FF}}{V}\right]_{\min}$$

The point on the P_r curve in Fig. 8.9 where a straight line drawn from the origin is tangent to the curve will indicate the maximum specific-range velocity.

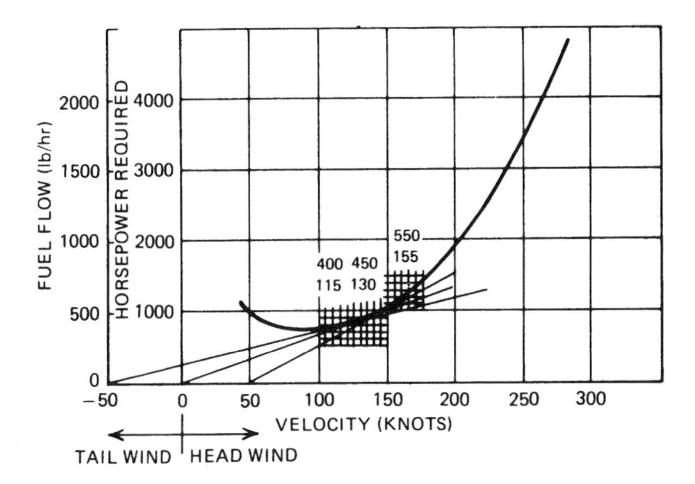

Fig. 8.10. Effect of wind on range.

For a power-producing aircraft, this is the $(L/D)_{max}$ point. Two items of aircraft performance occur at $(L/D)_{max}$ for power-producing aircraft:

1. Maximum engine-out glide ratio (minimum drag)
2. Maximum range

Every pilot should realize the importance of the $(L/D)_{max}$ and should know its airspeed.

Wind Effect on Specific Range

An aircraft flying into a headwind is in an unfavorable environment. Therefore, the airspeed should be increased to reduce the time that the aircraft is exposed to the headwind. Conversely, an aircraft flying with a tailwind is being helped on its way and should slow down to take advantage of this favorable environment. The amount that the airspeed should be changed is shown in Fig. 8.10.

To obtain the values of best airspeed, find the corrected airspeed for a headwind, by plotting the wind on the velocity scale to the right of the origin. Draw a new tangent line from this point to the P_r curve. Correcting for a tailwind is done similarly. The tailwind is laid off to the left of the origin and a new tangent line is drawn to the curve.

Total Range

The total range for power-producing aircraft is similar to that of thrust-producing aircraft (see Fig. 6.22).

SYMBOLS

c_b	Brake specific fuel consumption (lb fuel/BHP)
P_a	Power available (horsepower units)
P_r	Power required
P_{ri}	Power required, induced
P_{r_0}	Power required, profile
P_{rp}	Power required, parasite
BHP	Brake horsepower
SHP	Shaft horsepower
THP	Thrust horsepower
ESHP	Equivalent SHP
η	(eta) Propeller efficiency (%/100)

EQUATIONS

8.1 $$\frac{P_{ri_2}}{P_{ri_1}} = \frac{V_1}{V_2}$$

8.2 $$\frac{P_{r_{p2}}}{P_{r_{r1}}} = \left(\frac{V_2}{V_1}\right)^3$$

8.3 $$\text{ESHP} = \text{SHP} + \frac{TV}{325\eta}$$

8.4 $$\eta = \frac{\text{THP}}{\text{SHP}}$$

8.5 $$c_b = \frac{\text{FF}}{\text{BHP}}$$

8.6 $$\sin\gamma = \frac{325(P_a - P_r)}{V_k W}$$

8.7 $$\text{ROC} = \frac{33{,}000(P_a - P_r)}{W}$$

PROBLEMS

1. In the formula $\text{HP} = \dfrac{TV_k}{325}$

a. power required is in horsepower units.

b. power required is in ft-lb/sec.

2. Power is

a. (Force × Velocity)/Time.

b. Work/Time.

c. (Force × Distance)/Time.

d. Both (b) and (c)

3. Power required to overcome induced drag varies

a. inversely with V^2.

b. inversely with V^3.

c. inversely with V.

d. directly with V.

4. Power required to overcome parasite drag varies

a. directly with V^2.

b. directly with V^3.

c. directly with V.

d. inversely with V^2.

5. Maximum rate of climb for a propeller airplane occurs

a. at $(L/D)_{max}$.

b. at $P_{r(min)}$.

c. at $C_{L(max)}$.

d. at $(P_a - P_r)_{max}$.

6. The lowest point on the P_r curve is $(L/D)_{max}$.

a. True

b. False

7. Propeller aircraft are more efficient than jet aircraft because

a. they don't go so fast.

b. they process more air and don't accelerate it as much.

c. they use gasoline instead of JP fuel.

d. V_1 is less, so propulsive efficiency is greater.

8. Turboprop aircraft are classified as power producers because

a. nearly all of the engine output goes to the propeller.

b. the engine is a turbine engine.

c. the fuel flow is proportional to the power produced.

d. Both (a) and (c)

9. Propeller aircraft use a "zoom" climb

a. when a climb to avoid obstacles at take off is required.

b. when a large excess of power is available.

c. when flying at $(L/D)_{max}$.

10. Propeller aircraft get the highest angle of climb at $(L/D)_{max}$.

a. True

b. False

11. In Chapter 5, Problem 14, you calculated the drag for a 10,000-lb turbojet airplane for sea level standard day conditions. The values that you found are repeated below. Calculate the values for horsepower required and plot them on graph paper.

V_k	Drag	P_r
125	1233	
150	1058	
172	1020	
200	1067	
300	1720	
400	2860	

12. P_a–P_r curves for a propeller airplane are shown in Fig. 8.8. Find

a. V_{max} at full power

b. Sine of climb angle at V_s

c. Sine of climb angle at $(L/D)_{max}$

d. Rate of climb at V_s

e. Rate of climb at 150 knots

f. Velocity for maximum endurance

g. Velocity for maximum range

h. Velocity for maximum range with 50-knot headwind

9 Propeller Aircraft—Applied Performance

The aircraft performance items that we discussed in Chapter 8 were for one weight, clean configuration, and sea level standard day conditions. In this chapter we alter these conditions and examine the resulting performance. To simplify the explanation, only one condition is altered at a time.

VARIATIONS IN THE POWER-REQUIRED CURVE

The basic power-required curve (Fig. 8.2) was drawn for a 20,000-lb airplane in the clean configuration, at sea level on a standard day. We now show how the curve changes for variations in weight, configuration, and altitude.

Weight Changes

To find out how the P_r varies with weight changes, we go back to the basic relationship between P_r and T_r(drag) that we developed in Eq. 1.13:

$$P_r = \frac{DV}{325}$$

Adding subscripts and forming a ratio gives

$$\frac{P_{r_2}}{P_{r_1}} = \frac{D_2 V_2}{D_1 V_1}$$

From Eq. 7.1,

$$\frac{D_2}{D_1} = \frac{W_2}{W_1}$$

And from Eq. 4.2,

$$\frac{V_2}{V_1} = \sqrt{\frac{W_2}{W_1}}$$

Substituting gives

$$\frac{P_{r2}}{P_{r1}} = \left(\frac{W_2}{W_1}\right)^{3/2} \tag{9.1}$$

In addition to the increase in P_r, with an increase in weight there also must be an increase in velocity. This is the same as was found in Eq. 4.2:

$$\frac{V_2}{V_1} = \sqrt{\frac{W_2}{W_1}}$$

In plotting new curves for a change in weight, both Eqs. 9.1 and 4.2 must be applied. Figure 9.1 shows P_r curves for 20,000 and 30,000 lb.

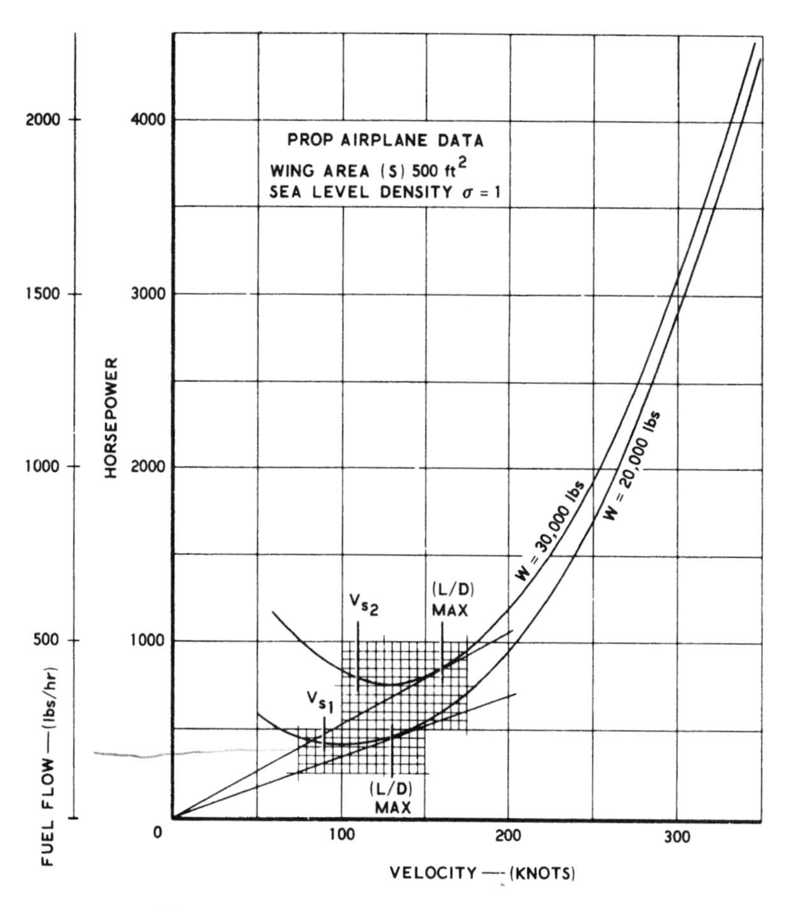

Fig. 9.1. Effect of weight change on a P_r curve.

Configuration Change

In going from the clean configuration to the gear and flaps down (dirty) configuration, the equivalent parasite area is greatly increased. For the airplane shown in Fig. 9.2, it increases by 50% (from 9.92 to 14.9 ft^2). This increases the parasite drag, and thus the parasite power required, but has little effect on the induced drag or induced power required.

There is no short-cut method of calculating the P_r curve for the dirty condition. You must calculate the parasite drag and then the total drag for the dirty condition, and then convert this to power required by Eq. 1.13. The effect of configuration change is shown on the P_r curves in Fig. 9.2.

Altitude Changes

In our discussion of the effect of altitude on the drag of an aircraft, we saw that the drag of the aircraft was unaffected by altitude, but that the true airspeed at

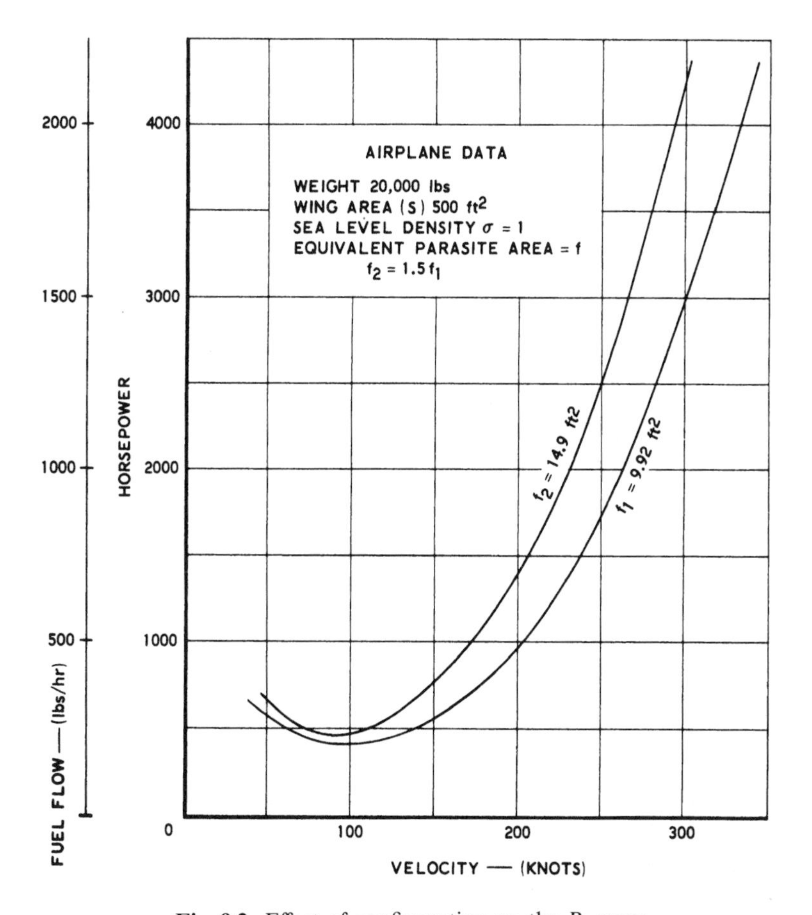

Fig. 9.2. Effect of configuration on the P_r curve.

which the drag occurred did change:

$$D_2 = D_1 \quad \text{and} \quad V_2 = \frac{V_1}{\sqrt{\sigma_2}}$$

Subscript 1 is sea level ($\sigma_1 = 1$) and subscript 2 is at altitude. Applying the above equations to Eq. 1.13 gives

$$\frac{P_{r_2}}{P_{r_1}} = \frac{1}{\sqrt{\sigma_2}} \tag{9.2}$$

The drag does not change with altitude but the P_r does. The velocity changes by the same amount:

$$\frac{V_2}{V_1} = \frac{1}{\sqrt{\sigma_2}} \tag{9.3}$$

In plotting new curves for a change in altitude, note that both Eq. 9.2 and Eq. 9.3 must be applied. All points on the sea level curve move to the right and upward, by the same amount, when correcting for altitude changes. This is shown in Fig. 9.3. Note that the line drawn from the origin, which is tangent to the curve and locates the $(L/D)_{max}$ point, remains tangent to the curve at all altitudes. The angle remains the same at all altitudes. The significance of this is explained later in this chapter.

Another point of interest in the altitude curves is that the curves move farther apart at higher velocities. Each point on the sea level curve is moved to the right and also moved upward by the same multiplier, $1/\sqrt{\sigma_2}$, so the change is greater when the velocity or power required is greater; thus, points on the right are changed more than those on the left.

VARIATIONS IN AIRCRAFT PERFORMANCE

Straight and Level Flight

Weight Change An increase in weight means an increase in P_r, but P_a is not changed; V_{max} is decreased. A reduction in weight has the opposite effect and V_{max} will be increased.

Configuration Change V_{max} in the dirty configuration is usually limited by structural strength of gear and flaps, so V_{max} is reduced.

Altitude Increase The power available at altitude is affected by the type of engine and/or propeller. If turbo supercharged reciprocating engines with constant speed props are flown below their critical altitude they can develop as much power as at sea level. In this case the V_{max} TAS will be increased significantly. Nonsupercharged reciprocating engines with fixed props, on the other hand, will suffer a power loss above about 5000 ft and may experience

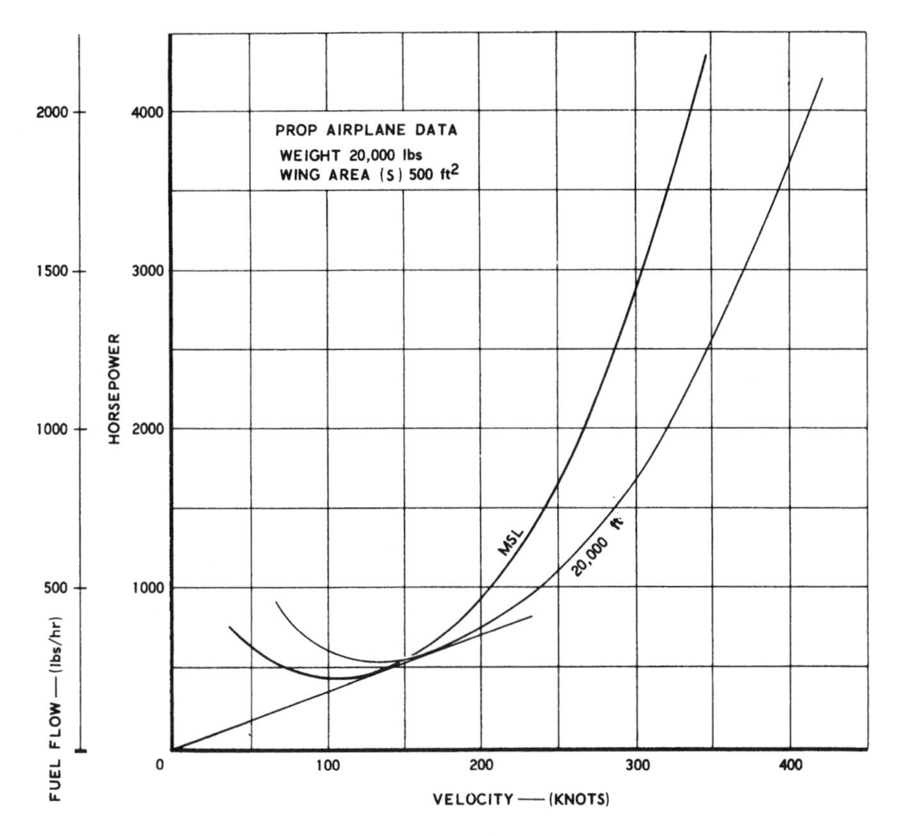

Fig. 9.3. Effect of altitude on a P_r curve.

either an increase or a decrease in V_{max}, depending on the intersection point of the P_a and P_r curves. Similar reasoning can be applied to turboprop engines. The engine loses power with increased altitude, but the P_r curve moves to the right, so the intersection of the curves will determine V_{max}. The IAS will be lower in all cases.

Climb Performance

Angle of Climb

Weight Change Increased weight results in increased power required, while power available is not changed. Equation 8.6 shows:

$$\sin\gamma = \frac{325(P_a - P_r)}{V_k W}$$

Both the excess power reduction and the increased weight cause a reduction in angle of climb.

Configuration Change Power required increases in the dirty configuration; thus, the angle of climb is reduced. However, the increase in P_r is much less than was the increase in T_r for the jet aircraft (see Figs. 9.2 and 7.4). Again the superior performance, at low speeds, of propeller aircraft is demonstrated.

Altitude Increase From Eq. 8.6, it can also be seen that the increase in P_r at altitude will reduce the angle of climb, even if P_a is not reduced.

Rate of Climb

Weight Change Rate of climb is calculated by Eq. 8.7:

$$\text{ROC} = \frac{33{,}000(P_a - P_r)}{W}$$

With an increase in weight there is an increase in P_r. Both of these reduce the rate of climb.

Configuration Change Dirty aircraft reduce climb performance. Clean up the aircraft as soon as practical.

Endurance

Weight Change More power is required with increased weight. Fuel flow is proportional to P_r, so endurance is also reduced. The opposite is true when weight is decreased.

Configuration Change Pull up the gear and flaps to improve endurance. If it is impossible to clean up the airplane, slow down to reduce parasite drag.

Altitude Increase With a turboprop, the reduction in fuel flow at altitude will more than compensate for the increase in P_r but if it is necessary to climb to altitude, more fuel may be required than if the airplane had remained at the lower altitude. For a reciprocating engine the endurance at altitude is reduced.

Specific Range

Weight Change As fuel is burned, weight is reduced, and the power-required curve moves downward and to the left. This is shown in Fig. 9.4. The tangent lines intersect the P_r curves at $(L/D)_{max}$ and indicate the maximum specific range velocity and fuel flow at this point. At a gross weight of 30,000 lb, the velocity must be 160 knots for maximum range and the fuel flow will be 425 lb/hr. The SR here is 160/425 = 0.3765 nmi/lb.

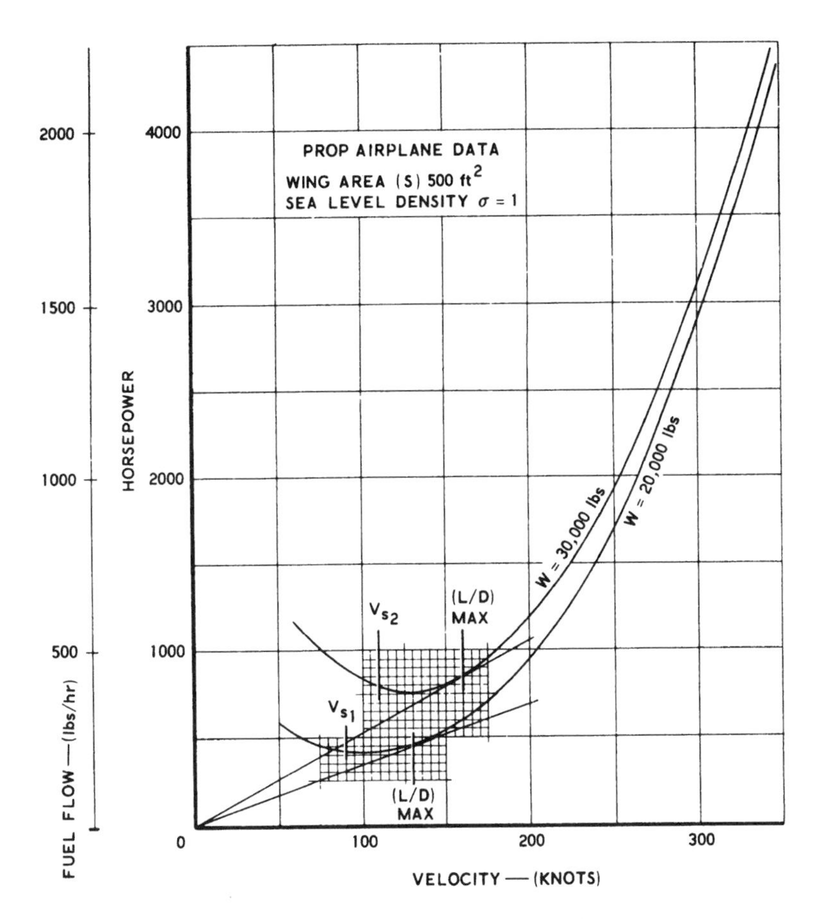

Fig. 9.4. Effect of weight change on specific range.

When the weight of the aircraft has been reduced to 20,000 lb, the aircraft must fly at 130 knots to attain maximum range. The fuel flow here will be 225 lb/hr and the SR = 130/225 = 0.5778 nmi/lb. This is a 53.5% improvement in specific range. The pilot must reduce the power output of the engine and reduce the airspeed to achieve this range. It is only necessary for the pilot to reduce the power setting because the airspeed will automatically be reduced when this is done.

The natural tendency of a pilot is to keep the power setting the same, as fuel is burned, and to allow the velocity to increase. This is wrong. It will decrease the range by an appreciable amount. For maximum range, as weight is reduced, *reduce airspeed*.

Configuration Change If it is necessary to fly in the dirty configuration, slow the airplane to reduce drag.

Altitude Increase In our earlier discussion we pointed out that an altitude increase changed both the P_r and the velocity by the same amount. We also saw that the tangent line to the curves remained the same for all altitudes. In Fig. 8.10 we saw that the intersection of the tangent line and the curve indicated the maximum specific range point. Thus, as far as the airframe is concerned, the altitude has no effect on the specific range. The ratio of fuel flow to velocity is constant with altitude. Reviewing Chapter 5, propeller-driven airplane do not show large increases in range as jet airplanes do.

We must consider at what altitude the engine–propeller combination is most efficient. First, let's look at the nonsupercharged reciprocating engine with a fixed-pitch propeller. This engine can develop sea level power to about 5000 ft. This seems to be about the altitude where maximum range is obtained. For the turbine supercharged reciprocating engine with constant speed propeller about 10,000 ft is best. The turbine engine of the turboprop aircraft likes high altitude and operates best at about 25,000 ft.

EQUATIONS

9.1 $$\frac{P_{r_2}}{P_{r_1}} = \left(\frac{W_2}{W_1}\right)^{3/2}$$

9.2 $$\frac{P_{r_2}}{P_{r_1}} = \frac{1}{\sqrt{\sigma_2}}$$

9.3 $$\frac{V_2}{V_1} = \frac{1}{\sqrt{\sigma_2}}$$

PROBLEMS

1. A lightly loaded propeller airplane will be able to glide ______ when it is heavily loaded.

a. farther than

b. less far than

c. the same distance as

2. To obtain maximum glide distance, a heavily loaded airplane must be flown at a higher airspeed than if it is lightly loaded.

a. True

b. False

3. In order to maximize range on a propeller-driven airplane at high altitude, true airspeed should be ______.

a. decreased

b. increased

c. not changed

4. In Fig. 9.3, the specific range for a propeller airplane is
   a. less at altitude than at sea level.
   b. more at altitude than at sea level.
   c. the same at altitude as at sea level.

5. The pilot of a propeller airplane is flying at the speed for best range under no-wind conditions. A head wind is encountered. To obtain best range the pilot now must
   a. speed up by the amount of the wind speed.
   b. slow down by the amount of the wind speed.
   c. speed up by less than the wind speed.
   d. slow down by more than the wind speed.

6. As a propeller airplane burns up fuel, to fly for maximum range, the airspeed must
   a. remain the same.
   b. be allowed to speed up.
   c. be slowed down.

7. The turboprop aircraft has its lowest specific fuel consumption at about 25,000 ft altitude because
   a. the turbine engine wants to fly at high altitudes.
   b. the propeller has higher efficiency at lower altitudes.
   c. This is a compromise between (a) and (b) above.

8. The power-required curves for an increase in altitude show that the
   a. P_r remains the same as altitude increases.
   b. P_r increases by the same amount as the velocity.
   c. P_r increases but the velocity does not.

9. A propeller aircraft in the dirty condition shows that the P_r moves up and to the left over the clean configuration. This is because
   a. the increase in the induced P_r is more at low speed.
   b. the increase is mostly due to parasite P_r.
   c. both induced P_r and parasite P_r increase by the same amount.

10. The increase in the P_r curves for a weight increase is greater at low speeds than at high speeds because the increase in
    a. induced P_r is greatest.
    b. parasite P_r is greatest.
    c. profile P_r is greatest.

11. Using Fig. 9.1, find the velocity for the best range for the airplane at 30,000 lb and 20,000 lb gross weight (GW).

12. Using Fig. 9.1, calculate the specific range for the airplane at both weights if it is flying at the best range airspeed.

13. Using Fig. 9.1, calculate the specific range for the 20,000-lb airplane if the throttle is not retarded and the airspeed is allowed to increase as weight is reduced. Compare your answer to that of Problem 12 and state your conclusions.

14. Using Fig. 9.3, calculate the specific range for this airplane if it is flying at best range airspeed at mean sea level (MSL) and also at 20,000 ft altitude.

15. A 20,000-lb propeller airplane at sea level standard day has the following data. Calculate its P_r and plot the curve on the graph paper used for Problem 17.

V	Drag	P_r
150	2074	
182.7	1923	
200	1954	
300	2948	
400	4809	

16. The airplane in Problem 15 now has itgs GW increased to 30,000 lb. Calculate the new values and plot the curve on the graph paper used for Problem 17.

$V_{20,000\ lb}$	$P_{r(20,000\ lb)}$	$V_{30,000\ lb}$	$P_{r(30,000\ lb)}$
150			
182.7			
200			
300			
400			

17. Calculate new values for the 20,000-lb airplane in Problem 15, but now flying at 20,000 ft density altitude. Plot the values and draw the curve on the graph paper below.

V_{SL}	$P_{r(SL)}$	$V_{(20,000\ ft)}$	$P_{r(20,000\ ft)}$
150			
182.7			
200			
300			
400			

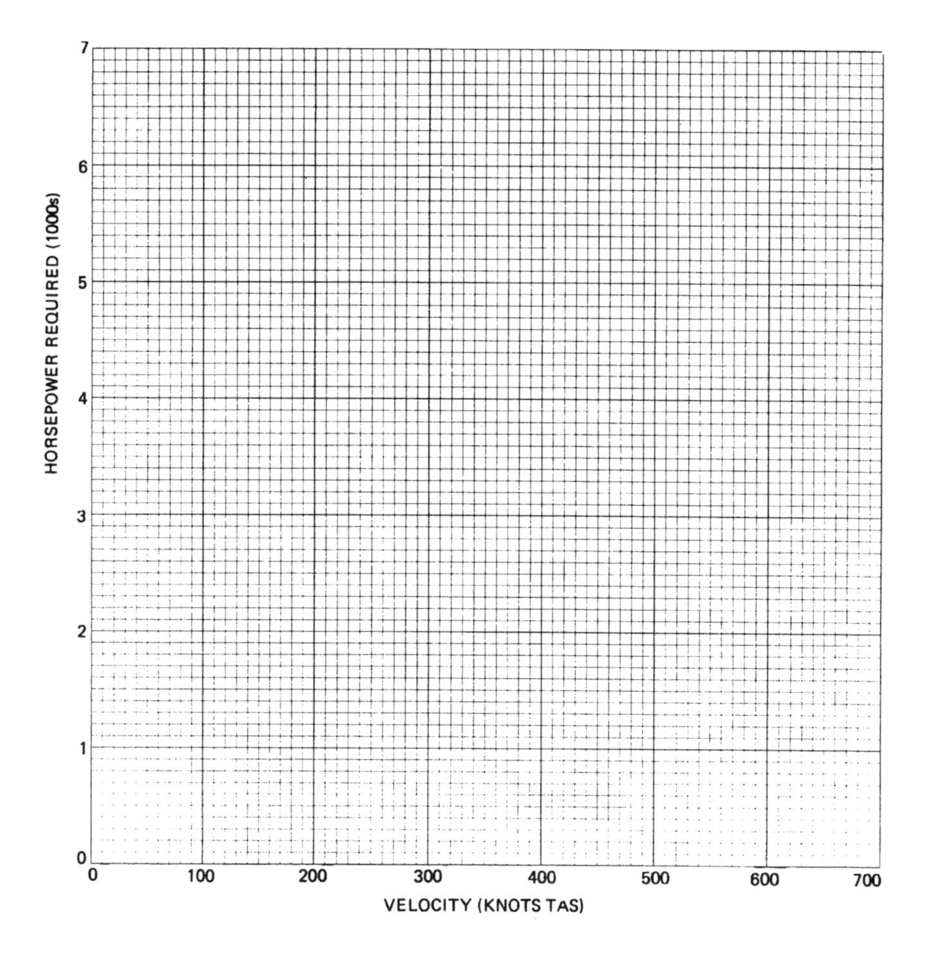
HORSEPOWER REQUIRED (1000s)
7
6
5
4
3
2
1
0
0
100
200
300
400
500
600
700
VELOCITY (KNOTS TAS)

10 Helicopter Flight Theory

A helicopter looks strange compared to a conventional fixed-wing aircraft, but the same principles of flight apply to both. In Chapter 3 we learned that Leonardo da Vinci stated that air passing around an airfoil produces the same aerodynamic forces when the airfoil is moving through still air as it does when the air is moving past a stationary airfoil. We can carry this principle further in saying that an airfoil will produce the same aerodynamic force whether the airfoil is being moved through the air by moving the entire fixed wing aircraft or by moving the airfoil while the rotary wing aircraft remains stationary.

MOMENTUM THEORY OF LIFT

Bernoulli's equation of lift is still a valid theory of how a helicopter produces lift, but, because the velocity of each rotor blade segment varies as the distance from the rotor hub, it is simpler to discuss the lift by using the momentum theory. The momentum theory is based upon Newton's laws. The second of these states that an unbalanced force acting on a body produces an acceleration of the body in the direction of the force. The "body" in the case of the flow of air through a helicopter rotor system is the mass of air. The unbalanced force is produced by the rotor blades being driven by the engine. The result is an acceleration of the air down through the rotor disk.

Newton's third law stated that for every action force there is an equal and opposite reaction force. Thus, by causing the air to be forced downward, lift was developed upward. This is quite similar to the production of thrust from an aircraft propeller or jet engine. Figure 10.1 shows a schematic of this theory, where

V_o = original velocity at about two rotor diameters above the rotor
V_i = induced velocity at the rotor
V_f = final velocity below the rotor

It can be seen from the figure that there is lower pressure above the rotor disk and higher pressure below the disk, with a sudden increase at the disk. The velocity increases gradually from approximately zero at about two rotor diameters above the disk to a maximum at about the same distance below the disk.

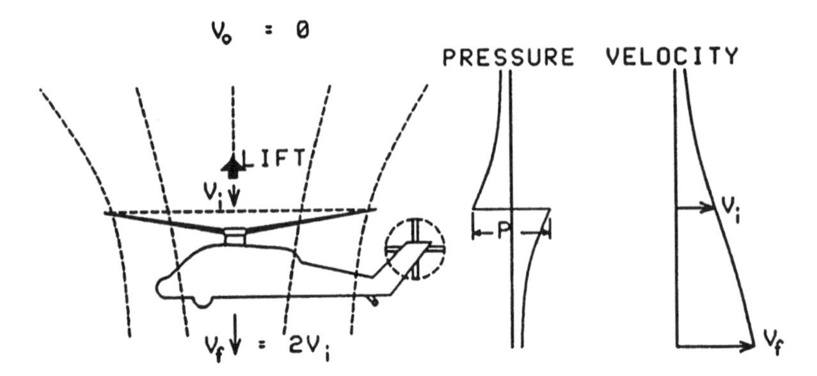

Fig. 10.1. Momentum theory airflow: (a) schematic, (b) pressure and velocity distribution.

AIRFOIL SELECTION

It was shown in Chapter 3 that cambered airfoils develop a nose-down pitching moment when subjected to an airflow, whereas symmetrical airfoils do not (see Figs. 3.9 and 3.10). These pitching moments change in value when the angle of attack changes. Later we discuss the fact that during the rotation of the blades the angle of attack is constantly changing. Rotor blades generally lack the stiffness to resist this moment and will twist if they are cambered. Twisting of the blades causes changes in AOA and lift, which complicates the problem, so it is avoided by using symmetrical airfoil sections. Another advantage of the symmetrical airfoil is that both the center of pressure and the aerodynamic center are at the same location and do not move with AOA changes. So most rotor blades have symmetrical airfoil sections. Figure 10.2 shows a typical helicopter rotor airfoil.

To reduce pitching moments on rotor blades caused by such factors as center of gravity location and axis of feathering (axis of rotation) location, it is desirable that they coincide with the center of pressure/aerodynamic center location. This is shown in Fig. 10.3.

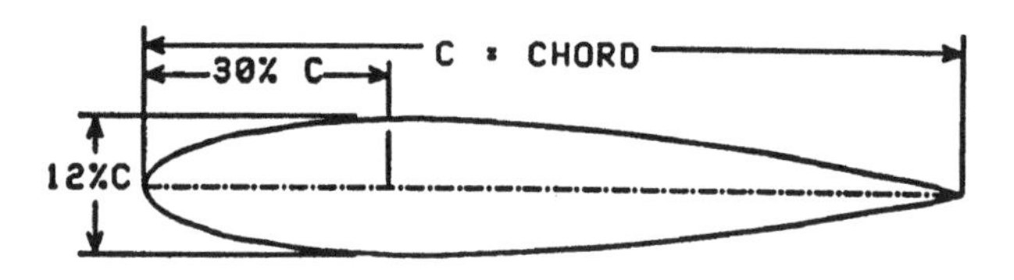

Fig. 10.2. NACA 0012 airfoil.

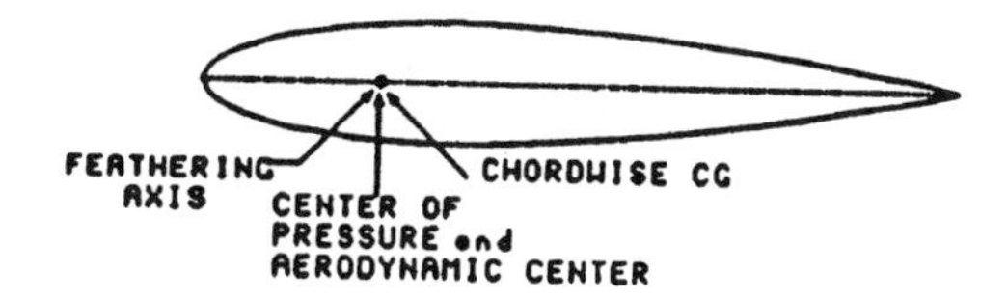

Fig. 10.3. Location of critical forces on an airfoil.

FORCES ON ROTOR SYSTEM

When the blades of a helicopter are not turning there is a noticeable droop to them. They do not have the inherent stiffness of fixed wings to prevent this droop. When the blades are rotated, however, centrifugal force (equal and opposite to the centripetal force that holds the rotating blades on the rotor shaft) causes them to straighten, as in Fig. 10.4. The force on only one blade is shown.

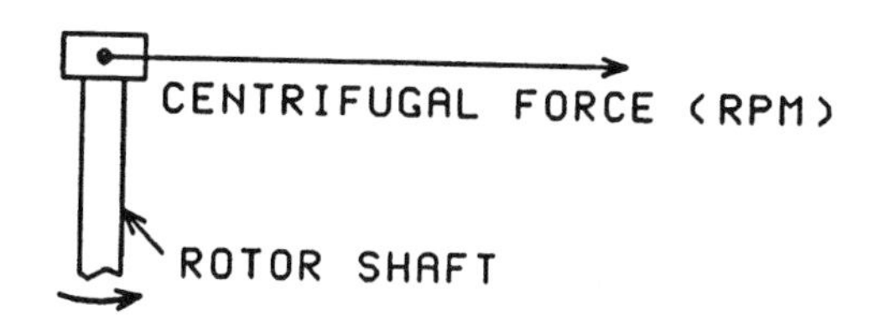

Fig. 10.4. Centrifugal force straightens rotor blade.

To produce the lift required to fly the helicopter a vertical vector must be exerted upon the rotor shaft. To produce the vertical lift vector shown in Fig. 10.5 the angle of attack of the symmetrical rotor blades must be set at a positive angle of attack. The blades are now being acted upon by two force vectors: (1) the centrifugal force vector and (2) the lift vector. The blades must assume the position of the resultant of these two vectors, as shown in Fig. 10.6, where the lift vector has been exaggerated for clarity.

The centrifugal force is much larger than the lift force. In fact, the lift force is only about 6% of the centrifugal force. To illustrate this, assume that the

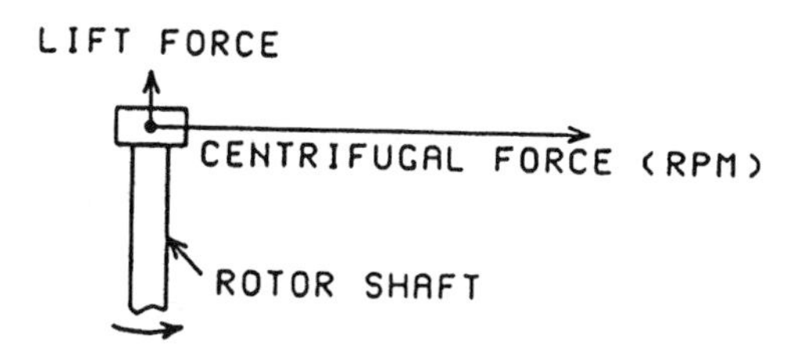

Fig. 10.5. Lift force and centrifugal force.

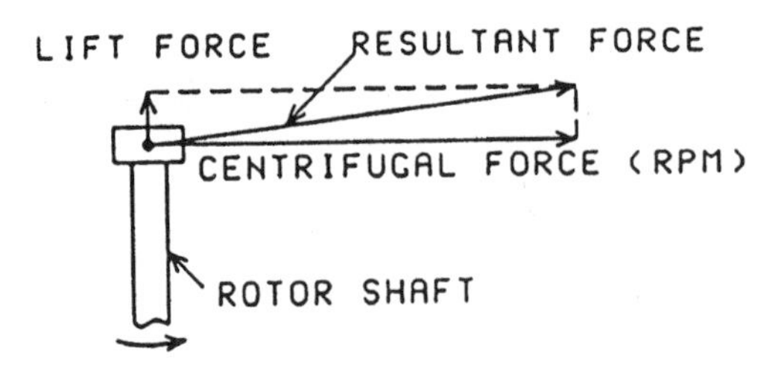

Fig. 10.6. Resultant of lift and centrifugal forces.

helicopter weighs 6000 lb and has two blades. The lift force on each blade would be 3000 lb and the centrifugal force would be about 50,000 lb. This is shown in Fig. 10.7. If the rotor blades are hinged to allow upward rotation, the lifting rotor system is as shown in Fig. 10.8.

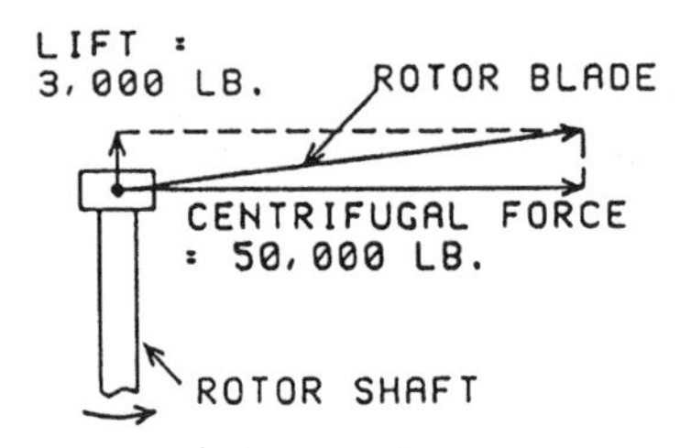

Fig. 10.7. Forces acting on a lifting blade.

Of particular interest is the coning angle shown between the two blades and the horizontal. When viewed from the side the rotating blades form a visual disk. A line drawn between the blade tips is called the tip-path plane. Figure 10.9 shows a hovering helicopter at a light weight. Under light weight conditions the coning angle is small because the lift force is relatively small as compared to the centrifugal force. In Fig. 10.10 the heavy weight requires more lift forces and, if the rpm is the same, the centrifugal forces remain the same as in the light weight condition. Thus, the coning angle is larger.

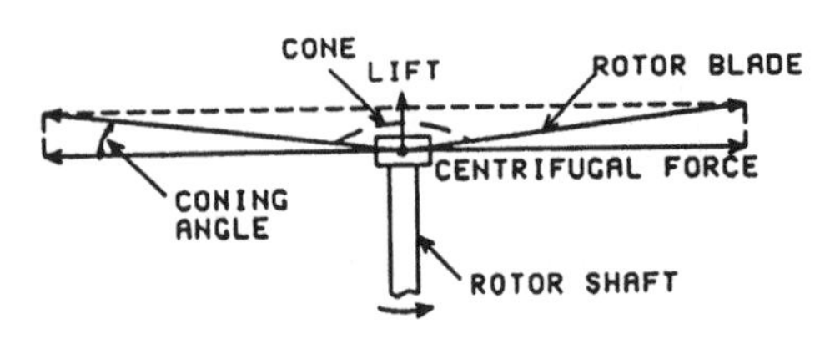

Fig. 10.8. Entire lifting rotor system.

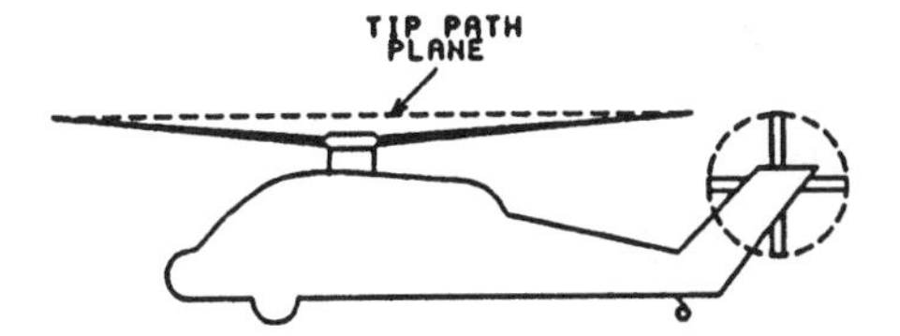

Fig. 10.9. Hovering helicopter at light weight.

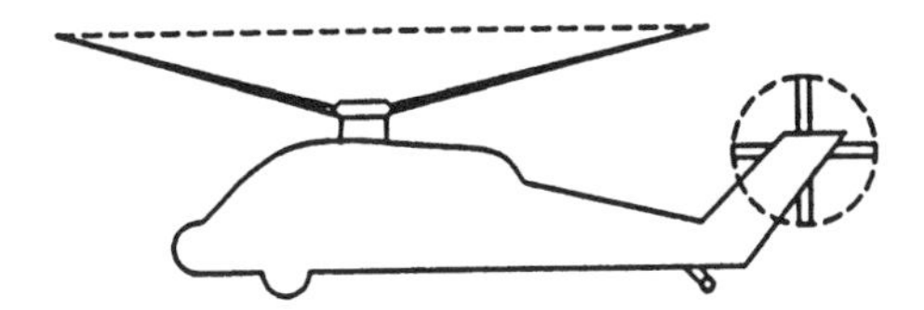

Fig. 10.10. Hovering helicopter at heavy weight.

THRUST DEVELOPMENT

Up to this point we have been discussing the forces on the rotor system while the helicopter was in a hover. To move the aircraft in any horizontal direction we must tilt the entire rotor disk in the desired direction. We have been calling the vertical force lift. Now we will see that in addition to the force required to overcome the weight, we also obtain a component force, which we call directional thrust. Therefore, we rename the lift force: We call it *total thrust.*

Figure 10.11 shows the total thrust vector resolved into the vertical lift component and the forward thrust component. Figure 10.12 shows that one of the first reactions that results from translating from a hover to flight in any direction is a loss of altitude. This is because the total thrust is no longer acting opposite to the weight of the helicopter as it was in the hover. Consequently, the effective lift is reduced and, if corrective action is not taken, the helicopter

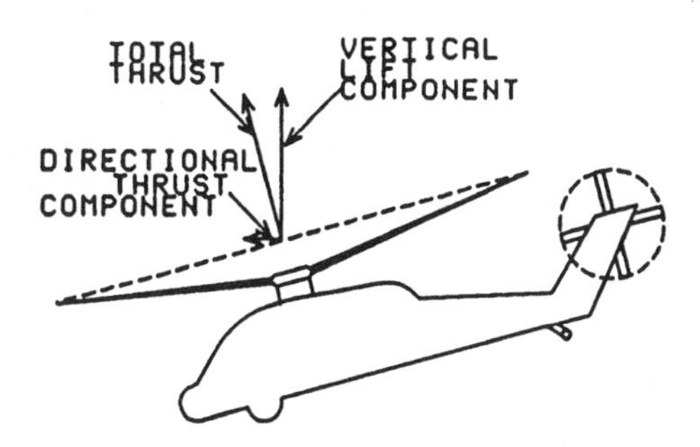

Fig. 10.11. Forward flight forces.

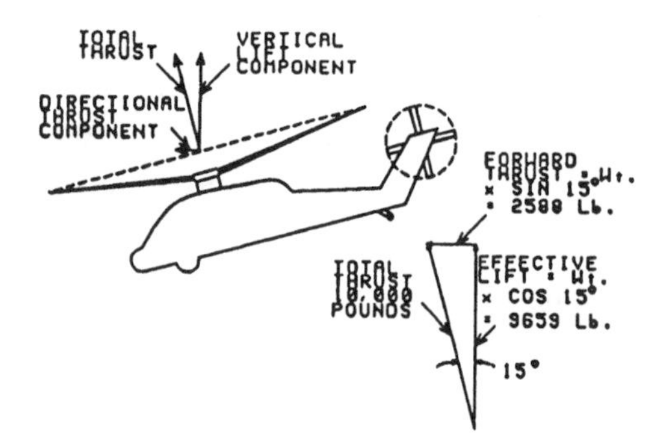

Fig. 10.12. Lift component of 10,000-lb total thrust at 15°.

will hit the ground. The effective lift of 9659 lb will not support the weight of 10,000 lb so the helicopter will lose altitude. The pilot must increase total thrust to 10,353 lb to have the effective lift equal the weight ($10{,}353 \cos 15^\circ = 10{,}000$).

HOVERING FLIGHT

Hovering means maintaining a constant position, usually a few feet above the ground. To hover, the helicopter must supply lift equal to the weight of the helicopter. To accomplish this, the blades must be rotating at high rpm and the angle of attack of all the blades is increased by increasing the angle of attack with the collective stick. We discuss the control devices in more detail later.

The lift and weight forces are equal and opposite in a no-wind hover condition and the tip-path plane remains horizontal. If the angle of attack is increased by the pilot lifting the collective, the helicopter will climb vertically. This also requires more power or the rpm will decrease.

In some helicopters there is an interlink between the collective and the throttle that automatically increases the throttle when the collective is raised. In others, the pilot must add throttle when collective is raised. In fixed-wing aircraft the speed of the airflow over the wings is determined primarily by the airspeed of the airplane. This is not so with rotary-wing aircraft. The speed of the airflow over the rotors is determined not only by the airspeed of the helicopter, but also by the speed of rotation of the rotor blades. Let us first examine the case where the helicopter is in a hover and thus the airspeed is zero.

Hovering Blade Velocity

The velocity of each blade section depends on its location with respect to the rotor shaft. The local velocity is a function of the distance from the shaft and

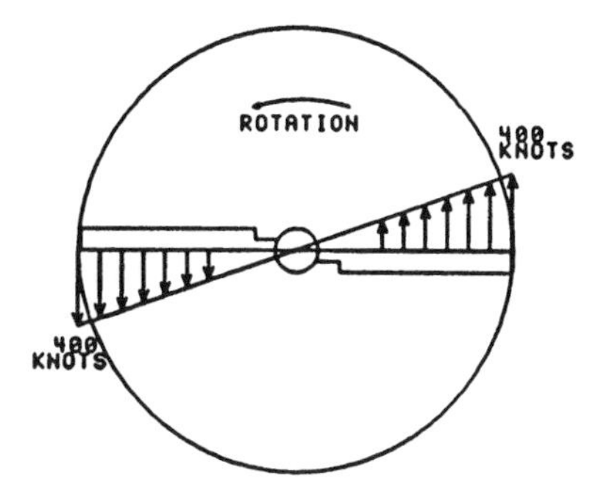

Fig. 10.13. Rotor velocity distribution in hover.

the rpm of the shaft. Figure 10.13 shows the velocity vectors of a typical two-blade rotor system having 20-ft rotors and operating at about 322 rpm. The rotor tip speed is about 400 knots.

Blade Twist

The relative wind of each blade is equal and opposite to the blade velocity. The speed of the blade increases as its distance from the shaft and, if each section operated at the same AOA, the lift would increase as the speed squared. (Remember the lift equation?) The blades are designed with a twist to prevent this uneven lift distribution. They have higher AOA at the root, which gradually decreases toward the tip.

Figure 10.14 shows the lift distribution on an untwisted blade and on an ideally twisted blade. Note that the twisted blade develops more lift near the root and less lift at the tip than the untwisted blade. The linear increase in lift of the twisted blade is desired.

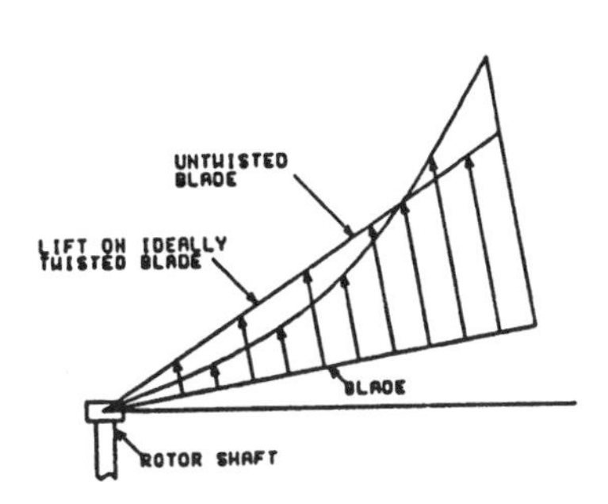

Fig. 10.14. Lift distribution on a twisted/untwisted blade.

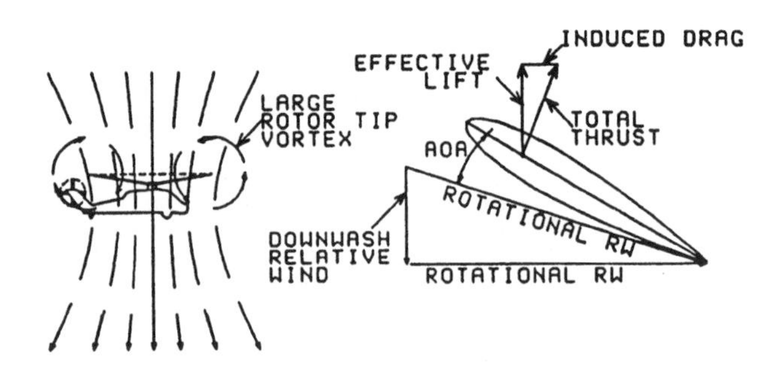

Fig. 10.15. Hovering out of ground effect.

GROUND EFFECT

In the early days of helicopter development it was found that it took much less power to hover a helicopter close to the ground than at a distance above the ground. Early theory attributed this to a build-up of a high-pressure "bubble" beneath the helicopter upon which the helicopter balanced. This also seemed to explain why the helicopter lost altitude as it moved in any direction. It "slid off the bubble." We have already disclaimed this theory as to the loss of lift (see Fig. 10.12); now we explain the reduction of power required to hover in *ground effect.*

In Chapter 5 we discussed induced drag and ground effect for fixed-wing aircraft. In case you missed it, please review those sections of the chapter. They apply to rotary-wing aircraft just as much as they do to fixed-wing aircraft.

Figures 10.15 shows a helicopter hovering out of ground effect and Fig. 10.16 shows one hovering in ground effect. Rotor tip vortices form when the rotors are producing lift. This is quite similar to the wingtip vortices discussed in Chapter 5 (see Fig. 5.5). The vortices cause the air to be accelerated downward causing a downwash behind the wing or rotor. In Fig. 10.15 the downwash relative wind affects the rotational relative wind, and the resultant

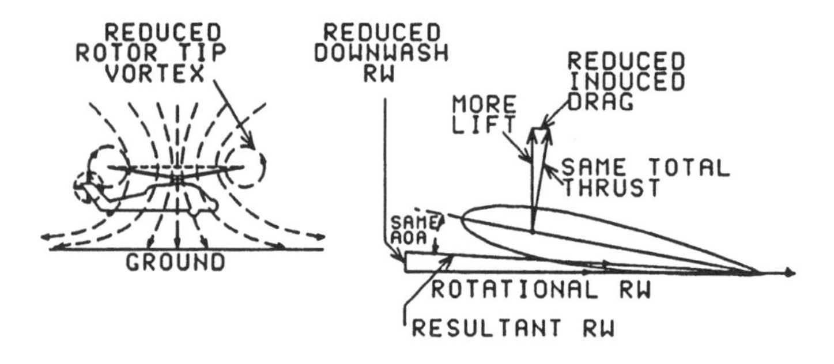

Fig. 10.16. Hovering in ground effect.

relative wind is as shown. The total thrust vector is 90° of the resultant relative wind and is tilted backward from the vertical. The effective lift is vertical and has a lower value than the total thrust. The rearward induced drag vector is also large. To summarize, downwash results in loss of effective lift and an increase in induced drag.

If the helicopter is hovered close to the ground (or water) the downwash is physically reduced, as shown in Fig. 10.16. The downwash relative wind is reduced and the angle that the resultant relative wind makes with the rotational relative wind is also reduced. Thus, the effective lift is increased and the induced drag is reduced. This reduction in drag results in a corresponding reduction in power required while in ground effect.

Torque

In American helicopters the rotor blades rotate in a counterclockwise direction, as viewed from above. According to Newton's third law of action and reaction, the fuselage tends to rotate in the opposite direction. This is called torque and must be counteracted before flight is possible. There are several ways of doing this. One method is to use a dual rotor system. These systems can be either tandem or side-by-side intermeshing rotors that rotate in opposite directions.

The most common form of canceling the torque effects is the antitorque rotor shown in Fig. 10.17. In this figure the helicopter is not in equilibrium. There is a balance of moments, but not a balance of forces. The sideward force of the antitorque rotor will cause the helicopter to drift to the right unless it is balanced by a force to the left. This is called "translating tendency." Figure 10.18 shows a method of compensating for this by tilting the main rotor slightly to the left.

ROTOR SYSTEMS

There are three primary types of rotor systems in use Each must accomplish the same task but do it in different ways.

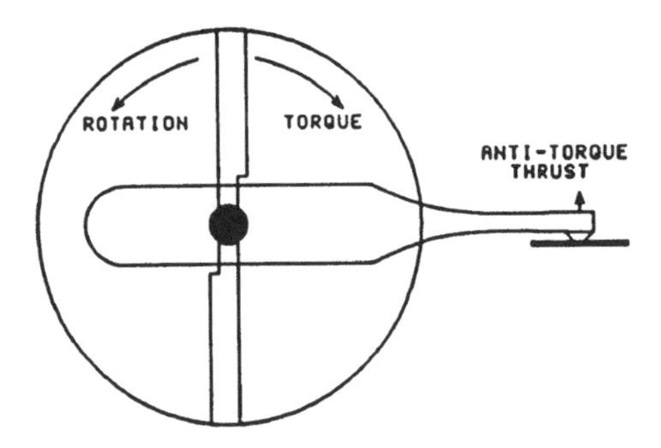

Fig. 10.17. Antitorque rotor.

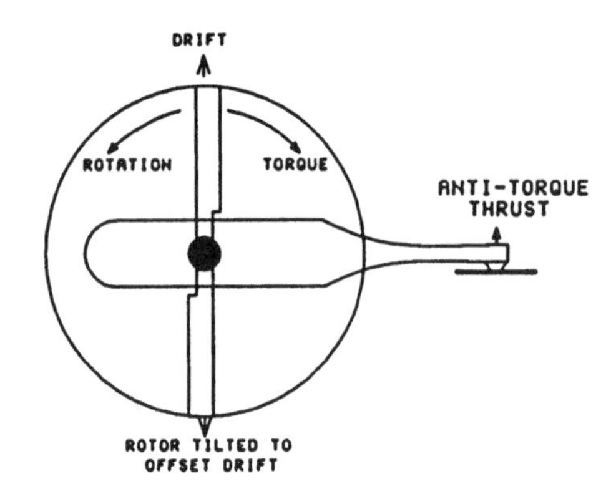

Fig. 10.18. Correction for antitorque rotor drift.

Rigid Rotor

The simplest system is the rigid rotor. It has but one degree of freedom. The angle of attack of the blades can be changed or *feathered*, but no other movement is allowed. Figure 10.19 is a schematic of the rigid rotor. If the blade is required to move up or down (flap) or to move forward or backward (hunt), the blade must be bent (flexed). The rotor itself absorbs these movements.

Semirigid Rotor (Seesaw)

The semirigid rotor system has two degrees of freedom. In addition to the feathering axis there is also freedom of flapping. No provision for hunting is available. Figure 10.20 shows a schematic of the semirigid system.

Articulated Rotor

Three degrees of freedom are provided by the articulated rotor system (Fig. 10.21). The vertical hinges allow hunting, the horizontal hinges allow flapping, and rotation of the blades allows them to change the angle of attack.

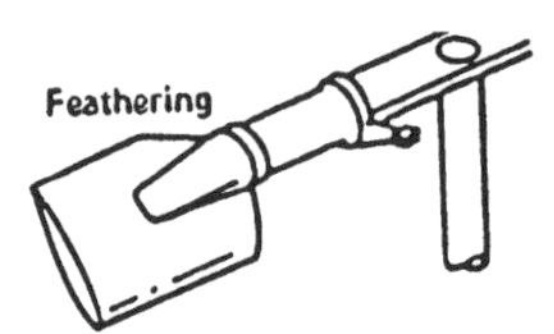

Fig. 10.19. Rigid rotor system.

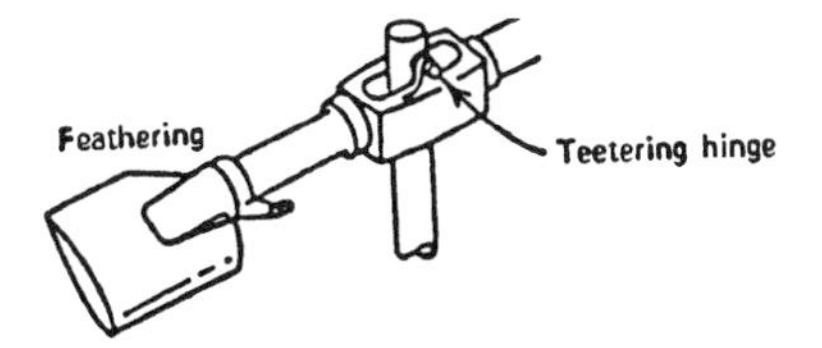

Fig. 10.20. Semirigid rotor system.

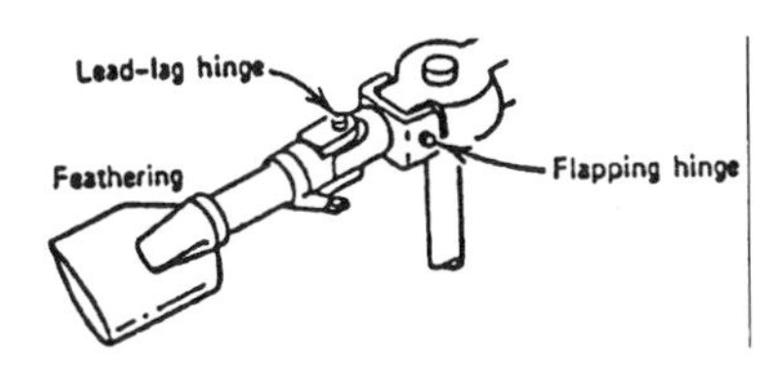

Fig. 10.21. Articulated rotor system.

DISSYMMETRY OF LIFT IN FORWARD FLIGHT

Early experiments in helicopter development produced an aircraft that could hover, but failed to achieve forward flight. The helicopter would roll over when it moved from a hover in any direction. To understand what caused this we must investigate *dissymmetry of lift*.

The airflow at the rotor tips that was caused by the rotation in a hover is shown in Fig. 10.22. As the helicopter starts to move in a horizontal direction, the airflow over the rotors is changed. For simplicity, consider that the helicopter is moving forward at 100 knots. The blade that is moving toward the front of the helicopter is called the *advancing blade* and the blade that is moving toward the rear is called the *retreating blade*. The advancing blade tip has an airspeed of 500 knots (Fig. 10.23), while the retreating blade-tip speed

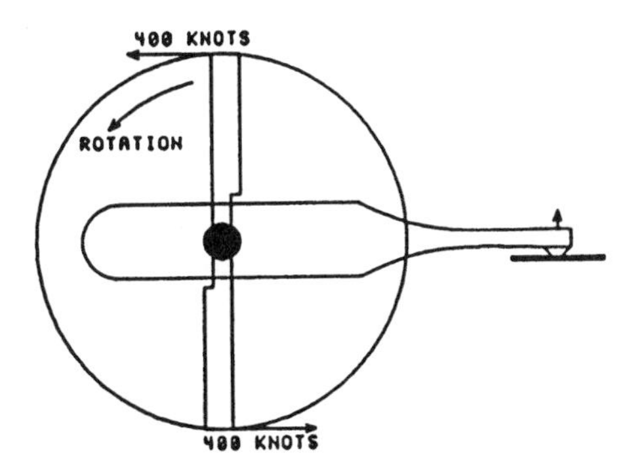

Fig. 10.22. Rotor tip velocities in a hover.

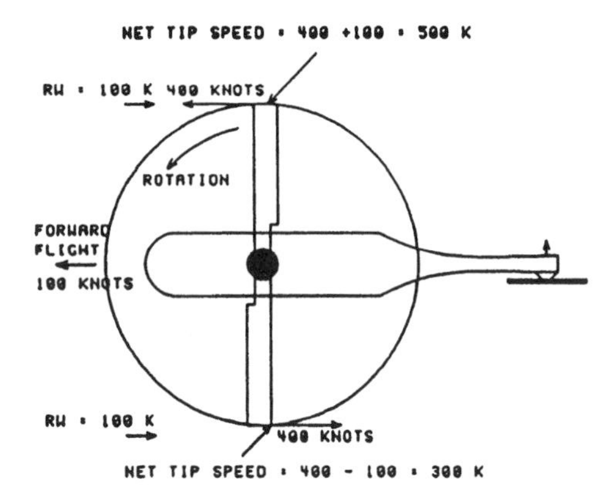

Fig. 10.23. Blade-tip velocity in forward flight.

is only 300 knots. If both blades are operating at the same AOA, the advancing blade will develop more lift than the retreating blade. This condition is dissymmetry of lift and cannot be tolerated. It is quite similar to flying a fixed-wing airplane whose ailerons are jammed in the full left wing down/right wing up position. The rolling moment that is produced is shown in Fig. 10.24.

Blade Flapping

The first successful solution to this problem was made by Juan de la Cierva, after he crashed his first prototype in his experimental autogyros from 1923 to 1935. He adapted the principle of designing rotors so that they could pivot up or down as they rotated. This motion is called *blade flapping*. As a blade leaves the tail position and advances around the right side of the helicopter, it is influenced by the increasing airspeed and produces more lift and climbs (flaps) upward. This climbing action changes the relative wind and reduces the angle

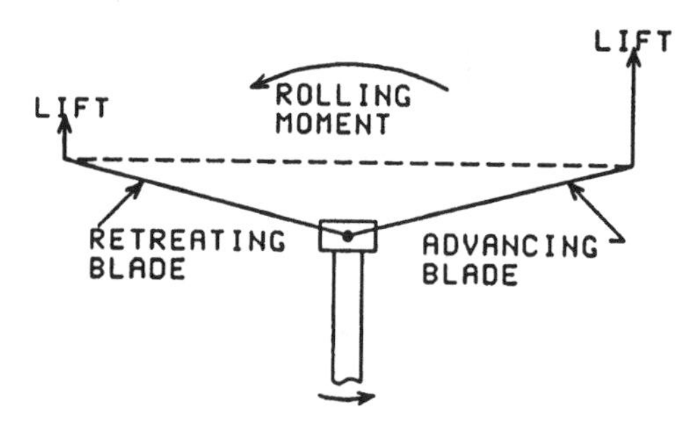

Fig. 10.24. Rigid rotor rolling moment in forward flight.

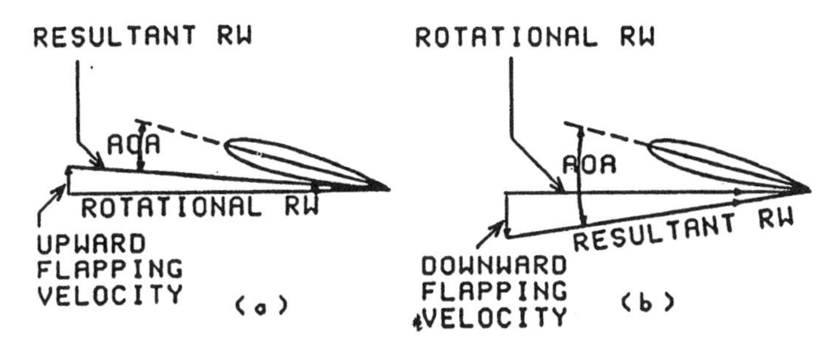

Fig. 10.25. Angle of attack and flight path changes: (a) advancing blade, (b) retreating blade.

of attack and thus the lift. Similarly, the retreating blade, as it moves from the nose of the helicopter toward the tail, experiences a reduction in airspeed, loses lift and flaps downward. This results in an increase in angle of attack and an increase in lift. The final result is that the lift is equalized, the rolling moment never materializes, and dissymmetry of lift does not exist.

Both the semirigid (seesaw) and fully articulated rotor systems have flapping hinges that automatically react to the changes in relative airspeed during blade rotation. Figure 10.25a shows the angle of attack changes on the advancing blade and Fig. 10.25b shows these changes on the retreating blade.

Figure 10.25a shows that the increasing speed of the advancing blade causes the blade to flap up. Since relative wind is always equal and opposite to the flight path, this upward movement causes the resultant relative wind to incline downward. This effectively reduces the angle of attack of the advancing blade.

Figure 10.25b shows that the decreasing speed of the retreating blade causes a loss of lift and the blade flaps down. This downward movement causes the resultant relative wind to be inclined upward, which increases the angle of attack of the retreating blade.

Blade Lead and Lag

The motion of blade lead and lag is called *hunting*. The hinges allow this action without requiring the blades to develop large bending stresses. The rigid and semirigid systems do not have these hinges. When the helicopter is moving horizontally, the blade pitch angles are constantly changing as the blades rotate to eliminate the dissymmetry of lift. Fully articulated rotor systems have vertical hinges that allow the blades to move forward and backward. As the pitch changes, so does the drag on the blades which causes the hunting action.

Another force, called Coriolis force, also causes the blades to lead and lag. A law of physics called the law of conservation of angular momentum states that the motion of a rotating body will continue to rotate with the same rotational velocity unless acted upon by some external force. If the mass is moved farther from the center of rotation it will decelerate. If the mass is

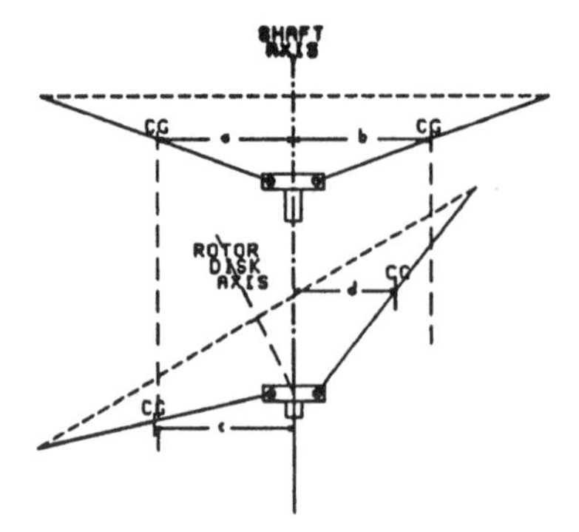

Fig. 10.26. CG radius change with flapping motion.

moved closer to the center of rotation it will accelerate. This means that as a rotor blade flaps up, its center of gravity moves closer to the drive shaft and it will speed up. When the blade flaps down, the opposite occurs.

Figure 10.26 shows this movement of the CG. The right rotor blade of the bottom sketch has flapped up and distance d is less than b. The blade will speed up (lead). The left blade has flapped down and the distance c is greater than a. The blade will slow down (lag). Figure 10.27 shows hunting.

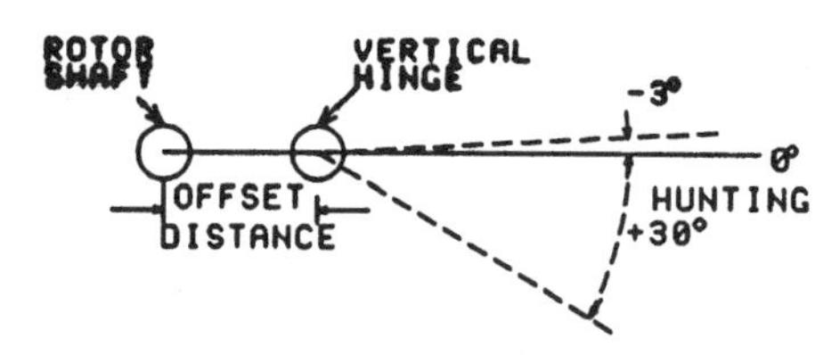

Fig. 10.27. Hunting motion of a fully articulated blade.

HIGH FORWARD SPEED PROBLEMS

As the forward speed of the helicopter is increased the relative wind speed over the advancing blade tip increases, while the relative wind speed over the retreating blade tip decreases. One of two possible problems arises at some critical forward speed: Either (1) the advancing blade tip approaches supersonic speed or (2) the retreating blade approaches stall speed.

Advancing Blade Compressibility

If the airspeed over an airfoil exceeds a certain value, called the critical Mach number, shock waves form on the upper surface of the airfoil. These shock waves cause the air to separate from the airfoil and a high-speed stall results. This causes a loss of lift and an increase in drag. A more detailed discussion of high-speed problems is found in Chapter 17.

Retreating Blade Stall

As the helicopter's forward speed increases, the relative wind over the retreating blade decreases. The resulting loss in lift causes the blade to flap down further and the effective angle of attack increases. At some high angle of attack the blades begin to become stalled. The tip stalls first and, if no corrective action is taken, the stall progresses inward. When approximately 25% of the disk is stalled, control of the helicopter is lost. Figure 10.28 shows the stall region and some typical angles of attack that the blades attain as they rotate.

Both advancing blade compressibility and retreating blade occur under similar conditions:

1. High forward speed
2. Heavy gross weight
3. Turbulent air
4. High-density altitude
5. Steep or abrupt turns

Both give warning by exhibiting abnormal vibrations under one or more of the above conditions. They differ as to rotor rpm: Advancing blade compressibility occurs at high rotor rpm, whereas retreating blade stall occurs at low rotor rpm. If corrective action is not taken, then advancing blade compressibility will pitch nose down and retreating blade stall will pitch nose up and roll the

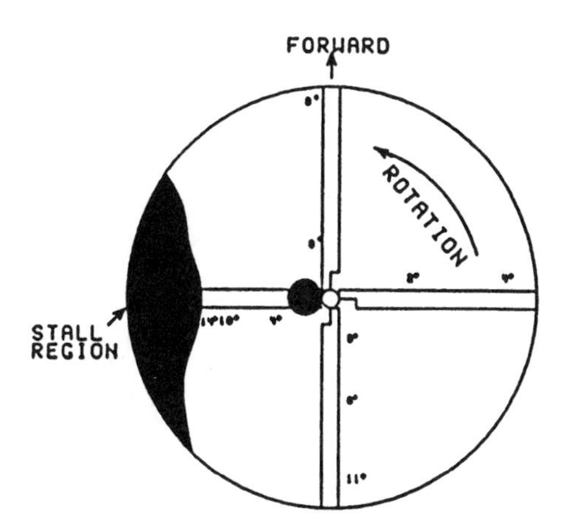

Fig. 10.28. AOA distribution during a retreating blade stall.

helicopter to the left. Corrective action for both advancing blade compressibility and retreating blade stall consists of

1. Checking and adjusting rotor rpm
2. Slowing the helicopter by reducing power, not by flaring the helicopter
3. Lowering blade AOA by using down collective
4. Leveling the helicopter if in a turn
5. Reducing *G*'s with down collective

The key is to unload the rotor blade.

Gyroscopic Precession

The rigid rotor helicopter acts more like a true gyroscope than either the seesaw or fully articulated types. The flapping hinges make the difference. However, all three types exhibit some of the characteristics of a gyroscope, as shown in Fig. 10.29.

A force applied to a rotating disk causes the disk to react 90° later in the direction of rotation. This explains the pitch-up and pitch-down reactions to advancing blade compressibility and retreating blade stall mentioned above. In either case, the blade stalls and lift is lost.

It is easy to see why the retreating blade stalls at high angle of attack, so we discuss that condition first. The blade is at the nine o'clock position when maximum angle of attack is encountered (see Fig. 10.28). Gyroscopic precession (Fig. 10.29) shows that the reaction occurs 90° later at the six o'clock or tail position, which causes a down load at the tail and results in pitch-up. Although it may seem strange that low angle of attack can produce stall, that is exactly what happens when critical Mach number is exceeded. Lift is lost at

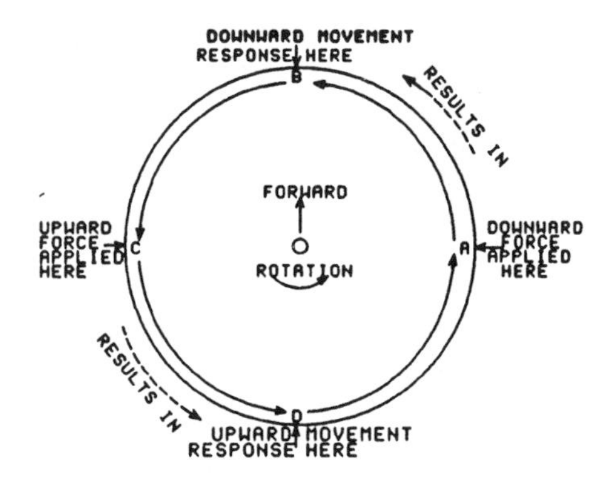

Fig. 10.29. Gyroscopic precession.

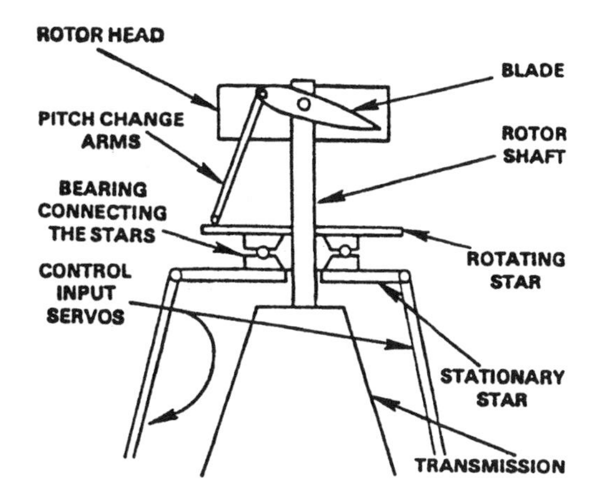

Fig. 10.30. Swash plate schematic.

the three o'clock position and the reaction occurs at the twelve o'clock position, causing pitch-down.

HELICOPTER CONTROL

The control of a helicopter consists of movement of the aircraft along the three principal axes and rotation about these axes. The rotation movements are called pitch, yaw, and roll and are discussed in some detail in Chapters 15 and 16. In the helicopter the movement along the longitudinal axis (forward and rearward flight) and along the lateral axis (sideward flight) is controlled by the cyclic stick. Movement along the vertical axis (vertical flight) is controlled by the collective stick. Pitch and roll are also controlled by the cyclic, while yaw is controlled by the foot pedals. This book is not intended to be a flight manual, so we will describe the action of these controls, rather than how to use them.

Rotor Head Control

Rotor control can be accomplished by direct tilting of the hub or by changing blade pitch. Blade pitch changing can be done by the swash plate, aerodynamic servo tabs, auxiliary rotors, jet flaps, and pitch links attached to a control gyro. The most common method is the swash plate (Fig. 10.30).

The stationary star and the rotating star are connected together by the bearing. This allows the rotating star, pitch change arms, blades, and rotor head to be rotated by the rotor head. The stationary star, control input servos, and transmission do not rotate. Both the rotating star and stationary star can move up and down on the rotor shaft. They both can also be tilted at various angles, but they always remain parallel to each other.

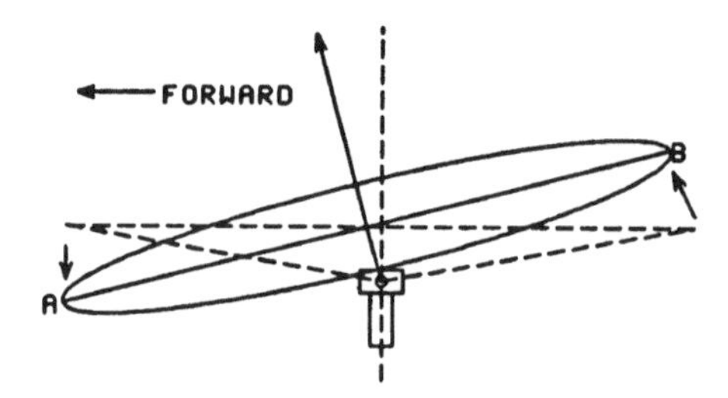

Fig. 10.31. Rotor flapping caused by cyclic stick movement.

Control of the Path

Vertical movement of the swash plate is controlled by the collective stick. If the pilot raises the collective, the pitch angle of all blades is increased and the helicopter climbs vertically. Down collective has the opposite effect. The cyclic stick tilts the stationary star through the control input servos. The rotating star remains parallel to the stationary star and changes the pitch angle of each blade by a different amount. This causes flapping of the blades and results in forces that change the tip path plane as shown in Fig. 10.31. Forward cyclic movement is shown.

Both cyclic and collective inputs can be combined so that horizontal and vertical motions can be coupled. Power is controlled by a motorcycle-type twisting hand grip on the collective stick.

Foot pedals change the blade angles of the antitorque rotor, changing its trust output and thus changing the heading of the helicopter. Figure 10.32 shows that the tail rotor is subject to dissymmetry of lift, similar to the main rotor system. A seesaw rotor hub allows blade flapping to eliminate this.

HELICOPTER POWER-REQUIRED CURVES

In the case of fixed-wing aircraft, we have seen that the power required consists of the power required to overcome induced drag and the power required to overcome the parasite drag. Rotary-wing aircraft have another kind of drag: the drag caused by the rotation of the rotor system. The power required to turn the rotors is given a new name—*profile power required* (P_{ro}). It is the power required to overcome the profile drag of the rotor system. Unlike the parasite drag of the helicopter, it occurs without any forward motion of the aircraft. The total power required is shown in Fig. 10.33. It is the sum of the power required to overcome the induced drag, parasite drag, and profile drag.

Power available has several small variations with helicopter speed, but for the sake of this discussion it is assumed to be constant. One other difference between rotary- and fixed-wing aircraft is the ability of the helicopter to fly at zero airspeed. The power-required curves do not terminate at stall speed, but continue down to zero airspeed.

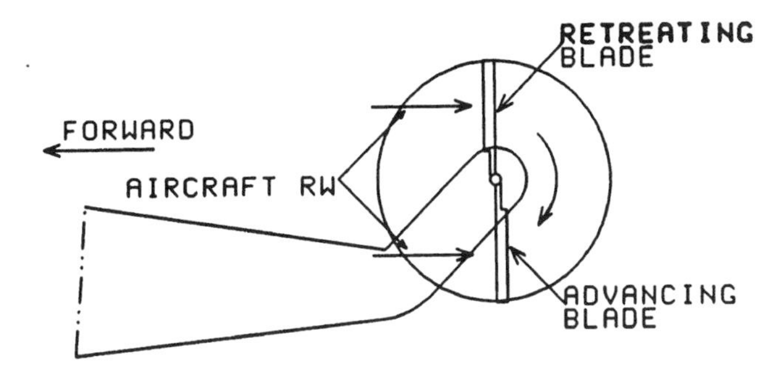

Fig. 10.32. Tail rotor dissymmetry of lift.

Notice that more power is required to hover than in intermediate-speed forward flight. This is true in or out of ground effect.

Translational Lift

It can be seen from Fig. 10.33 that there is a noticeable reduction in the total power-required curve as airspeed increases from zero (hover). This is caused by translational lift. The efficiency of the hovering rotor system is improved as the helicopter moves forward. In a hover there is much turbulence caused by the tip vortices. As the aircraft moves into a region of undisturbed air and the vortices are left behind, the airflow becomes more horizontal and the efficiency of the rotor system is improved.

SETTLING WITH POWER

One of the most dangerous flight conditions that the helicopter pilot can encounter is known as settling with power or, technically, the vortex ring state.

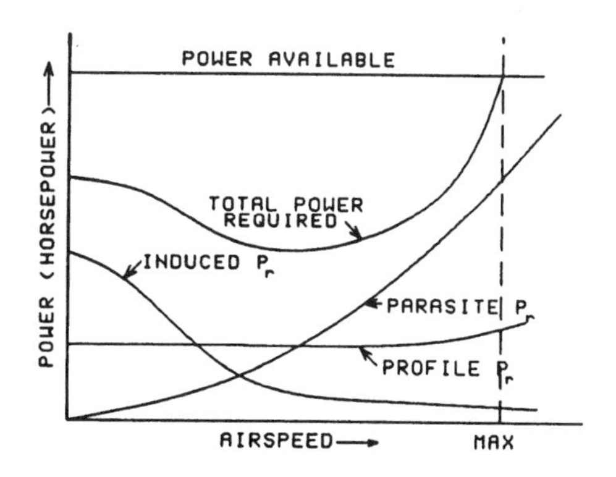

Fig. 10.33. Helicopter power available and power required.

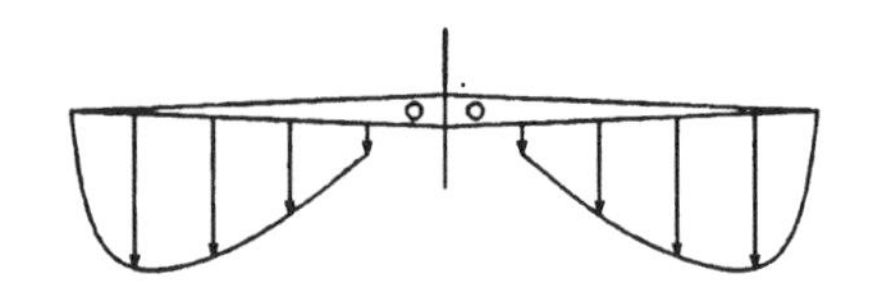

Fig. 10.34. Induced flow velocity in a hover.

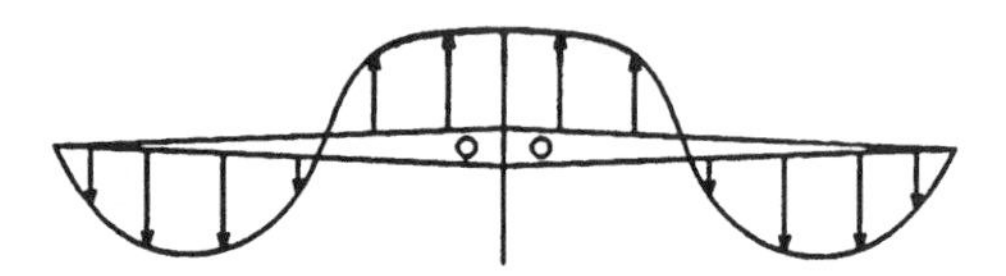

Fig. 10.35. Induced flow velocity during vortex ring state.

In this condition the helicopter is descending into its own downwash. This can happen when the aircraft is making a vertical or near vertical descent with low forward speed. The rotor system is under some power from the engine, not in an autorotation. A helicopter pilot attempting to land at a site surrounded by high trees or other obstructions cannot make a normal approach. Forward speed cannot be maintained so the pilot brings the helicopter to a hover at some height above the ground.

The airflow through the rotor system is as shown in Fig. 10.34. The maximum downward velocity is at the blade tips where the blade airspeed is highest and decreases nearer the rotor shaft. As the helicopter descends it is acted upon by an upward relative wind that counteracts the induced flow. The resulting induced flow velocity may be as shown in Fig. 10.35.

With upward and downward air flows in opposite directions, there is no lift on the helicopter and it is a free-falling body. Corrective action consists of increasing power (if available) to reestablish down flow. An alternative solution is to lower collective and increase forward speed. The normal tendency to increase collective pitch while applying power is wrong. This action can aggravate the power settling. If it is necessary to make a landing in a site where a forward speed of at least 15 knots cannot be maintained and there is sufficient power available, the aircraft could be brought to a hover and a very low rate of descent used.

AUTOROTATION

In forward flight the airflow through the rotor system and the force vectors are as shown in Fig. 10.36. To maintain constant rotor rpm, the drag vector must be counteracted by an equal and opposite force supplied by the engine. If the engine fails, or is deliberately disengaged from the rotor system, some other force must be applied to sustain the rotor rpm. This unpowered flight is similar

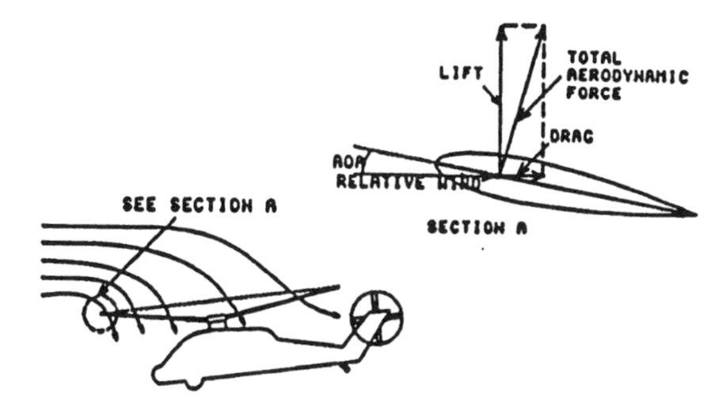

Fig. 10.36. Airflow and force vectors in forward flight.

to the glide of a fixed-wing aircraft and is called autorotation. The force that keeps the blades rotating is generated by lowering the collective pitch and thus reducing the angle of attack of all blades.

As the helicopter loses altitude, the airflow through the rotor disk changes, as shown in Fig. 10.37. The airflow during descent provides the energy to overcome the blade drag and turn the rotor. The helicopter is trading the potential energy that it has due to its altitude into kinetic energy used to turn the blades. Later we see how this kinetic energy is used to cushion the landing.

Note that the total aerodynamic force is now vertical. The forward component of the lift vector equals the backward component of the drag vector. There are no unbalanced forces in the horizontal direction and rpm will remain constant. Thus, forward speed is important for successful autorotation from altitude. To make a successful landing, the pilot must stop the forward airspeed and reduce the rate of descent just before landing. Both of these actions can be started by flaring the aircraft. This is done by moving the cyclic stick to the rear, which tilts the rotor disk to the rear and slows the forward speed. This is shown in Fig. 10.38. There is also an increase in rotor rpm, which will help to

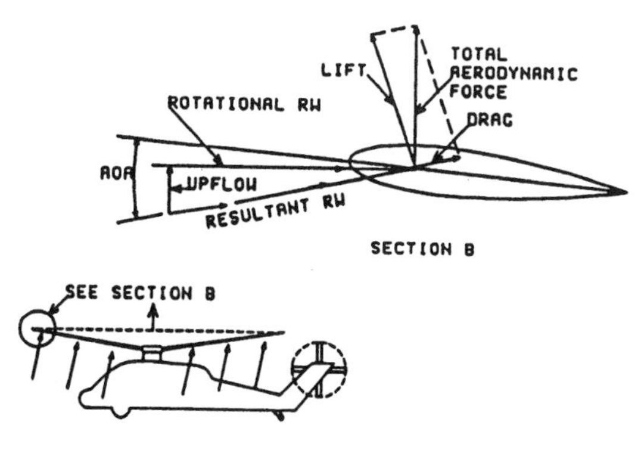

Fig. 10.37. Airflow and forces in steady-state descent.

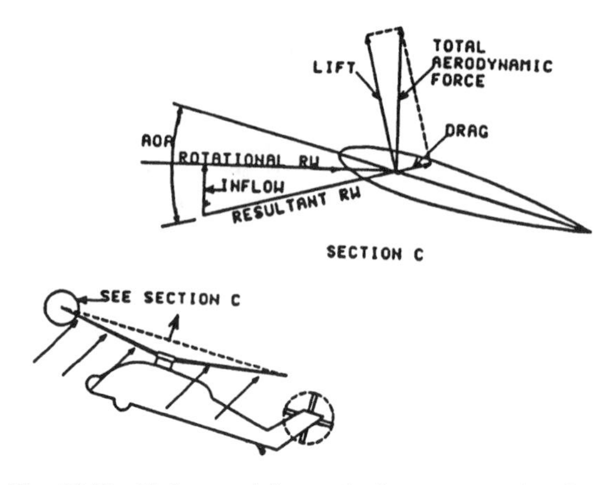

Fig. 10.38. Airflow and forces during autorotative flare.

cushion the landing. After forward speed stops, the cyclic is returned to neutral and the final vertical velocity is further reduced by using up collective. The collect pitch is also used to avoid rotor overspeed.

Vertical autorotations are seldom attempted except during engine failure at hover. The pilot has much better visibility of the landing zone and better directional control if forward velocity is maintained.

PROBLEMS

1. Most helicopter rotors have symmetrical airfoils because
   a. they produce more lift than cambered airfoils.
   b. they can produce both upward and downward lift.
   c. they do not develop pitching moments.
   d. All of the above

2. The lift theory most often used in helicopter aerodynamics is
   a. the gyroscopic precession theory.
   b. the momentum theory.
   c. Bernoulli's theory.
   d. the settling with power theory.

3. A helicopter can hover near the ground with less power than it can hover away from the ground because
   a. the air is denser near the ground.
   b. a high-pressure air "bubble" is produced below it.
   c. the rotor blades have less induced drag.
   d. None of the above

4. The primary stress in rotor blades in flight is

a. tension.

b. bending.

c. torsion.

d. shear.

5. As a helicopter moves forward from a hover to a speed of about 15 knots, the power required is reduced because

a. the efficiency of the rotor is improved.

b. the helicopter is moving into a region of undisturbed air.

c. the tip vortices are left behind.

d. All of the above

6. If a helicopter is flying at high speed and the engine fails, the pilot should enter an autorotation by slowing the helicopter by using back cyclic then reducing blade angles with the collective.

a. True

b. False

11 Slow-Speed Flight

Throughout the history of aviation, the slow-speed region of flight has been the most hazardous. The reasons for this are quite different for straight-wing aircraft and for swept-wing aircraft. Examination of the coefficient of lift curves for each type of aircraft will help explain some of these differences.

Figure 11.1 shows the lift vs. AOA curves for a straight-wing, propeller-driven aircraft and a swept-wing, turbojet aircraft. Four differences can be seen from these curves:

1. The straight-wing aircraft has a higher value of C_L. Recall the basic lift equation

$$L = \frac{C_L \sigma V^2 S}{295}$$

It can be seen that if the lift equals the aircraft weight, stall speed, V_S, will be a minimum when the value of the lift coefficient is a maximum, $C_{L(max)}$. The higher the value of $C_{L(max)}$, the lower the stall speed will be. This means that the straight-wing aircraft has a lower stall speed for the same weight and wing area.

2. The swept-wing aircraft must fly at a higher AOA to achieve maximum lift.
3. There is a sudden reduction of C_L for the straight-wing aircraft at stall, but not for the swept-wing aircraft.
4. The straight-wing aircraft is more sensitive to AOA changes.

For purposes of this discussion, let us compare fighter aircraft in each category. The straight-wing, propeller fighter aircraft of World War II was susceptible to slow-speed stalling. The basic principles of the stall were discussed earlier. We now discuss stall patterns.

STALL PATTERNS

The lift distribution in the spanwise direction is shown in Fig. 11.2. It is in the shape of an ellipse. The lift is zero at the wingtips and increases elliptically, with the maximum lift occurring at the wing root. If the wing area is exactly

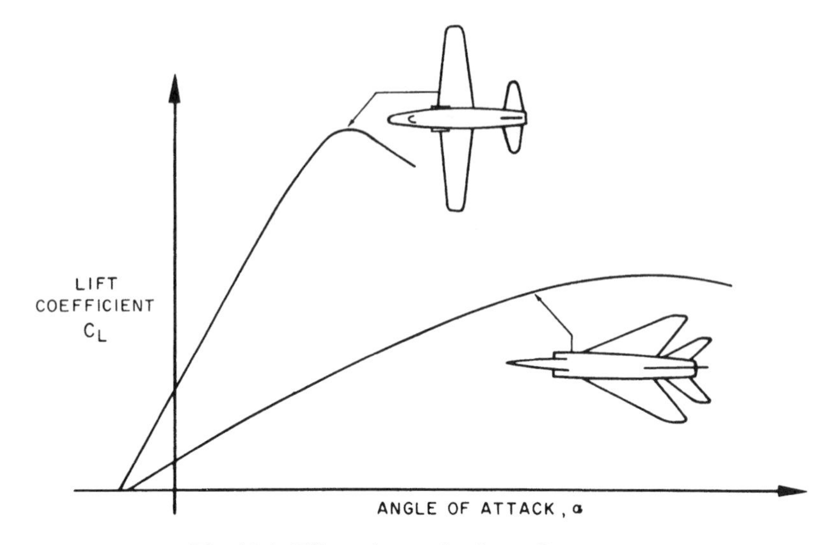

Fig. 11.1. Effect of sweepback on C_L–α curves.

proportional to the lift at all points on the wing, the local wing loading will be a constant value everywhere on the wing. Stall of a wing will occur where the ratio of local lift coefficient, C_l, to the wing lift coefficient, C_L, is highest. If the wing planform area is also elliptical, the local wing loading will be constant. This is shown by the value of $C_l/C_L = 1.0$ in Fig. 11.3. Thus, the entire trailing edge of an elliptical planform wing will stall at the same time. The elliptical wing is considered to be the most efficient planform and was used with much success on the British Spitfire. Structurally, the elliptical wing is difficult to manufacture, so it is not widely used.

The straight wing is usually rectangular or it has a slight taper. The wing area near the wingtip is large compared to the lift in that area, and the wing loading is, therefore, low. Figure 11.3 shows that the value of C_l/C_L approaches zero near the wingtip for the rectangular wing. The maximum wing loading is at the wing root. Stall will, therefore, start at the trailing edge of the wing root and then spread outward and forward as the stall progresses, as shown in Fig. 11.4.

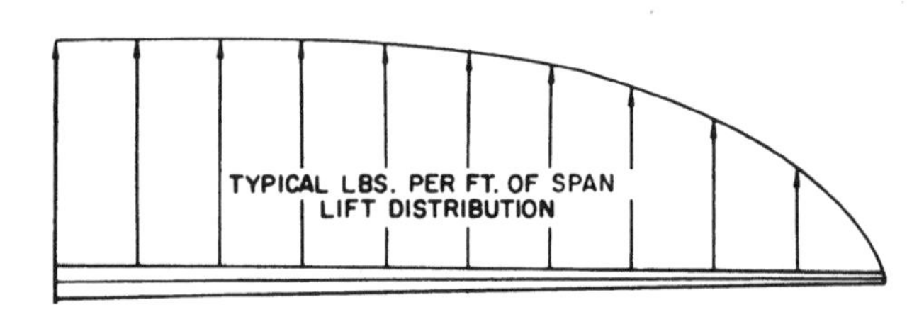

Fig. 11.2. Spanwise lift distribution.

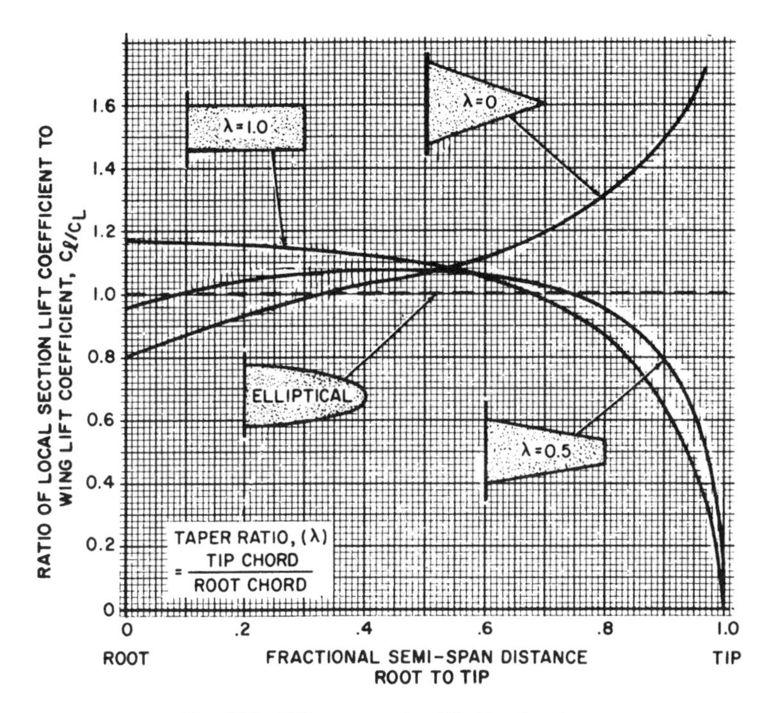

Fig. 11.3. Wing spanwise lift distribution.

Swept wings have more taper, with the extreme being the delta wing, which has a taper ratio of zero. The delta wing's area at the tip is very small, and the wing loading is very high. Stall starts where the wing loading is highest, so the stall starts at the wingtip and progresses inboard and forward. The high tip wing loading can be seen from the values of C_l/C_L in Fig. 11.3. The stall pattern is shown in Fig. 11.4.

Straight-wing aircraft, with stall starting at the roots, seem to have advantages over swept-wing aircraft, as far as stall characteristics are concerned. First, there is more adequate stall warning, caused by the separated air buffeting the fuselage. Second, the ailerons of the straight-wing aircraft are not enveloped in the stalled air until the stall has progressed outward from the wing roots, compared to the relatively early envelopment from tip stalling of the swept wing. Swept-wing aircraft are usually equipped with stall warning devices to compensate for the lack of buffet warning.

The lack of a sharp stall point on the C_L curve of a swept-wing aircraft makes it nearly impossible to stall it under normal low-speed flight conditions. However, this does not mean that the low-speed region is safe for swept-wing aircraft. In Chapter 5 we saw that induced drag was very high at low airspeeds. Induced drag is also greater with aircraft having low aspect ratios. Sweeping the wings reduces the span and effectively increases the chord. Aspect ratio is the span divided by the average chord, so swept-wing aircraft, with the same

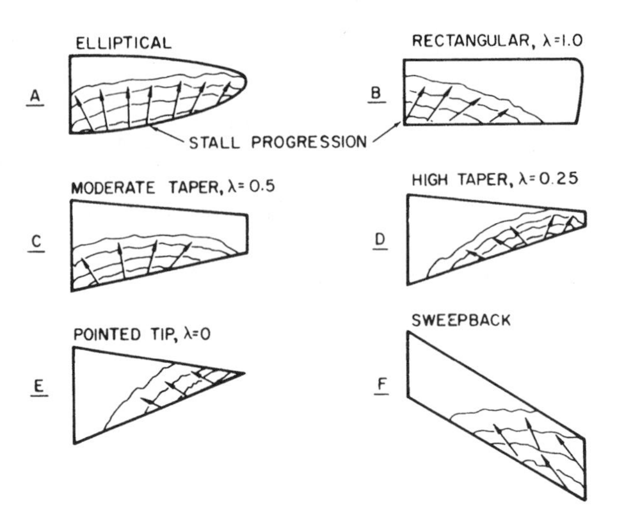

Fig. 11.4. Stall patterns.

wing area, have lower values of aspect ratio. The high drag existing at low airspeeds was evident in the comparison of thrust-required curves of the swept-wing jet aircraft to the power-required curves of propeller-driven straight-wing aircraft.

Because the high drag on jet aircraft at low airspeeds is so important, a discussion of the *region of reversed command* is warranted. Most pilots are familiar with this flight region as the "backside of the thrust curve" or, incorrectly, as the "backside of the power curve." This last expression is correct only when propeller aircraft are being discussed. Propeller aircraft are not hindered by a lack of power in the low-speed region, so only jet aircraft and thrust-required curves are considered here.

REGION OF REVERSED COMMAND

A typical thrust-required curve for a turbojet aircraft in the dirty configuration is shown in Fig. 11.5. Minimum drag occurs at the $(L/D)_{max}$ point. For this aircraft these figures are 2400 lb thrust required at 160 knots TAS. All airspeeds greater than that for minimum drag are said to be in the region of normal command. In this region added thrust must be supplied if greater airspeed is desired. To increase the airspeed to 200 knots, the pilot must add throttle until the engine(s) produce 2750 lb of thrust.

In the region of normal command, thrust is directly proportional to velocity. All airspeeds below that for minimum drag are said to be in the region of reversed command. In this region, the thrust required to fly the aircraft is inversely related to the airspeed. The slower the aircraft flies, the greater the

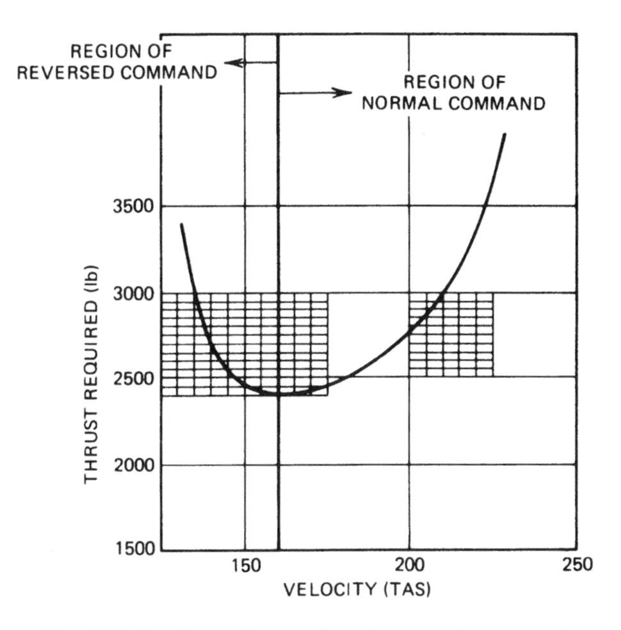

Fig. 11.5. Regions of normal and reversed command.

thrust required. At an airspeed of 140 knots, for instance, the T_r is 2750 lb, the same as was required at 200 knots.

It can be reasoned that "if more thrust is required to fly slower, then to slow down the pilot should add throttle." This, of course, is not true. The fallacy of this statement lies in the fact that, at any stabilized airspeed, the thrust available equals the thrust required, and adding throttle produces a thrust excess. This excess will cause the aircraft either to climb or to accelerate. The correct statement should be, *If, for any reason, the aircraft should be slowed down, more thrust is required to maintain altitude at that airspeed.*

Take the case of an aircraft flying at minimum drag. This is 160 knots for the aircraft in Fig. 11.5. The T_r is 2400 lb, and the T_a must also be 2400 lb for stabilized flight. There is no excess thrust, so the aircraft cannot climb under these conditions. To slow the aircraft, the pilot has several alternatives: (1) increase the drag by using speed brakes or by lowering the landing gear; (2) retard the throttle thus reducing thrust; (3) increase the AOA by back pressure on the control stick.

Let us examine these alternatives in more detail. Increasing the drag is certainly a viable method of slowing the aircraft. That is why speed brakes were invented. However, speed brakes and lowering the gear usually result in rapid changes in airspeed, so let's examine how we could slowly and gradually reduce the airspeed.

Reducing the throttle results in a thrust deficiency. At any stabilized flight airspeed, a thrust deficiency results in a rate of sink. This can be seen by an adaptation of Eq. 6.4:

$$\text{Rate of sink} = 101.3V\left(\frac{D - T}{W}\right) \quad \text{(fpm)}$$

The pilot who elects to slow the aircraft by this method cannot maintain altitude. Any attempt to increase AOA to compensate for less thrust will merely decrease the airspeed and increase the drag. This results in an even greater thrust deficiency and greater rate of sink.

The third method of slowing the aircraft is to increase the AOA. In our study of the basic lift equation, we made a point of the fact that the AOA controlled the value of the lift coefficient, and this in turn controlled the airspeed. We said, *Angle of attack is the primary control of airspeed in steady flight.*

Let's see how this works. The pilot pulls back on the control stick and thus increases AOA. Lift coefficient is increased, airspeed is reduced, and drag is increased. A deficiency now exists in thrust and a sink rate will develop. So, the pilot must add throttle to maintain altitude. This can be a smooth coordinated method of slowing the aircraft without loss of altitude and without excessive "throttle jockeying."

Another example of using AOA to control airspeed is illustrated in Fig. 11.6. Suppose a pilot is flying an airplane having the T_r curve shown in Fig. 11.6 and

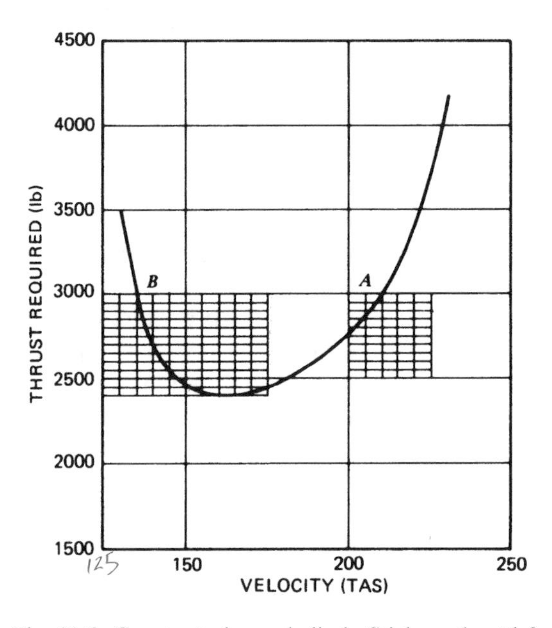

Fig. 11.6. Constant airspeed climb. Stick or throttle?

the throttle is set to produce 3000 lb of thrust. For stabilized flight, the airplane must be at point *A* or at point *B*. At point *A* the airspeed is 210 knots, and at point *B* it is 135 knots. To climb the airplane from point A, the pilot can pull back on the control stick and increase the AOA. This will increase the C_L and decrease airspeed. If this maneuver slows the aircraft to 200 knots, the thrust available is still 3000 lb, but the thrust required has dropped to 2750 lb. The aircraft now has an excess thrust of 250 lb and will be able to climb.

Now, suppose the plane is at point *B*. To climb the aircraft the pilot pulls back on the control stick and increases the AOA. The airspeed will be reduced again, but this time the thrust required will be increased. If the airspeed is decreased to 130 knots, the thrust required will be about 3500 lb, for a thrust deficiency of 500 lb. The climb method that worked in the region of normal command will not work in the region of reversed command.

There is a method of climbing the aircraft that works at all airspeeds, and it is strongly recommended. Always control the airspeed with the stick. If the airspeed is held constant, thrust required will be constant. Climb is then controlled by the excess thrust, and the thrust available is determined by the throttle setting. So, *throttle setting controls the rate of climb or descent in steady flight.*

Flight in the region of reversed command cannot be avoided. Every takeoff and landing requires flight in this region. Pilots must be careful not to paint themselves into a corner, where drag is so high that enough thrust may not be available to overcome it. When the drag (T_r) exceeds the thrust available (T_a) there is only one way to go—*down.*

LOW-LEVEL WIND SHEAR

Wind shear can be defined as a sudden and prolonged change in wind direction and/or speed, relative to an airplane's ability to accelerate. While the wind-shear phenomenon can occur at altitude, it is most hazardous during landing and takeoff operations, so this discussion is limited to low-level wind shear. Under unusual weather conditions, wind-direction changes of 180° and wind-speed changes as high as 50 knots have been observed.

We have been taught that wind cannot affect an aircraft in flight except for groundspeed and drift. This is true for steady winds or winds that change gradually, but it is not true when rapid wind changes occur. When the air velocity changes, the mass of the aircraft must be accelerated or decelerated, and this takes time. Slow-accelerating airplanes will have more trouble recovering from wind shear.

Wind shears are caused by thunderstorm activity, weather fronts, and low-level jet streams. Figure 11.7 illustrates a wind-shear situation caused by a thunderstorm. The downward air flow directly below the thunderstorm is called a *downburst.* Investigation of an airline crash revealed that vertical velocities in excess of 1800 fpm were encountered in such a downburst. The

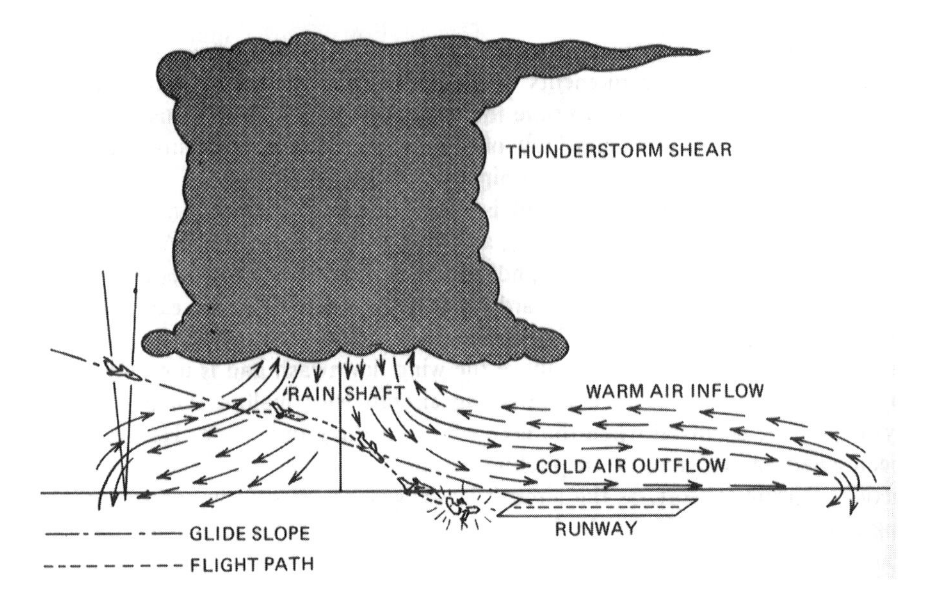

Fig. 11.7. Wind shear caused by a thunderstorm.

downburst is the most dangerous type of wind shear because it can force the airplane into the ground.

When a downburst hits the ground, it is diverted outward. This creates horizontal air velocities known as *headwind bursts*, *tailwind bursts*, and *cross-wind bursts*, depending on the position of the aircraft relative to the center of the thunderstorm. These are shown in Fig. 11.8, as viewed from above. The performance of an aircraft flying through these bursts is discussed in the next section.

Another hazard associated with thunderstorms is called first gust. This is a wind shift line or gust front that may precede a thunderstorm by as much as

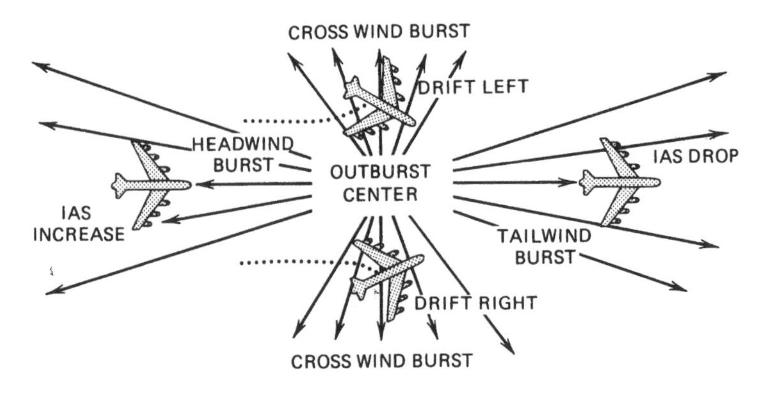

Fig. 11.8. "Bursts" caused by a thunderstorm.

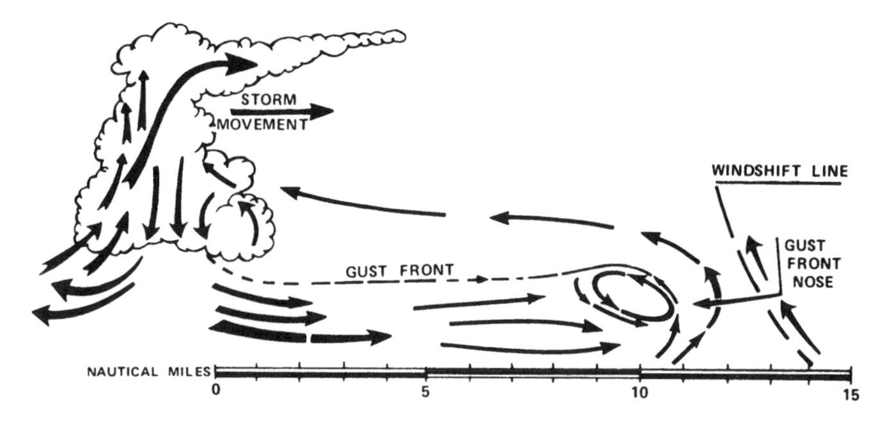

Fig. 11.9. Thunderstorm gust front.

15 miles. This is shown in Fig. 11.9. The slope of the gust front, makes it impossible to determine when an aircraft will fly through it.

Wind shear may also be present in certain types of cold and warm fronts. Turbulence may or may not exist in wind-shear conditions. Most fronts have shallow wind gradients; thus, the changes in wind direction and velocity are gradual. Other fronts have steep wind gradients and severe amounts of wind shear. A rule of thumb to determine the possibility of wind shear is this: If a temperature differential of 10°F or greater exists across the front and/or if the frontal speed is 30 knots or more, there is the possibility of significant low-level wind shear.

A third source of low-level wind shear is caused by temperature inversions. These are called low-level jet streams. They are caused by warm air moving above a pocket of cool, calm air, as shown in Fig. 11.10. This type of wind shear can be found on clear nights, when the ground is dry and smooth. Wide temperature variations during the day and night period help create this situation. The horizontal winds at 1000 ft above ground

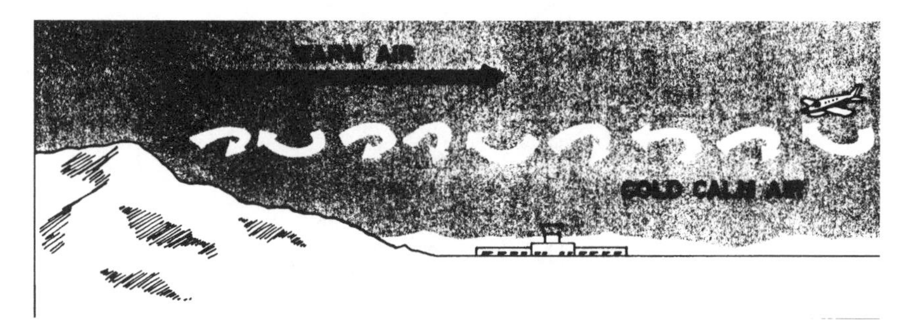

Fig. 11.10. Low-level jet stream.

level have been found to be as much as 70 knots, while the surface winds remained calm.

Little or no turbulence, or vertical wind velocity, is produced inside the low-level jet stream. When an aircraft is flying at a constant altitude, no significant wind shear will occur. Large variations in headwind, tailwind, and crosswind components can be experienced, however, when climbing or descending through a low-level jet stream.

AIRCRAFT PERFORMANCE IN LOW-LEVEL WIND SHEAR

During Takeoff and Departure

The most hazardous condition caused by a wind shear during takeoff is when an increasing tailwind is encountered shortly after takeoff, as shown in Fig. 11.11. The sequence of events is as follows:

1. Takeoff appears normal.
2. Tailwind wind shear encountered just after liftoff.
3. Airspeed decreases, resulting in pitch down moment.
4. Airplane crashes.

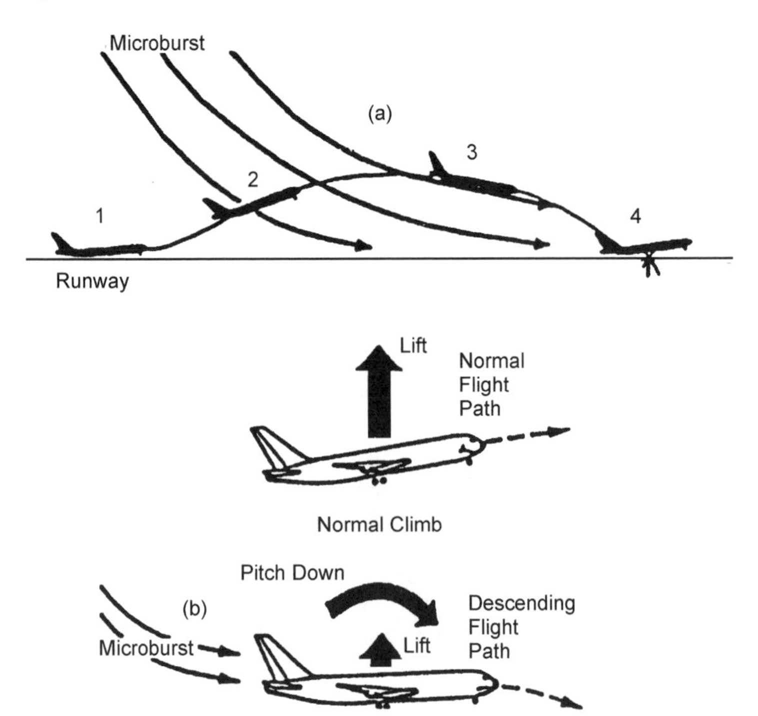

Fig. 11.11. Tailwind wind shear encountered on takeoff: (a) flight path, (b) forces and moments.

During Approach to a Landing

Wind shear encountered during final approach can lead to dangerous situations. The two most likely conditions are discussed here.

1. If an aircraft passes through a front where a wind shift from a headwind to either a tailwind or a reduced headwind occurs, then the shearing situation illustrated in Fig. 11.12 may exist. The sequence of events is
   a. Normal approach.
   b. Increasing downdraft and tailwind encountered.
   c. Airspeed decreases and pitch down results.
   d. Aircraft crashes short of runway.

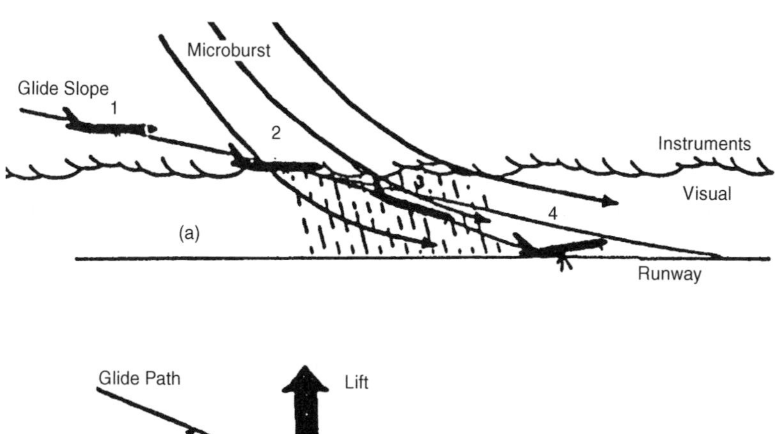

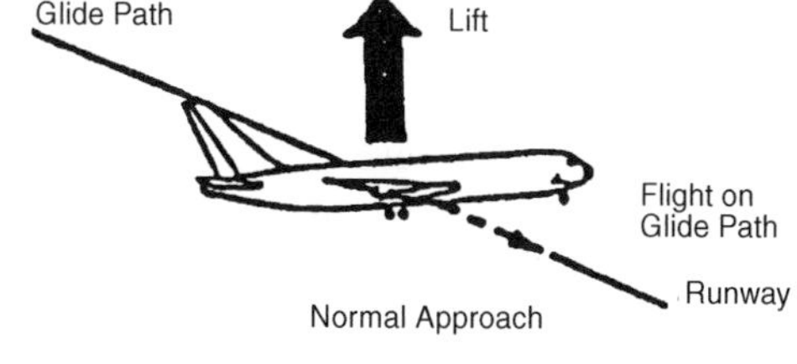

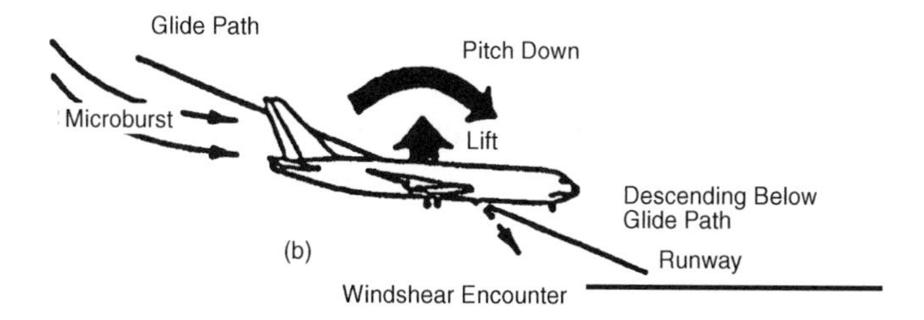

Fig. 11.12. Tailwind shear encountered in landing approach: (a) flight path, (b) forces and moments.

2. When an aircraft makes an approach in a tailwind condition that changes to a headwind or calm condition, the reverse situation occurs. When the wind shear is encountered, the IAS increases, the aircraft pitches up, and it climbs above the glide slope. If no thrust correction is made the aircraft climbs above the flight path and a high and fast approach results with the possibility of overshooting the runway. If the pilot reduces thrust after going high on the glide slope, then thrust must be reestablished as the aircraft returns to the glide slope or the airplane will go below the flight path and a short, hard landing will result.

Crosswind Burst Response

A crosswind shear burst causes the airplane to roll and/or yaw. Large crosswind shears may require large and rapid aileron inputs. Wind shears may increase the pilot workload and cause increased distraction. Horizontal vortices may also be encountered, which can cause severe roll forces and require full aileron movement to maintain aircraft control.

Turbulence Effects

Turbulence differs from wind shear in that it is sporadic. Airspeed may fluctuate in an unpredictable pattern. This can make changes in airspeed difficult to detect and thus delay the recognition of the presence of wind shear.

Heavy Rain

Wind shears are often accompanied by heavy rain. Earlier studies of rain effects at altitude concluded that rain had a significant effect on performance, but, due to sufficient altitude, would not force the aircraft to the ground. In the approach or post takeoff flight phase, however, conditions are much more critical and the performance margin of the aircraft is considerably reduced.

Rain affects the aircraft in several ways:

1. The momentum of the raindrops impact the aircraft in a downward and backward direction
2. The increased weight of the rain water film affects the gross weight of the aircraft.
3. The water film is roughened by the impact of the raindrops and the aerodynamic properties of the wing are adversely affected.
4. The raindrops hit the airplane unevenly, causing possible pitching or rolling moments.

Of these effects, the roughened airfoil and resulting loss of aerodynamic properties is the most important. Recent studies have indicated that as much as a 30% increase in C_D and a loss of lift of more than 30% can occur due to this roughness. The AOA for $C_{L(max)}$ can also be reduced by 2 to 6°. At least nine major commercial aircraft accidents investigated and reported by the National Transportation Safety Board (NTSB) have involved heavy rain during takeoffs and landings.

EFFECT OF ICE AND FROST

The formation of ice or frost on the lifting surfaces of an airplane will cause deterioration of the lifting ability of the surfaces. The effects of ice and frost on the lift coefficient curves are shown in Fig. 11.13. Ice is likely to accumulate near the leading edge if it is deposited during flight. This changes the contour of the airfoil and results in a considerable reduction of lift. The smoothness of the skin is also reduced, and a large increase in drag results. As a result of the

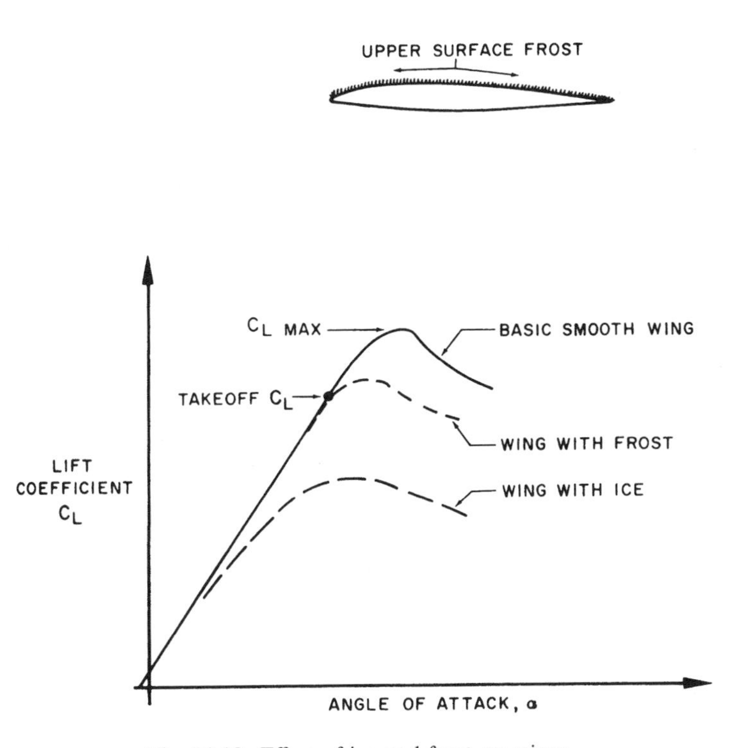

Fig. 11.13. Effect of ice and frost on wings.

reduced lift and the increased drag, thrust or power required will be greater and the stall speed will increase. A third effect will be that the weight of the ice will increase the gross weight of the aircraft.

Obviously, all ice must be removed before flight, and any ice formed during flight must be removed using the aircraft's de-icing equipment. If the plane does not have de-icing equipment, a 180° turn is strongly recommended.

The effect of frost on the wings is not so obvious. A layer of evenly distributed frost will not appreciably change the contour of the wing but it will increase the roughness. Drag will be increased, and the kinetic energy of the air will be reduced so that stall will occur at a lower AOA and at a higher airspeed. One study showed that 0.1 in. of evenly distributed frost on the aircraft's wings will increase the stalling speed by 35%. This roughly doubles the required takeoff run.

WAKE TURBULENCE

Turbulence caused by aircraft in flight was once attributed to "propwash." However, this phenomenon has become better understood and is now classified in two categories. *Thrust stream turbulence* is one of these types. It is caused by high-velocity air from propeller blades or from jet exhausts. This type of disturbance is of primary concern during ground operations, where it can be dangerous to ground personnel and to loose gear and can cause foreign object damage (FOD) to aircraft and engines. It should not be a flight hazard except in close formation flying, or during takeoff and landing behind an aircraft that is making a ground runup.

Thrust stream turbulence dissipates fairly rapidly compared to the second category of turbulence, *wake turbulence*. Wake turbulence, is generated during flight and is caused by wingtip vortices. Wingtip vortices were discussed in Chapter 5, in connection with induced drag.

We will discuss wake turbulence as a hazard to other aircraft. A *vortex* is a highly developed rotational mass of disturbed, high-energy air created by the wing of an aircraft as it produces lift. An aircraft creates two such rotating vortices, as shown in Fig. 11.14. As an aircraft that is producing lift passes through the air, there is a differential pressure on the bottom and the top of the wing. This causes a movement of air around the wingtips. The air rolls into two distinct vortices, one at each wingtip. The rotating process is normally complete in about 2 to 4 times the wingspan (200 to 600 ft) behind the aircraft. Vortices are also developed by helicopters and trail along behind the aircraft in the same manner as those of fixed-wing aircraft. This rotational energy is directly related to the weight, lift being generated, and wingspan of the aircraft. It is inversely related to the airspeed and, therefore, the vortices are a maximum during takeoff and landing at high gross weights.

The force of the air in wingtip trailing vortices can easily exceed the aileron control capacity or climb rate of light aircraft. These forces can cause some

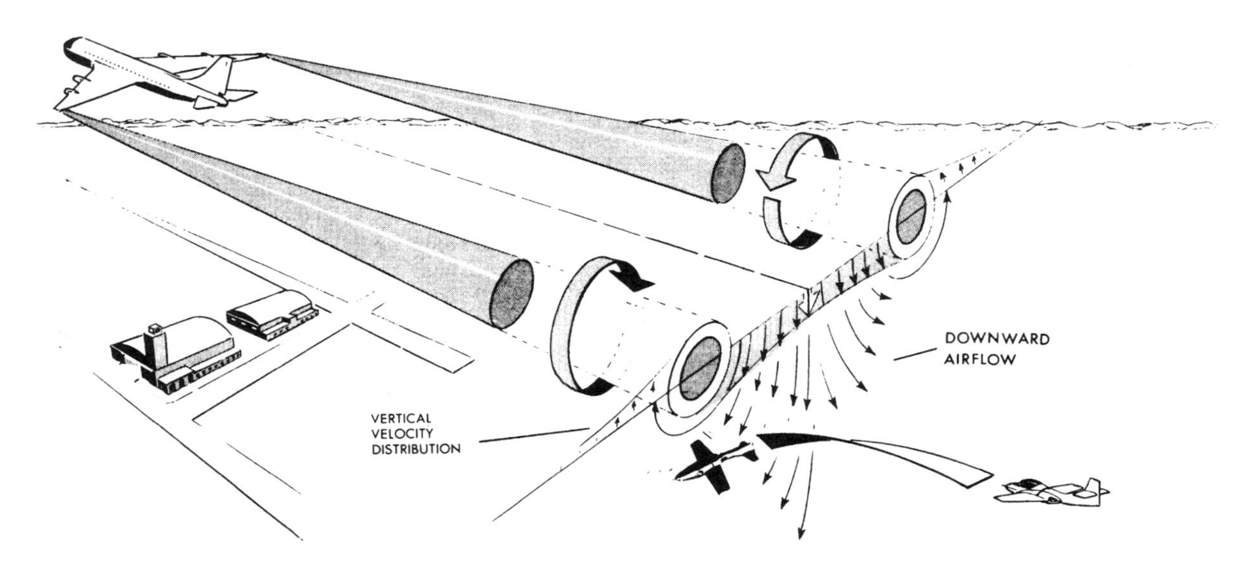

Fig. 11.14. Wingtip vortices behind an aircraft.

light aircraft to roll completely over and to be forced into the ground. A typical mishap is the case where, just before touchdown, a wing drops and contact is made with the runway while the aircraft is in a wing-down attitude. Another possible result, at altitude, is the inadvertent entry into an unusual attitude and resultant spin.

Current avoidance of wake turbulence problems during takeoffs and landings is accomplished by requiring time intervals between aircraft. Figure 11.15 shows both the takeoff and landing space clearance patterns. In the takeoff pattern the following aircraft should become airborne before the lead aircraft and should climb above the turbulence. In the landing pattern the following aircraft should touch down past the point where the lead aircraft has touched down.

Lateral movement of tip vortices due to crosswinds can present a hazard to takeoffs and landings when aircraft are operating from fields with parallel runways that are less than 2500 ft apart. This is illustrated in Fig. 11.16. The no-wind movement of the vortices is away from the centerline of the airplane at about 5 knots, as shown in Fig. 11.16a. If a 5-knot crosswind from the right is present, the left vortex will move to the left at a speed of 10 knots (Fig. 11.16b) and will present a hazard to aircraft operating on a parallel runway to the left.

Much work is currently being undertaken to develop splines, winglets, and other devices to limit the amount of tip vortices. This is primarily done to reduce induced drag, but reduced wake turbulence is a spinoff advantage. In the meantime, strict adherence to existing regulations as to time and space separation is imperative.

SPINS

Intentional spins are prohibited in most light aircraft, but spins and spin recovery should be taught (in an unrestricted airplane) as part of pilot training. Spins are much more difficult to control in a jet aircraft, so they are not practiced. Actual spin recovery techniques vary for different aircraft, thus the discussion presented here is general in nature.

Spin Warning

For an aircraft to spin it must be near aerodynamic stall. Therefore, stall warnings must be heeded, if spins are to be avoided. In our discussion of stalls earlier in this chapter we mentioned the advantages of the straight wing in providing stall warning to the pilot. When pilots of straight-wing aircraft feel the buffeting, they merely have to reduce back pressure to avoid the possibility of a spin.

Figure 11.17 shows how a straight-wing aircraft stalls at the wing root first and how the stalled air envelops the tail and provides buffet warning to the

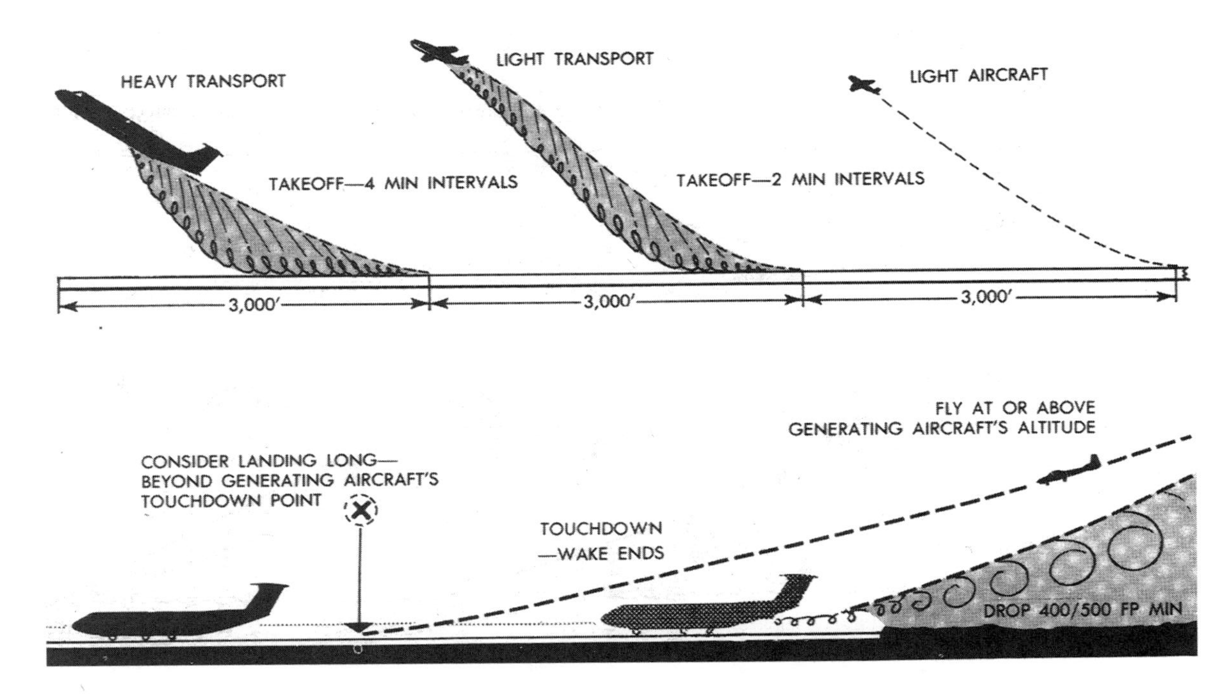

Fig. 11.15. (a) Takeoff and (b) landing clearance to avoid wake turbulence.

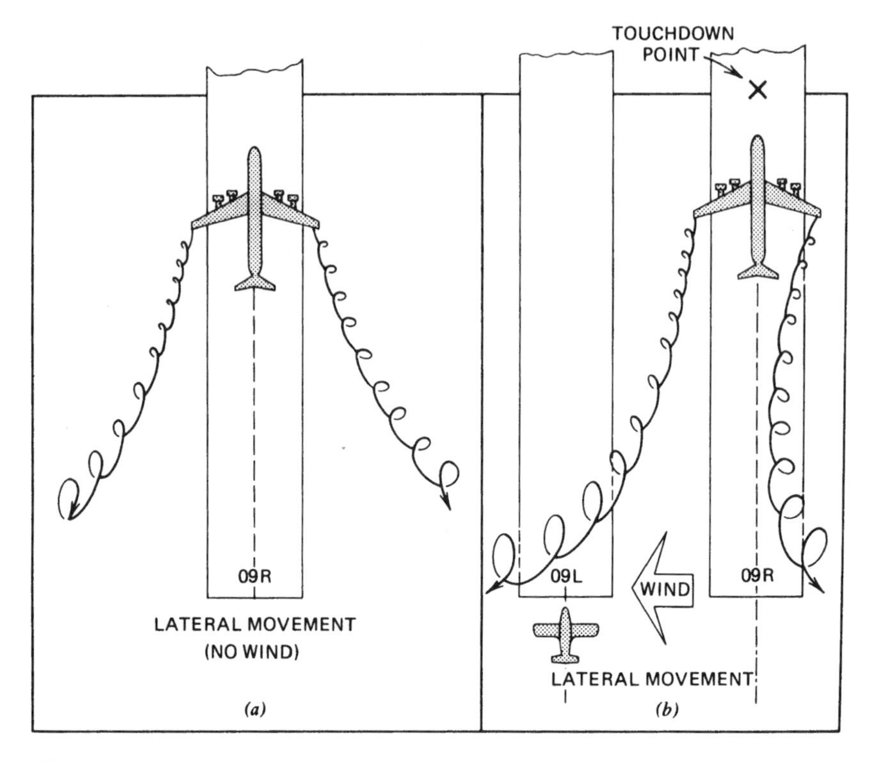

Fig. 11.16. Lateral movement of tip vortices with (a) no wind and (b) crosswind.

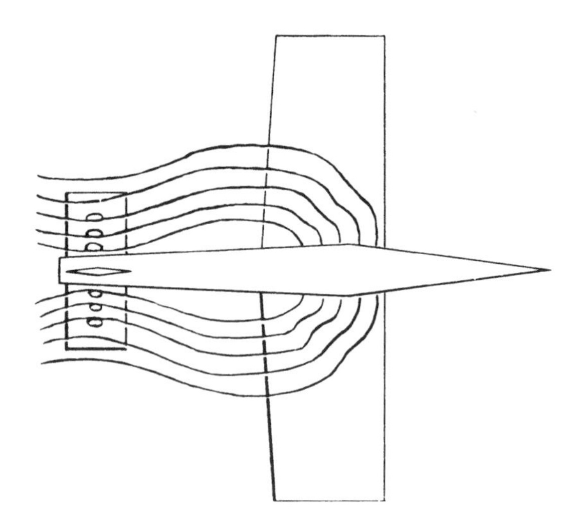

Fig. 11.17. Stall hitting the horizontal tail.

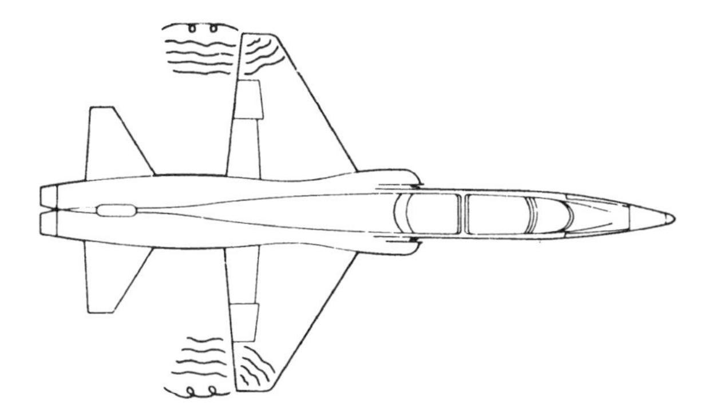

Fig. 11.18. Swept wings stall at tips first.

pilot. At this point, the ailerons are not blanked out and the pilot still has good lateral (roll) control.

Swept-wing aircraft stall at the wingtips first, rather than at the roots. This is shown in Fig. 11.18. The stall warning here is much less than for the straight-wing aircraft. Also, the ailerons are enveloped in the stalled air and become ineffective early in the stall–spin sequence. Stall warning devices are helpful in anticipating an incipient spin.

Aerodynamic Characteristics of a Spin

The aerodynamic characteristics of a spin for a straight-wing aircraft and those for a swept-wing aircraft are quite different. The straight-wing aircraft has C_L and C_D curves as shown in Fig. 11.19. If this type of aircraft is flying at an AOA at or above that for $C_{L(max)}$ and the aircraft is rolled to the right (and/or the

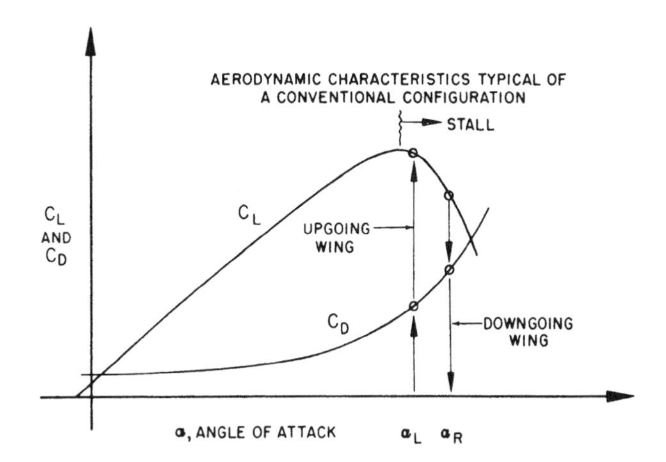

Fig. 11.19. Aerodynamics of spin for straight-wing aircraft.

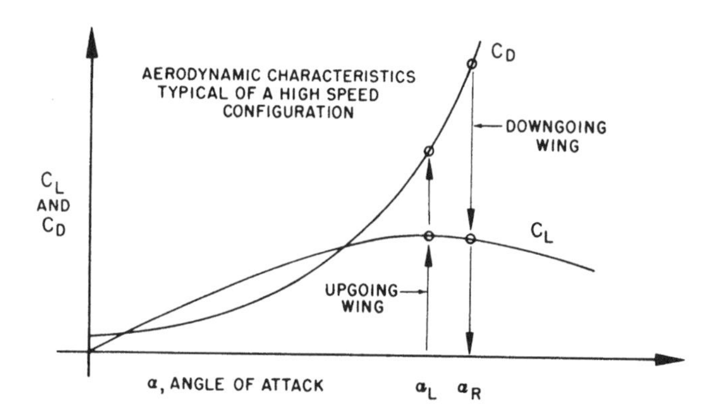

Fig. 11.20. Aerodynamics of spin for swept-wing aircraft.

right wing drops), the right wing is flying with a higher local AOA and becomes even more stalled. The left wing, on the other hand, moves upward and is flying with a lower local AOA and becomes less stalled, thus producing more lift. The right rolling motion of the aircraft is aided by the relative increase in lift of the left wing, and a yawing moment to the right is produced. These moments result in autorotation. This means that the aircraft will continue to roll and yaw without any input from the pilot. In addition to the aerodynamic forces acting on the aircraft, there are also inertia and gyroscopic effects that complicate the problem.

The straight-wing aircraft's spin is characterized primarily by the rolling motion with a moderate yaw. The attitude of the spin is about 40° or more, nose down. In straight-wing aircraft, a stalled condition must exist before a spin can develop, but this is not true for a swept-wing aircraft. The swept-wing aircraft has C_L and C_D curves as shown in Fig. 11.20. Note that the C_L curve does not have a well-defined maximum lift point. When this type of aircraft is rolled at high angles of attack, only small changes in C_L take place. In the swept-wing aircraft there is no definite stall, and the wing autorotation contribution will be quite weak. The change in C_D that occurs between the two wings, however, is substantial and a strong yawing moment is developed.

Swept-wing aircraft have low aspect ratios. The mass of the aircraft is distributed along the longitudinal axis of the plane rather than in the wings. As the yaw develops, during the spin, this mass distribution contributes to the inertial moments and tends to flatten the spin. This results in very high angles of attack and high sink rates.

This explanation is extremely simplified. Each part of the airplane affects pro-spin or anti-spin characteristics separately. Each model of airplane spins differently, and each will spin differently when the configuration of the aircraft is changed. Pilots must be familiar with the spin characteristics of their airplanes as described in the operator's handbook.

Spin Recovery

The most effective spin-recovery technique for straight-wing aircraft is to stop the spin rotation by use of opposite rudder and to lower the AOA with forward stick. Care must be used during pullout from the resulting dive to prevent an accelerated stall, which could result in an entry into another spin. In addition to the control positions described for straight-wing recovery, the ailerons are often used for swept-wing spin recovery. If ailerons are used into the direction of the spin they create drag forces that oppose rotation. Proper use of ailerons vary from airplane to airplane. Consult the flight manual for proper spin-recovery procedures for your aircraft.

PROBLEMS

1. The main difference between the C_L–α curves for straight-wing aircraft is that
   - **a.** the swept-wing aircraft has a lower value of $C_{L(max)}$.
   - **b.** the straight-wing aircraft does not fly at as high an AOA for $C_{L(max)}$.
   - **c.** the swept-wing aircraft does not have an abrupt loss of lift at $C_{L(max)}$.
   - **d.** All of the above
2. Low-speed stall will start at the wing's trailing edge ________
   - **a.** where C_l/C_L is a minimum.
   - **b.** where C_l/C_L is a maximum.
   - **c.** at the root for a swept wing.
   - **d.** at the wingtip for a straight rectangular wing.
3. The region of reversed command for a jet aircraft is also correctly known as
   - **a.** the backside of the thrust required curve.
   - **b.** the backside of the power curve.
   - **c.** the backside of the drag curve.
   - **d.** Both (a) and (c) above
4. An aircraft should never be flown in the region of reversed command.
   - **a.** True
   - **b.** False
5. Pulling back on the control stick (or yoke) will cause the airplane to climb if the plane is flying at
   - **a.** low speed.
   - **b.** high speed.
   - **c.** any speed.

6. A better way to climb, which will work at any speed, is to control the airspeed with the stick and add throttle to climb.

a. True

b. False

7. An airplane is making a final approach for a landing and encounters a horizontal wind shear. Which of the types of bursts below is the most dangerous?

a. Headwind burst

b. Crosswind burst

c. Tailwind burst

8. An airplane in flight encounters wing icing. The greatest danger is that

a. weight is increased.

b. C_D is increased.

c. C_L is decreased.

9. Wake turbulence can cause an airplane to be turned completely upside down. To escape wake turbulence a pilot should avoid

a. flying behind and below a large airplane.

b. taking off or landing behind a large airplane.

c. taking off or landing on a parallel runway that is downwind from one being used by large airplanes.

d. All of the above

10. Spin recovery consists of

a. stopping spin rotation with rudder.

b. lowering AOA with forward stick.

c. applying ailerons into the spin direction (swept-wing airplanes).

d. All of the above

11. A 10,000-lb airplane has the T_r curve shown in Fig. 11.6. The T_a is 3000 lb and it is the same for all airspeeds. The airplane is flying at 210 knots and the pilot decides to climb and pulls back on the stick, thus reducing the speed to 200 knots. Find the ROC.

12. Is the same situation as in Problem 11, except the airspeed at the start of the maneuver is 135 knots and after the pilot pulls back on the stick, the speed is 130 knots. Find the ROC.

12 Takeoff Performance

Takeoff performance involves accelerated motion. During takeoff, the aircraft starts from zero velocity, accelerates to takeoff velocity, and becomes airborne. The relationships among acceleration, velocity, takeoff distance, and time are not familiar to the average pilot and should be understood as well as experienced. This chapter discusses the factors involved in attaining takeoff velocity.

The following are important factors in takeoff performance:

- Takeoff velocity, which could be a function of stall speed, minimum control speed, or the thrust (or power) available-thrust (or power) required relationship
- The acceleration during takeoff
- The distance required to complete the takeoff

Takeoff and landing distance computation is a complex subject, and the takeoff charts in the pilot's handbook should be closely studied. This book makes no effort to analyze the problems in detail but simply explains how changes in weight, altitude, runway slope, headwind/tailwind, and other factors influence the aircraft's performance. The problem is simplified if we assume that the aircraft's acceleration is a constant during the takeoff roll.

LINEAR MOTION

Newton's laws of motion express relationships among force, mass, and acceleration, but they stop short of discussing velocity, time, and distance. These are covered here. The formulas that are derived here are valid only for constant acceleration.

$$\text{Acceleration, } a = \frac{\text{Change in velocity}}{\text{Change in time}} = \frac{V - V_0}{t - t_0}$$

where

V = velocity at time t

V_0 = velocity at time t_0

If we start at time $t_0 = 0$ and rearrange the above equation we get

$$V = V_0 + at \tag{12.1}$$

The distance, s, traveled in a certain time is

$$s = V_{av} t$$

The average velocity, V_{av}, is

$$V_{av} = \tfrac{1}{2}(V + V_0)$$

So

$$s = \tfrac{1}{2}(V_0 + at + V_0)t \quad \text{or} \quad s = V_0 t + \tfrac{1}{2}at^2 \tag{12.2}$$

Solving Eqs. 12.1 and 12.2 simultaneously and eliminating t, we can derive a third equation:

$$s = \frac{V^2 - V_0^2}{2a} \tag{12.3}$$

In applying these equations to the takeoff problem where the aircraft starts from a "brakes locked" position on the runway, V_0 is zero. The equations are then simplified and Eq. 11.3 becomes

$$s = \frac{V^2}{2a} \tag{12.4}$$

where

s = distance (ft)
V = takeoff velocity (fps)
a = acceleration (fps^2)

The acceleration of an aircraft during takeoff is determined by applying Newton's second law: $F = ma$. The force that provides the acceleration for takeoff is the unbalanced force acting on the aircraft during the takeoff roll. This is called the net accelerating force and is illustrated in Fig. 12.1.

Two assumptions are made in constructing Fig. 12.1. The first is that no lift is being developed as the aircraft gathers speed. This is true for many of the higher-performance aircraft that have the wing positioned on the aircraft at an angle of incidence that produces the least drag rather than one that produces lift. These aircraft must "rotate" at takeoff speed to develop lift. Most transport

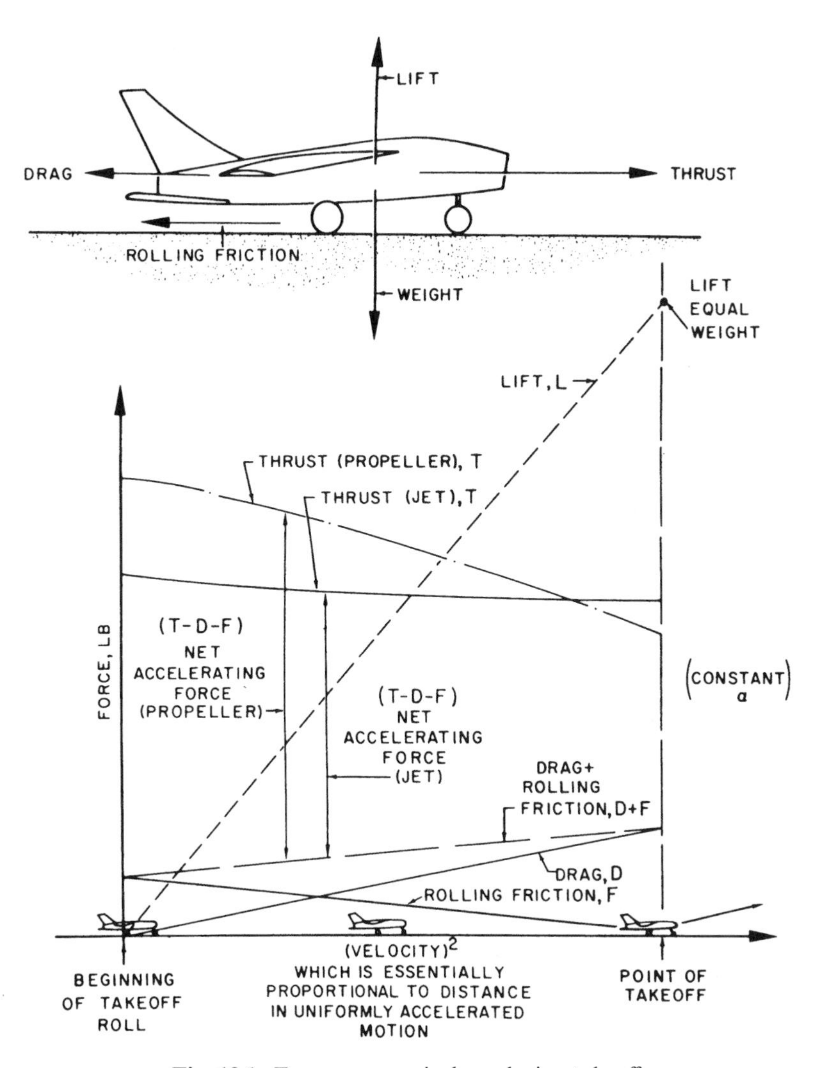

Fig. 12.1. Forces on an airplane during takeoff.

aircraft and military aircraft fit into this category. If no lift is developed during the takeoff run, rolling friction remains constant until rotation. The second assumption is that the thrust increases during acceleration and that net accelerating force is constant throughout the takeoff run and thus Eq. 12.4 is valid.

Figure 12.1 shows that the net accelerating force on the aircraft, F_N, equals the thrust T, minus the drag, D, minus the rolling friction, F:

$$F_N = T - D - F$$

By substituting in Newton's second law equation and rewriting we obtain

$$a = \frac{F_N}{m} = \frac{g(T - D - F)}{W} \tag{12.5}$$

where

a = acceleration (fps^2)
F_N = net accelerating force (lb)
W = weight (lb)
g = gravitational acceleration (= 32 fps^2)
m = mass (slugs) (= W/g)

Equation 12.5 shows that the acceleration of the aircraft is affected by the thrust, the drag, the rolling friction, and the weight of the aircraft.

FACTORS AFFECTING TAKEOFF PERFORMANCE

The factors that affect takeoff performance are

1. Aircraft gross weight
2. Thrust on the aircraft
3. Temperature
4. Pressure altitude
5. Wind direction and velocity
6. Runway slope
7. Runway surface

To see how these factors affect takeoff distance, we use Eq. 12.4 to make a ratio. A subscript 1 denotes a known set of conditions, and a subscript 2 applies to a new set of conditions:

$$\frac{s_2}{s_1} = \left(\frac{V_2}{V_1}\right)^2 \left(\frac{a_1}{a_2}\right) \tag{12.6}$$

The effects of weight change, altitude change, and wind conditions on takeoff distance can be found by applying Eq. 12.6.

Weight Change

Increasing the gross weight of an aircraft has a threefold effect on takeoff performance: (1) takeoff velocity is increased, (2) there is more mass to accelerate, and (3) there is more rolling friction. Takeoff velocity varies as the square of the weight. This was discussed in Chapter 4 and shown in Eq. 4.2:

$$\frac{V_2}{V_1} = \sqrt{\frac{W_2}{W_1}}$$

Squaring this equation gives

$$\left(\frac{V_2}{V_1}\right)^2 = \frac{W_2}{W_1}$$

The effect of increasing weight on an aircraft's acceleration is twofold: There is more mass to be accelerated and there is more rolling friction. For a dry concrete runway, the coefficient of rolling friction is about 0.02, so an increase in weight of 1000 lb means a loss in accelerating force of 20 lb. This change is small, so the principal effect on acceleration is due to the change in the mass of the aircraft.

Acceleration is inversely proportional to the mass (or weight) of the aircraft. Therefore, if the rolling friction increase is ignored, the acceleration term is

$$\frac{a_1}{a_2} = \frac{W_2}{W_1}$$

If we substitute values from these last two equations into Eq. 12.6, the effect of a weight change on takeoff distance is

$$\frac{s_2}{s_1} = \left(\frac{W_2}{W_1}\right)^2 \tag{12.7}$$

Altitude

Pressure altitude corrected for runway temperature results in runway density altitude. This may be considerably higher than the field density altitude, because runway temperatures can be higher than the official field temperature. An increase in density altitude has a twofold effect on takeoff performance: (1) A higher takeoff velocity (TAS) is required, and (2) for unsupercharged reciprocating engines and for turbine engines, less thrust is available.

To produce lift equal to the weight of the aircraft at a given AOA, the dynamic pressure must be the same, regardless of the altitude. Thus, the airplane will take off at the same equivalent airspeed (EAS) at altitude as at sea level. Because of the reduced density of the air, however, the true airspeed (TAS) will be greater at altitude.

From basic aerodynamics, we learned that the relationship between true and equivalent airspeed was expressed by Eq. 2.12:

$$\frac{\text{TAS}}{\text{EAS}} = \frac{1}{\sqrt{\sigma}}$$

Thus, the velocity for takeoff at an airfield with a density altitude above sea level (V_2), as compared to the velocity required at sea level on a standard day, (V_1) is

$$\frac{V_2}{V_1} = \frac{1}{\sqrt{\sigma_2}} \quad \text{or} \quad \left(\frac{V_2}{V_1}\right)^2 = \frac{1}{\sigma_2}$$

For unsupercharged reciprocating engines and turbine engines, the power or thrust available decreases approximately as the air density (or density ratio). As thrust is decreased, acceleration (a_2) is also decreased, as follows

$$\frac{a_1}{a_2} = \frac{1}{\sigma_2}$$

Substituting values from these last two equations into Eq. 12.6 gives us the effect of altitude on takeoff distance for unsupercharged reciprocating or turbine engined aircraft:

$$\frac{s_2}{s_1} = \left(\frac{1}{\sigma_2}\right)^2 \tag{12.8}$$

Supercharged reciprocating engines can deliver sea level power up to their critical altitude, so no decrease in thrust or acceleration occurs below that altitude. Thus,

$$\frac{s_2}{s_1} = \frac{1}{\sigma_2} \tag{12.9}$$

where

s_1 = standard sea level takeoff distance
s_2 = altitude takeoff distance
σ_2 = altitude density ratio

Wind

Takeoff into a headwind allows the aircraft to reach takeoff velocity at a lower groundspeed than for a no-wind condition. The takeoff velocity, V_2, with a headwind, in terms of the no-wind takeoff velocity, V_1, and the velocity of the headwind V_W, is

$$V_2 = V_1 = V_W$$

Dividing both sides of this equation by V_1 and squaring gives

$$\left(\frac{V_2}{V_1}\right)^2 = \left(\frac{V_1 - V_W}{V_1}\right)^2 = \left(1 - \frac{V_W}{V_1}\right)^2$$

Although the acceleration over the ground is reduced by a headwind, the acceleration through the air mass is not affected. Therefore,

$$a_2 = a_1 \quad \text{or} \quad \frac{a_1}{a_2} = 1$$

Substituting the values of these last two equations into Eq. 12.6, we find that the effect of a headwind on the takeoff distance is

$$\frac{s_2}{s_1} = \left(1 - \frac{V_W}{V_1}\right)^2 \tag{12.10}$$

Takeoff with a tailwind simply means that the groundspeed must be increased over the no-wind groundspeed by the amount of the tailwind:

$$\frac{s_2}{s_1} = \left(1 + \frac{V_W}{V_1}\right)^2 \tag{12.11}$$

Runway Slope

If the runway has a slope to it, the component of the weight that is parallel to the runway surface will reduce the net accelerating force, in case of an upslope, or increase the accelerating force, in case of a downslope:

$$\text{Accelerating force} = \pm W \text{ sine slope}$$

The takeoff distance charts in the pilot's handbook usually include runway slope calculations. A ruleof thumb is *five percent increase in takeoff distance for each percent of uphill slope.*

DEFINITIONS THAT ARE IMPORTANT RELATIVE TO TAKEOFF PLANNING

Runway Available Actual runway length (less overrun) less the airplane lineup distance (200 ft).

Takeoff Safety Speed, V_2 The speed to which the airplane must be accelerated at or before reaching 35 ft above the end of the runway.

Takeoff Ground Run The distance through which the airplane must be accelerated, without loss of an engine, to reach takeoff speed.

Critical Engine Failure Speed, V_{CEF} The speed to which the airplane can be accelerated, lose an engine, and then either continue to the takeoff with the remaining engines or stop, in the same total runway distance. V_{CEF} does not apply to single-engine aircraft.

Critical Field Length (CFL) The total length of runway required to accelerate on all engines to critical engine failure speed, experience an engine failure, and then continue to takeoff or stop. This field length is commonly called "balanced field length (BFL)." For a safe takeoff the CFL must be no greater than the runway available. CFL does not apply to single-engine aircraft.

Ground Minimum Control Speed, V_{MGC} The minimum airspeed at which the airplane, while on the ground, can lose an outboard engine and maintain directional control. V_{MGC} does not apply to single-engine aircraft.

In-Flight Minimum Control Speed, V_{MCA} The minimum speed at which an engine can be lost and directional control maintained using full rudder deflection and not more than 5° of bank. V_{MCA} does not apply to single-engine aircraft.

Refusal Speed, V_{REF} The maximum speed that the aircraft can obtain under normal acceleration and then stop in the available runway. V_{REF} applies to single-engine as well as multiengine aircraft.

Maximum Braking Energy Speed, V_{MBE} The highest speed from which the airplane may be brought to a stop without exceeding the maximum energy absorption capacity of the brakes.

Takeoff Decision Speed, V_1 The minimum indicated airspeed at which an engine failure can be experienced and the takeoff safely continued. Safe abort capability is assured if the takeoff is aborted prior to reaching this speed. V_1 is the critical engine failure speed or ground minimum control speed, whichever is higher. However, it must never exceed refusal speed, maximum braking speed, or takeoff speed. V_1 does not apply to single-engine aircraft.

ABORTED TAKEOFFS

A graphic representation of the refused takeoff situation is shown in Fig. 12.2. At the brakes release point the aircraft starts to accelerate, and the variation of velocity and distance is as shown on the takeoff acceleration profile. The deceleration profile shows the variation of velocity with distance as the airplane is brought to a stop at the end of the runway. The refusal speed is determined by the intersection of the acceleration and deceleration profiles. This speed is commonly called decision speed (V_1). An allowance for pilot reaction time is included in the calculation of refusal speeds.

Once the refusal speed is exceeded, the aircraft cannot be brought to a stop in the runway remaining. If a single-engined airplane loses its engine above refusal speed it will end up in the overrun. A multiengined airplane should continue to make an engine-out takeoff if an engine fails above this airspeed. This is shown in Fig. 12.3.

The refusal speed for multiengined aircraft is called the critical engine failure speed, V_{CEF}. If the airplane has not reached this speed, the pilot can retard all engines and brake the aircraft to a stop before the end of the runway in the

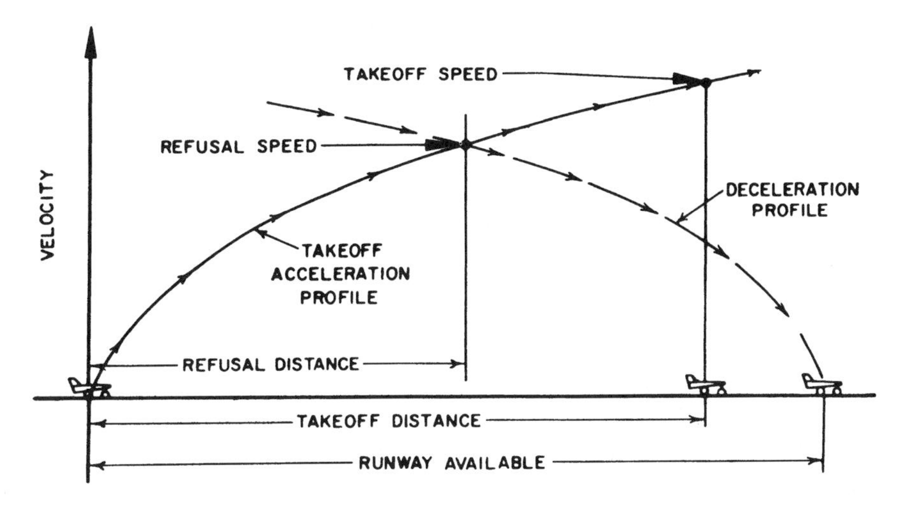

Fig. 12.2. Single-engine velocity–distance profiles.

same manner as for the single-engined airplane. At or above V_1 the pilot of the multiengined airplane must continue the takeoff after losing an engine. As can be seen in Fig. 12.3, acceleration will be less with the loss of an engine, but takeoff velocity, V_2, will be the same. Thus, takeoff distance will be greater. In some cases, the remaining runway may not be long enough to allow the airplane to accelerate to V_2. Thus, critical field length becomes important since

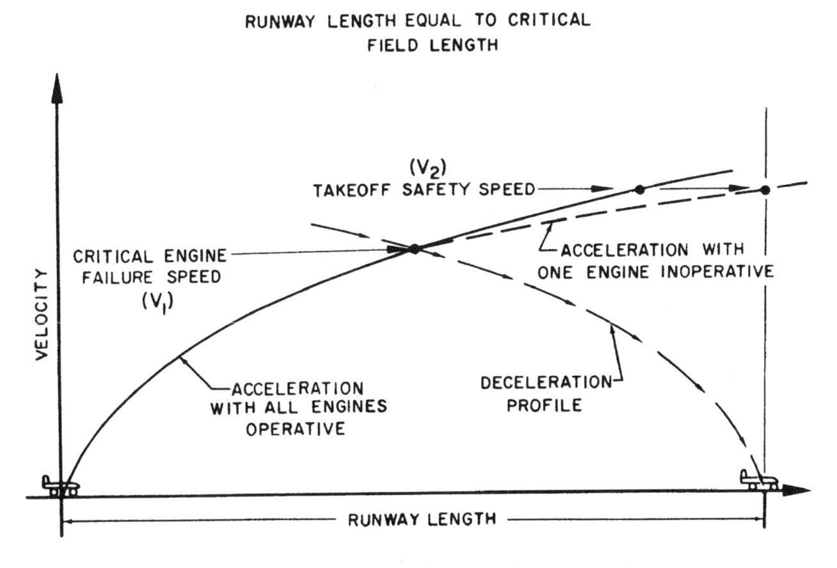

Fig. 12.3. Multiengine velocity–distance profiles.

it assures that this will not happen. Figure 12.3 shows the condition where actual field length is exactly equal to critical field length and V_2 is reached at the exact end of the runway.

DANGERS OF OVERROTATION

The dangers of overrotation during takeoff are twofold. First, high AOA results in high induced drag and possible loss of lift (if $C_{L(max)}$ is exceeded). Second, once the aircraft is airborne, it may be impossible to climb the aircraft out of ground effect. This was discussed in Chapter 5. When the aircraft departs the vicinity of the runway, the downwash is unhampered by the ground and is noticeably increased and a large increase in induced drag occurs. The increased downwash may envelop the horizontal tail and cause pitch-up moments, which add to the high AOA condition.

The most photographed aircraft accident in history occurred when an F-100 fighter attempted an emergency landing. This accident has been nicknamed the "Sabre Dance." It is a classic example of the danger of overrotation. The aircraft had a misaligned nose wheel and the pilot was attempting an emergency landing. The runway was partially foamed to lessen the chance of fire. The pilot thought he was overrunning the foamed portion and decided to go around for another approach. To prevent the damaged nose wheel from contacting the runway, the pilot purposely held the nose high as he added throttle for the go-around. The airplane remained airborne but could not accelerate or fly away from the runway (out of ground effect) because of the high induced drag. Several thousand feet down the runway the aircraft lost directional control and crashed at about 90° heading to the runway.

SYMBOLS

F	Rolling friction force (lb)
F_N	Net accelerating force
V_1	Takeoff decision speed (knots)
V_2	Takeoff speed (knots)
V_{CEF}	Critical engine failure speed (V_1 or V_{MGC}) (knots)
V_{MB}	Maximum braking speed (knots)
V_{MCA}	In-flight minimum control speed (knots)
V_{MGC}	Ground minimum control speed (knots)
V_{REF}	Refusal speed (knots)
V_W	Wind speed (knots)

EQUATIONS

12.1 $V = V_0 + at$

12.2 $s = \frac{1}{2}(V_0 + at + V_0)t$ or $s = V_0 t + \frac{1}{2}at^2$

12.3 $s = \dfrac{V^2 - V_0^2}{2a}$

12.4 $s = \dfrac{V^2}{2a}$

12.5 $a = \dfrac{F_N}{m} = \dfrac{g(T - D - F)}{W}$

12.6 $\dfrac{s_2}{s_1} = \left(\dfrac{V_2}{V_1}\right)^2 \left(\dfrac{a_1}{a_2}\right)$

12.7 $\dfrac{s_2}{s_1} = \left(\dfrac{W_2}{W_1}\right)^2$

12.8 $\dfrac{s_2}{s_1} = \left(\dfrac{1}{\sigma_2}\right)^2$ (unsupercharged reciprocating or any turbine aircraft)

12.9 $\dfrac{s_2}{s_1} = \dfrac{1}{\sigma_2}$ (supercharged reciprocating aircraft below critical altitude)

12.10 $\dfrac{s_2}{s_1} = \left(1 - \dfrac{V_W}{V_1}\right)^2$ (headwind)

12.11 $\dfrac{s_2}{s_1} = \left(1 + \dfrac{V_W}{V_1}\right)^2$ (tailwind)

PROBLEMS

1. Takeoff velocity for a multiengined airplane is a function of:
  a. stall speed.
  b. minimum control speed.
  c. low-speed region where $T_a = T_r$ or $P_a = P_r$.
  d. The highest of the above

2. Takeoff distance is a function of
  a. takeoff velocity.
  b. acceleration.
  c. Both (a) and (b)

3. Takeoff distance is
   a. directly proportional to the acceleration.
   b. inversely proportional to the velocity squared.
   c. inversely proportional to the acceleration.
   d. directly proportional to the velocity squared.
   e. Both (c) and (d)

4. As weight increases the takeoff distance increases because
   a. there is less thrust and thus less acceleration.
   b. there is more mass and thus less acceleration.
   c. the takeoff velocity is higher.
   d. Both (b) and (c)

5. As weight (W_2) increases the takeoff distance increases
   a. directly as the square of W_2/W_1.
   b. directly as W_2/W_1.
   c. inversely as the square of W_2/W_1.
   d. inversely as W_2/W_1.

6. For jet aircraft, takeoff distance at altitude is greater than at sea level because
   a. the thrust available is less.
   b. the EAS for takeoff is greater.
   c. the TAS for takeoff is higher.
   d. Both (a) and (c)

7. Which of the aircraft below will pay the greatest penalty in high-altitude takeoff distance?
   a. One with a supercharged reciprocating engine
   b. One with a turbine engine
   c. One with an unsupercharged reciprocating engine
   d. Both (b) and (c)

8. Takeoff acceleration through the air mass is not reduced by a headwind.
   a. True
   b. False

9. Takeoff with a tailwind requires that the takeoff groundspeed be ______ by the amount of the tailwind.
   a. increased
   b. decreased
   c. not affected

Aircraft data for Problems 10–15:

Thrust available during takeoff = 3000 lb
Turbojet engines
Combined average drag and rolling friction = 500 lb
Gross weight = 10,000 lb
Takeoff speed = 120 knots (203 fps)
Sea level standard conditions

10. Calculate the airplane's acceleration.

11. What should the airspeed be at the 1000-ft runway marker?

12. Calculate the no-wind takeoff distance.

13. If the weight is increased to 15,000 lb, calculate the no-wind takeoff distance.

14. Calculate the takeoff distance for the 10,000-lb airplane if there is a 12-knot headwind

15. Calculate the no-wind takeoff distance for the 10,000 lb airplane operating from anairfield where the density ratio is 0.8.

Boeing 767200 (Courtesy of the Boeing Company).

13 Landing Performance

Landing performance involves decelerated motion. During the landing phase of flight, the aircraft touches down at a certain velocity, then decelerates to zero velocity. This chapter discusses the approach to a landing and factors in reducing speed once the aircraft has touched down. The following are important factors in landing performance:

1. Approach paths and approach speeds
2. Hazards of hydroplaning
3. Deceleration during landing
4. The distance required to stop the aircraft

As we discussed in the previous chapter, this is a complex subject and the simplified discussion presented here should not be used to supersede the information presented in the pilot's handbook.

PRELANDING PERFORMANCE

Gliding Flight

One of the most important concepts of aerodynamics is equilibrium of forces, which was discussed in Chapter 1. If you do not thoroughly understand this concept, you should review that material now. An airplane that is flying at a constant speed is in equilibrium, as shown in Fig. 1.2. If the engine(s) should fail, the thrust force is reduced to zero, and an unbalance of forces results. This means that the aircraft will not be able to maintain altitude and will start to descend. If this occurs the pilot would want to know several things:

1. How far can the airplane glide?
2. How long will the airplane remain airborne?
3. What will the sink rate be?
4. Can the plane glide to a suitable landing site?
5. Can a successful engine-out landing be made?

An understanding of the basic aerodynamics of gliding flight should help the pilot make an intelligent estimate of the performance of the airplane in the above situation. Once the engine quits and the aircraft starts to descend, the forces acting on it are as shown in Fig. 13.1. The figure shows that the forces

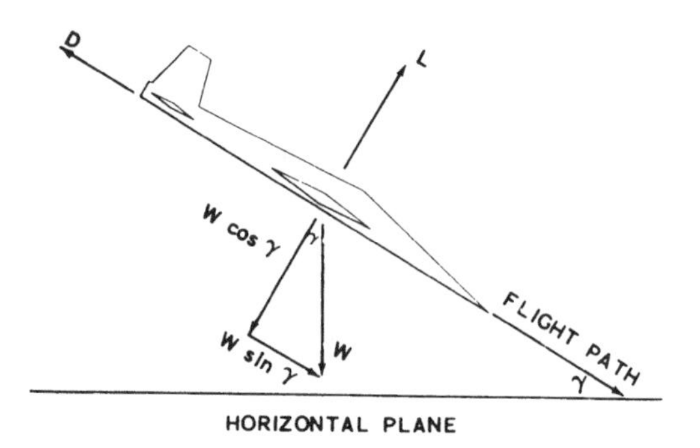

Fig. 13.1. Forces acting in a power-off glide.

of lift and drag act along the vertical and longitudinal axes of the aircraft, but the weight acts toward the center of the earth. To better analyze the reaction of these forces, it is desirable to replace the weight force by two component forces that act in the same rectangular coordinate system as the lift and drag forces.

The glide path makes a glide angle with the horizontal which we will call γ (gamma). The component of the weight that acts in the direction of the aircraft's vertical axis is equal to $W \cos \gamma$. It acts at the aircraft's center of gravity (CG), and opposes the lift, L. The component of weight that acts along the longitudinal axis is equal to $W \sin \gamma$ and opposes the drag D. In Fig. 13.1 the component $W \sin \gamma$ is shown at some distance below the CG. This component actually acts at the CG and is shown in the figure merely to explain the right triangle relationship of the vectors.

Once glide has been established and velocity is constant, equilibrium again exists and the force equations are

$$L = W \cos \gamma \tag{13.1}$$

$$D = W \sin \gamma \tag{13.2}$$

The pilot is concerned about how to achieve maximum glide ratio. Flying at maximum glide ratio means that maximum glide range will be attained. The ratio of horizontal distance to the vertical distance (altitude) is the glide ratio. To cover the most distance over the ground, the pilot wants to have the maximum glide ratio. The flight path angle must be a minimum to achieve this.

Dividing Eq. 13.1 by Eq. 13.2 gives

$$\frac{L}{D} = \frac{W \cos \gamma}{W \sin \gamma} = \frac{1}{\tan \gamma} \quad \text{or} \quad \tan \gamma = \frac{D}{L} = \frac{1}{L/D} \tag{13.3}$$

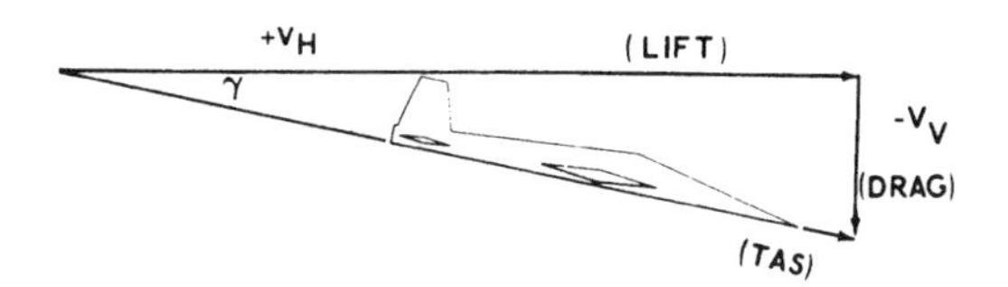

Fig. 13.2. Glide ratio vector diagram.

Equation 13.3 shows that the minimum glide angle is obtained at $(L/D)_{max}$ and, therefore, the airplane must be flown at the airspeed corresponding to the $(L/D)_{max}$ AOA. The glide ratio vector diagram is shown in Fig. 13.2. The tangent of the glide angle in Fig. 13.2 is equal to the opposite side divided by the adjacent side of the right angle shown in the figure. From Eq. 13.3, the tangent of the glide angle is D/L, so the opposite side of the triangle can be labeled as drag and the adjacent side can be labeled as lift. The vertical velocity $-V_V$ and the horizontal velocity $+V_H$ also represent the opposite side and adjacent side of the triangle, respectively; thus, the numerical value of the L/D ratio is equal to the glide ratio.

A pilot who tries to stretch the glide by flattening the glide angle will actually decrease the glide distance. Maximum glide distance is achieved only at a minimum glide angle, and this occurs when the plane flies at $(L/D)_{max}$.

The Landing Approach

Precision flying is required to ensure a stabilized steady flight path to touchdown. The approach speed specified for the weight of the aircraft is given in the pilot's handbook and provides a margin of safety above the minimum safe airspeed for the aircraft. The minimum airspeed may depend on the aircraft's stall speed, minimum control speed, or the speed at which $P_a = P_r$ (or $T_a = T_r$).

A fairly long approach path allows the pilot to stabilize the variable quantities such as airspeed, glide slope, and drift. Once these are brought under control, a smooth flight path can be maintained and a minimum of control forces will be needed to make the actual landing. Steep turns should be avoided during the alignment of the aircraft with the runway. Steep turns increase the induced drag and will bleed off the airspeed. Stall speed also increases in the turn. Once proper alignment to the runway has been established, only minor banking for drift corrections will be required. The major preoccupation now becomes the approach glide path.

Various approach glide paths are shown in Fig. 13.3. Approach path A shows a steep, low-power approach. Such an approach usually results from making the turn from the downwind leg too soon and not allowing for a long enough straightaway. The pilot finds the plane too close to the touchdown

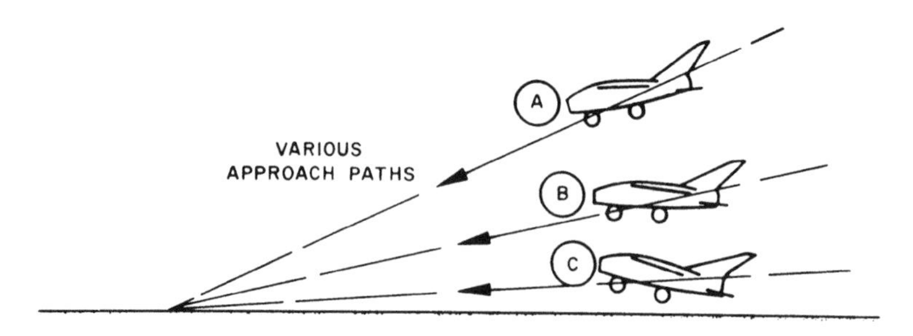

Fig. 13.3. Approach glide paths.

point and at too high an altitude. There are three ways to correct this:

1. Cut the throttle and attempt to lose altitude rapidly. If the approach is made at idle throttle, the pilot must dive (or sideslip) the airplane to lose altitude.
2. Continue the normal approach and thus land long on the runway.
3. Execute a go-around and make a new approach.

The first way results in a high rate of descent and/or an increase in airspeed. Pilots of light aircraft may not have difficulty in reducing the high rate of descent by either flaring the airplane or adding throttle. In either case the plane will probably make a hard landing or float for a distance. The high rate of descent is much more serious for heavier airplanes. Recovery involves more than merely pulling the nose up to flare the plane. Such a recovery technique will add greatly to the induced drag, but will provide little increase in lift. A large increase in thrust is required to overcome the high rate of descent. In some cases there is simply not enough thrust available and a hard, short landing results.

Landing long may succeed if the runway is long enough. However, the third way— making a go-around—is recommended. Although it may be hard on pilot ego, it is easy on the airplane. If the approach is not properly lined up on final approach, it is certainly better to "take it around" rather than to "press on" and hope to salvage a good landing from a poor approach.

Approach path C in Fig. 13.3 shows a long, shallow approach. This is a result of getting too low during the approach and then having to "drag it in" using high thrust and high AOA. Such approaches have been successful in propeller aircraft, but they must be avoided in turbojets. Propeller aircraft derive much more lift at high power settings due to the air being blown over the wing area behind the propellers, as shown in Fig. 13.4a.

The only advantage of making a high-thrust approach in a conventional jet aircraft is the generation of lift from the vertical component of thrust, as shown

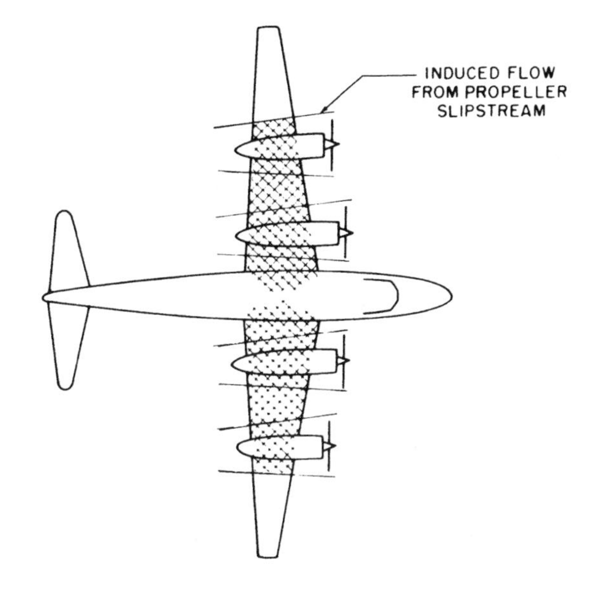

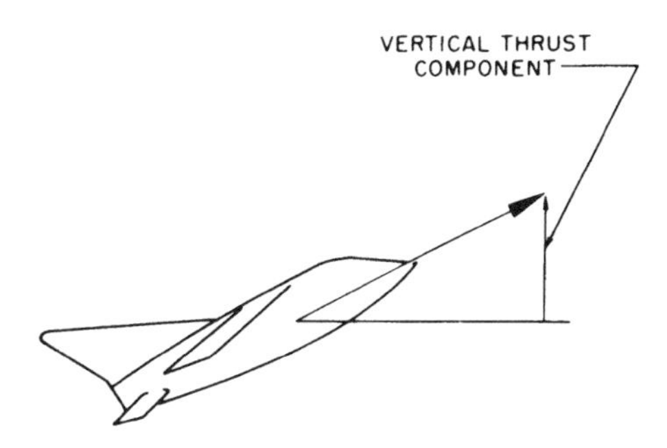

Fig. 13.4. Lift from (a) propellers and (b) turbojets.

in Fig. 13.4b. The large increase in drag that results from the high AOA for the jet aircraft far exceeds the benefits of the lift.

Propeller aircraft are power producers and, as shown from Eq. 1.6, this means that they produce high thrust at low airspeeds. Propeller aircraft do not suffer from a thrust deficiency at low airspeeds as do jet aircraft. Even with the advantages of blowing air and high thrust available, propeller aircraft should not be flown in a low, flat, high-powered approach. Engine failure under this type of approach can be disastrous. Thus, the intermediate approach B is desirable. It does not require a high rate of sink, excessive speed, or the extreme flare that the steep approach requires. The intermediate path also does not require the high thrust and AOA that the low path does.

Once established on the proper approach path and aircraft heading, the pilot must control the airspeed and altitude. The proper AOA will produce the desired airspeed. Lowering the AOA will increase the airspeed, and increasing the AOA will lower the airspeed. As we have emphasized several times, *the stick controls the airspeed.* Once proper airspeed has been established, *the primary control of the rate of descent will be the throttle.*

LANDING DECELERATION, VELOCITY, AND DISTANCE

Forces on the Aircraft During Landing

Transport and other large aircraft use a different landing technique than fighter and attack aircraft. Nearly all modern transports have nonskid brakes and thrust reversers. They land in a three-point attitude and apply brakes and thrust reversers soon after touchdown. The landing forces on these aircraft are shown in Fig. 13.5. Thrust reverser forces are not shown in the figure.

Fighter, attack, and similar high-performance aircraft use aerodynamic drag braking as well as brake friction to slow the airplane. These forces are shown in Fig. 13.6. It is assumed that

1. The aircraft has a tricycle landing gear and lands on the main gear and the pilot holds the nose up to take advantage of the aerodynamic drag.

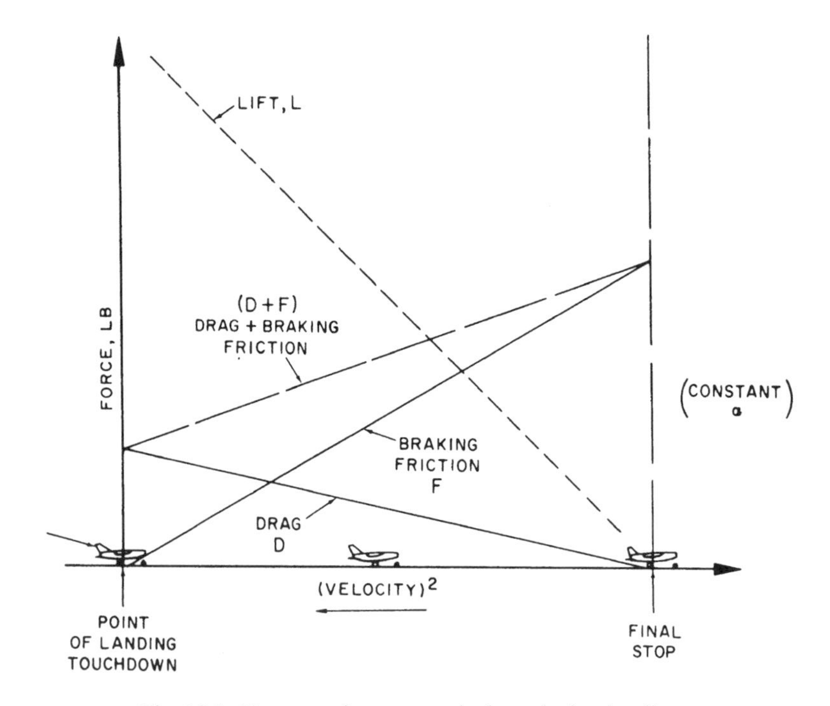

Fig. 13.5. Forces acting on an airplane during landing.

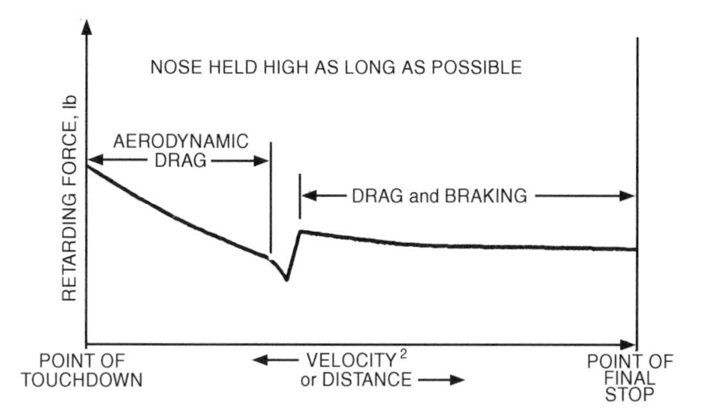

Fig. 13.6. Aerodynamic braking and wheel braking.

2. Foot brakes are not applied until the pilot can no longer keep the nose wheel off the runway.
3. The aircraft does not develop lift in the three-point position and the nose wheel is on the runway.

Braking action of the wheel brakes is not well understood by the average pilot. Many factors are involved, including

- Tire material
- Tread design and condition
- Runway surface material and condition
- Amount of braking applied (wheel slippage)
- Amount of normal (squeezing) force between tires and the runway

Friction is defined as a force that develops between two surfaces that are in contact with each other when an attempt is made to move them relative to each other. The friction force, F, is related to the force that is pressing the surfaces together, called the normal force, N, by a dimensionless factor called the coefficient of friction, μ (mu):

$$F = \mu N$$

Figure 13.7 illustrates these forces.

The coefficient of friction for a certain tire material, tread design, and tire wear operating on a dry concrete runway is a function of the amount of brakes being applied. This braking is defined as a percentage of slip between the tire and the runway. Zero percent slip is a rolling tire with no braking action. A locked wheel that skids without turning is defined as 100% slip.

A plot of the coefficient of friction versus wheel slippage is shown in Fig. 13.8. The maximum braking coefficient of friction can be seen to be about

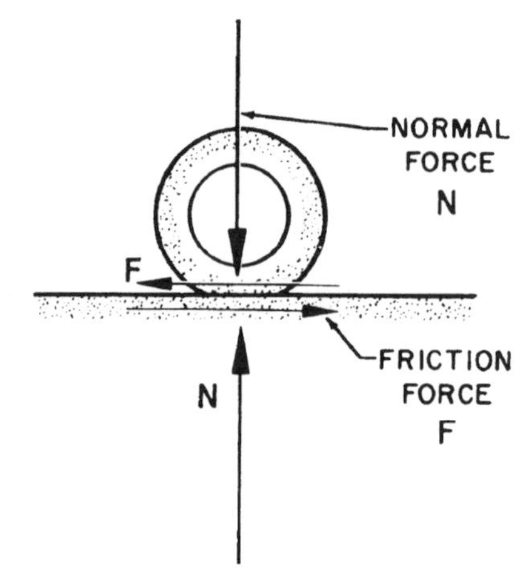

Fig. 13.7. Normal and friction forces.

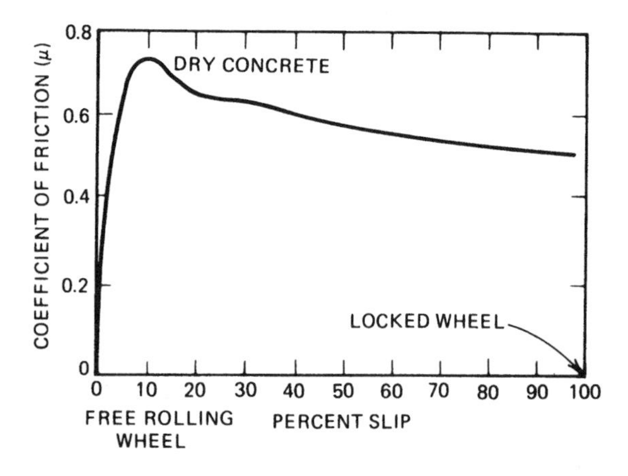

Fig. 13.8. Coefficient of friction versus wheel slippage.

0.7. It occurs when wheel slip is about 10%. Locking the wheels reduces the value to about 0.5, causes the tires to wear unevenly, and may cause tire blowout. Water, snow, or ice on the runway causes a large reduction of the coefficient of friction, as shown in Fig. 13.9.

Braking Techniques

The technique of using braking devices may vary in detail with aircraft type and model, but general principles can be stated for aircraft using aerodynamic braking. During the initial landing roll the velocity and dynamic pressure is high and aerodynamic braking will be most effective. Holding the nose of a

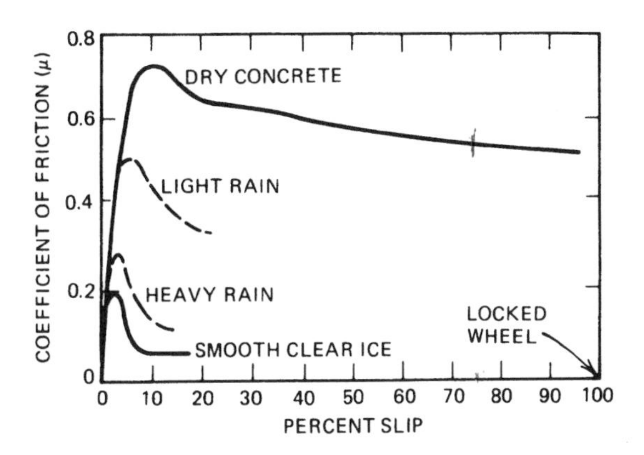

Fig. 13.9. Effect of runway condition on coefficient of friction.

tricycle landing gear airplane high will increase the parasite drag and is an effective method of slowing the aircraft. Use of speed brakes and full down flaps has the same effect.

At some point during the landing roll the reduced airspeed will cause the elevator to lose effectiveness, and the nose of the aircraft can no longer be held off the runway. Usually this is the point where wheel braking is initiated and becomes the principal method of stopping the aircraft. There is often a debate over whether it is better to retract the flaps, thus increasing the weight on the wheels, or to leave them down, thus preserving the aerodynamic braking effect. Inadvertent retraction of the landing gear instead of the flaps has occurred and may influence this decision.

Wheel braking force for all aircraft will be a maximum when the normal force on the braked wheels is greatest. For a tricycle landing gear aircraft, this will occur when the control stick is held in the back position. Many pilots have a tendency to release back pressure when the nose wheel contacts the runway. This should be avoided and the stick should be held full back as long as braking action is desired.

LANDING EQUATIONS

General Equation

The calculation of landing distance, S, in terms of touchdown velocity, V_0, and deceleration, $-a$, is similar to the takeoff case, which was shown in Eq. 12.4 to be

$$S = -\frac{V_0^2}{2(-a)} = \frac{V_0^2}{2a}$$

To determine how the landing distances are affected by variations in weight, altitude, and wind conditions, a ratio is established similar to that used for the takeoff case (Eq. 12.6):

$$\frac{s_2}{s_1} = \left(\frac{V_2}{V_1}\right)^2 \left(\frac{a_1}{a_2}\right)$$

Effect of Weight Change

Changing the gross weight of a landing aircraft affects the landing speed in the same way the takeoff speed was affected. The landing speed must be changed by using Eq. 4.2:

$$\frac{V_2}{V_1} = \sqrt{\frac{W_2}{W_1}} \quad \text{or} \quad \left(\frac{V_2}{V_1}\right)^2 = \frac{W_2}{W_1}$$

Changing the weight of the aircraft has no effect on the deceleration of the aircraft, caused by the braking action. At first, this statement sounds strange, but the heavier the aircraft, the more the weight on the braking wheels. So, more braking force is available to counteract the greater weight to be decelerated. For the braked phase of the landing,

$$\frac{a_1}{a_2} = 1$$

Substituting these last two equations into Eq. 12.6 gives the effect of a weight change on landing distance:

$$\frac{s_2}{s_1} = \frac{W_2}{W_1} \tag{13.4}$$

Effect of Altitude

Aircraft airspeeds are affected by the same factors at altitude whether the aircraft is taking off or landing. Thus, the equation showing the velocity change at altitude holds for the landing phase as well:

$$\left(\frac{V_2}{V_1}\right)^2 = \frac{1}{\sigma_2}$$

Unless the aircraft is equipped with thrust reversers, it does not rely on the engine performance to decelerate the aircraft as it did for takeoff acceleration. So,

$$\frac{a_1}{a_2} = 1$$

Substituting the values from these last two equations into Eq. 12.6 gives us the effect of altitude on landing distance for all non-thrust-reverser aircraft:

$$\frac{s_2}{s_1} = \frac{1}{\sigma_2} \tag{13.5}$$

Effect of Wind

Headwinds and tailwinds affect takeoff and landing performance in exactly the same way, so Eqs. 12.10 and 12.11 apply to the landing situation also. We rewrite them with new equation numbers:

$$\frac{s_2}{s_1} = \left(1 - \frac{V_W}{V_1}\right)^2 \quad \text{(headwind)} \tag{13.6}$$

$$\frac{s_2}{s_1} = \left(1 + \frac{V_W}{V_1}\right)^2 \quad \text{(Tailwind)} \tag{13.7}$$

HAZARDS OF HYDROPLANING

Hydroplaning occurs when a tire loses contact with the runway surface due to a buildup of water in the tire–ground contact area. NASA has researched this problem and identified three forms of hydroplaning: dynamic, viscous, and reverted rubber.

Dynamic hydroplaning is caused by the buildup of hydrodynamic pressure at the tire–pavement contact area. The pressure creates an upward force that effectively lifts the tire off the surface. When complete separation of the tire and pavement occurs, the condition is called total dynamic hydroplaning, and wheel rotation will stop.

Figure 13.10 shows the forces on a tire without the presence of water on the runway. Figure 13.10a shows a standing tire with the only forces being the weight of the aircraft and the ground reaction opposing it. As the aircraft moves to the left, as shown in Fig. 13.10b, the ground friction causes a spin-up moment and wheel rotation results. When the tire is rolling freely at a fixed speed on a dry runway, the vertical ground reaction shifts forward of the axle and a spin-down moment that offers resistance to the wheel rotation is developed. When these two moments are equal, the wheel is turning at a constant rpm.

The introduction of water on the runway leads to dynamic hydroplaning, which is shown in Fig. 13.11. Deep fluid on the runway creates additional drag on the tire when it is displaced from the tire path, and a high-spray pattern is produced, as shown in Fig. 13.11a. As the forward speed of the aircraft is increased (Fig. 13.11b), the spray pattern thrown up by the tire lowers and the wedge of water penetrates the tire–ground contact area and produces a hydrodynamic lift force on the tire. This is *partial hydroplaning*. As the speed

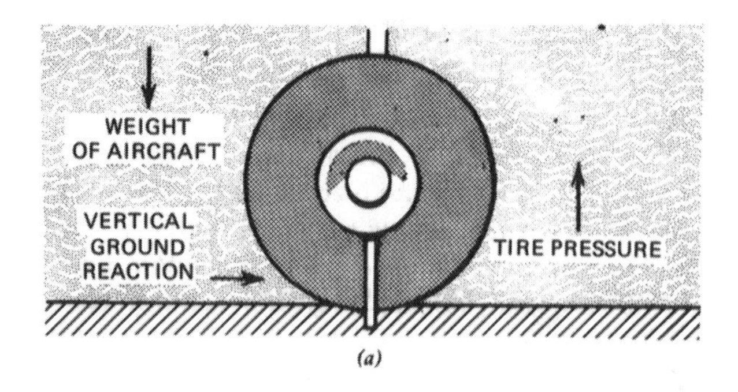

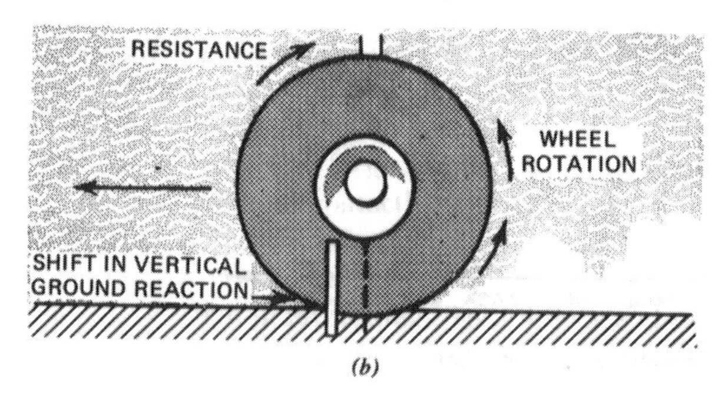

Fig. 13.10. Forces on tire: (a) static condition, (b) rolling tire.

increases the spray pattern becomes flatter, and the wedge of fluid penetrates farther into the ground contact area until at some high forward speed complete separation of the tire and runway takes place and total hydroplaning occurs, as shown in Fig. 13.11c.

The ground friction is progressively reduced as the wedge of water penetrates beneath the tire. It approaches zero at total hydroplaning, and the spin-down moment causes the tire to stop the wheel rotation. Obviously no braking action is available when the wheel is not making contact with the runway and has stopped rotating.

Total dynamic hydroplaning is more of a landing than a takeoff problem, but crosswind takeoffs are dangerous under these conditions. The approximate speed at which total dynamic hydroplaning occurs is

$$V_{\mathrm{H}} = 9\sqrt{P} \tag{13.8}$$

where

V_{H} = hydroplaning speed (knots)
P = tire inflation pressure (psi)

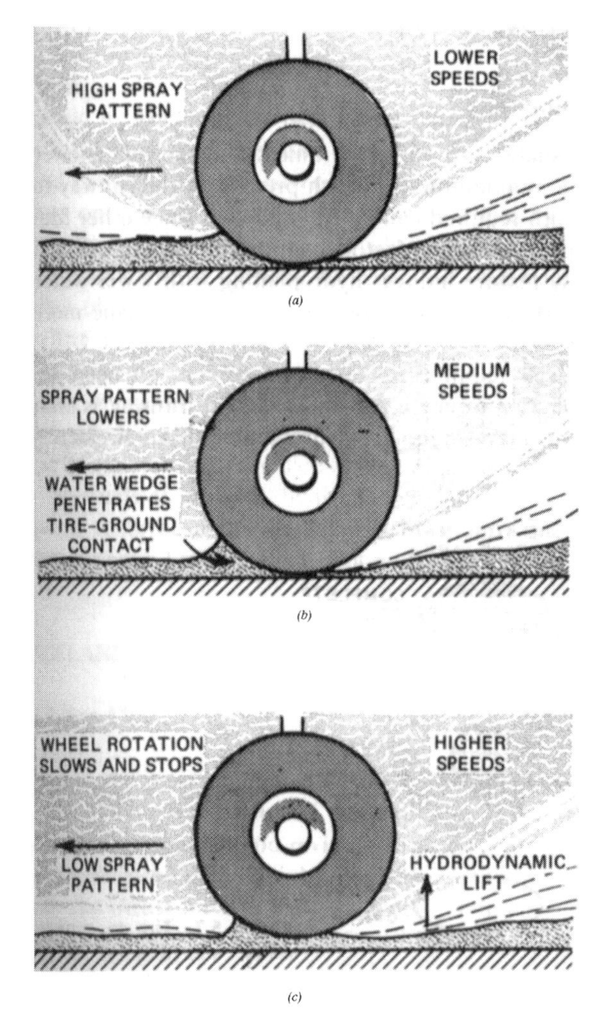

Fig. 13.11. Hydroplaning forces on tire: (a) low speed, (b) medium speed, (c) high speed.

Total dynamic hydroplaning usually does not occur unless a severe rain shower is in progress. There must be a minimum water depth present on the runway to support the tire. The exact depth cannot be predicted since other factors, such as runway smoothness and tire tread, influence dynamic hydroplaning. Both smooth runway surface and smooth tread tires will induce hydroplaning with lower water depths. While the exact depth of water required for hydroplaning has not been accurately determined, a conservative estimate for an "average" runway is that water depths in excess of 0.1 in. may induce full hydroplaning.

Viscous hydroplaning is more common than dynamic hydroplaning. Viscous hydroplaning may occur at lower speeds and at lower water depths than dynamic hydroplaning. Viscous hydroplaning occurs when the pavement surface is lubricated by a thin film of water. The tire is unable to penetrate this film, and contact with the pavement is partially lost. Viscous hydroplaning often occurs on smooth runway pavements or where rubber deposits are present, usually in the touchdown area where a thin water film can significantly reduce the coefficient of friction.

The third type of hydroplaning is known as *reverted rubber hydroplaning.* White streaks on the runway are an indication that this type of hydroplaning has occurred. Examination of the aircraft tire will show an elliptically shaped tacky or melted rubber condition. This condition occurs when the heat that is generated during a locked wheel skid reverts the rubber to a softer slippery state.

Methods to prevent hydroplaning include transverse runway grooving, frequent removal of rubber deposits from the touchdown areas, maximum use of aerodynamic braking, and pilot education.

SYMBOLS

N	Normal force (lb)
P	Tire inflation pressure (psi)
V_H	Hydroplaning speed (knots)
γ (gamma)	Glide angle (degrees)
μ (mu)	Coefficient of friction

EQUATIONS

13.1 $L = W \cos \gamma$

13.2 $D = W \sin \gamma$

13.3 $\tan \gamma = \dfrac{1}{L/D}$

13.4 $\dfrac{s_2}{s_1} = \dfrac{W_2}{W_1}$

13.5 $\dfrac{s_2}{s_1} = \dfrac{1}{\sigma_2}$

13.6 $\dfrac{s_2}{s_1} = \left(1 - \dfrac{V_W}{V_1}\right)^2$ (Headwind)

13.7 $\dfrac{s_2}{s_1} = \left(1 + \dfrac{V_W}{V_1}\right)^2$ (Tailwind)

13.8 $V_H = 9\sqrt{P}$

PROBLEMS

1. An airplane is in trimmed flight and its velocity is constant.

a. It has no unbalanced forces acting on it.

b. It has no unbalanced moments acting on it.

c. It is in a state of equilibrium.

d. All of the above are true.

2. An airplane is in a constant velocity engine out glide. Its altitude is decreasing. It is not in a state of equilibrium.

a. True

b. False

3. At $(L/D)_{max}$ an airplane will

a. be at minimum glide angle.

b. be achieving maximum glide distance.

c. have a glide ratio equal to the numerical value of $(L/D)_{max}$.

d. All of the above

4. A steep, low-power approach is more dangerous for heavy airplanes than light airplanes because

a. recovery from a high rate of descent involves a great increase in power (or thrust).

b. the aircraft will float down the runway and possibly overshoot the runway.

c. flaring the aircraft to decrease the rate of descent increases induced drag.

d. Both (a) and (c)

5. A low-angle, high-power (or thrust) approach

a. may be used for propeller aircraft if a short field landing is required.

b. is dangerous for propeller aircraft if an engine fails.

c. should be avoided at any cost for jet airplanes as high induced drag results.

d. All of the above

6. Aerodynamic drag is more effective than wheel braking during the first part of a landing.

a. True

b. False

7. As a general rule, when the nose of a tricycle landing gear airplane can no longer be held off the runway, it is time to start applying wheel brakes.

a. True

b. False

8. In applying wheel brakes, it is not a good idea to apply full brake pressure because:

a. you might cause a blow out.

b. you get lower coefficients of friction when wheel slippage exceeds about 15%.

c. Both (a) and (b)

9. In general, if maximum wheel braking of a tricycle landing gear airplane is desired, the pilot should keep the stick full back even after the nose is on the runway.

a. True

b. False

10. Headwinds and tailwinds affect the landing distance by the same amount as they affect the takeoff distance.

a. True

b. False

Aircraft data for Problems 11–15:

Landing speed = 100 knots (169 fps)

Gross weight = 8000 lb

Average retarding force = 2000 lb

Sea level standard conditions.

11. Calculate the deceleration.

12. Calculate the no-wind stopping distance.

13. If the weight is increased to 13,000 lb, calculate the no-wind stopping distance. Assume the same landing speed (100 knots).

14. If the 8000-lb airplane lands with a 10-knot tailwind, calculate the stopping distance.

15. If the aircraft is operating from an airfield where the density ratio is 0.8, calculate the no-wind stopping distance.

14 Maneuvering Performance

Maneuvering performance can be roughly divided into two main categories: *general turning performance* and *energy maneuverability*. General turning performance is common to all aircraft. It involves constant altitude turns and constant speed (but not constant velocity) conditions. Remember, velocity is a vector, having speed and direction. Energy maneuvering is relevant to tactical military missions, such as air-to-air combat maneuvers and air-to-ground maneuvers.

The discussion in this chapter does not cover energy maneuverability, but is confined to general turning flight and the flight envelope limitations on an aircraft.

GENERAL TURNING PERFORMANCE

In turning flight the airplane is not in a state of equilibrium, since there must be an unbalanced force to accelerate the plane into the turn. At this point let us review the subject of acceleration.

In Chapter 1 acceleration was defined as the change in velocity per unit of time. Velocity was defined as a vector quantity that involves both the speed of an object and the direction of the object's motion. Any change in the direction of a turning aircraft is therefore a change in its velocity, even though its speed remains constant. An aircraft in a turn is subjected to an unbalanced force acting toward the center of rotation. Newton's second law states that an unbalanced force will accelerate a body in the direction of that force. This unbalanced force is called the centripetal force, and it produces an acceleration toward the center of rotation known as radial acceleration.

The forces acting on an aircraft in a coordinated, constant altitude turn are shown in Fig. 14.1. In wings level, constant altitude flight, the lift equals the weight of the airplane and is opposite to it in direction. But in turning flight, the lift is not opposite, in direction, to the weight. Only the vertical component of the lift, called effective lift, is available to offset the weight. Thus, if constant altitude is to be maintained, the total lift must be increased until the effective lift equals weight.

Solving the triangle in Fig. 14.1 gives

$$\cos \phi = \frac{\text{Effective lift}}{\text{Total lift}} = \frac{W}{L} \tag{14.1}$$

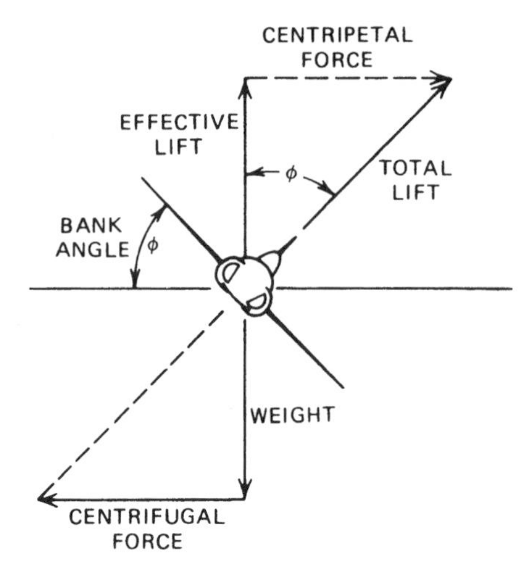

Fig. 14.1. Forces on an aircraft in a coordinated level turn.

The load factor, G, on the aircraft is defined as lift/weight:

$$G = \frac{L}{W} \tag{14.2}$$

Inverting Eq. 4.1 and substituting the value of L/W into Eq. 14.2, we obtain

$$G = \frac{1}{\cos \phi} \tag{14.3}$$

G is called the *load factor*. The centripetal force causing the airplane to turn is found by solving the triangle in triangle in Fig. 14.1:

$$\sin \phi = \frac{\text{Centrifugal force}}{\text{Total lift}}$$

So

$$\text{Centripetal force} = L \sin \phi \tag{14.4}$$

The equal and opposite reaction force described in Newton's third law is called the *centrifugal force*. This force is the airplane's inertia, which resists the turn.

A vector diagram of the forces acting on an airplane in a coordinated constant altitude turn is shown in Fig. 14.2. From Eq. 14.2 it can be seen that the total lift, L, can be replaced by its equivalent, GW.

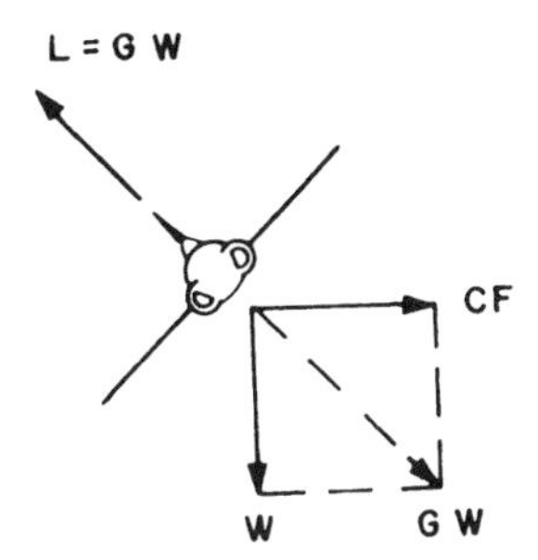

Fig. 14.2. Vector diagram of forces on an aircraft in a turn.

Load Factors on an Aircraft in a Coordinated Turn

Equation 14.3 tells us that the *G*'s required for an aircraft to maintain altitude in a coordinated turn are determined by the bank angle alone. Type of aircraft, airspeed, or other factors have no influence on the load factor. Figure 14.3 depicts the load factors required at various bank angles.

Note that the load factors required to fly at constant altitude in turning flight were determined by assuming that the wings of the aircraft provided all the lift of the aircraft. According to this analysis, the load factor at a 90° bank will be infinity and thus impossible to attain. But we have all seen aircraft fly an eight-point roll, with the roll stopped at the 90° position. The secret is, of course, that lift is provided by parts of the aircraft other than the wings. If the nose of the aircraft is held above the horizontal, the vertical component of the thrust will act as lift. The fuselage is also operating at some effective AOA and will develop lift. The vertical tail and rudder will also develop lift.

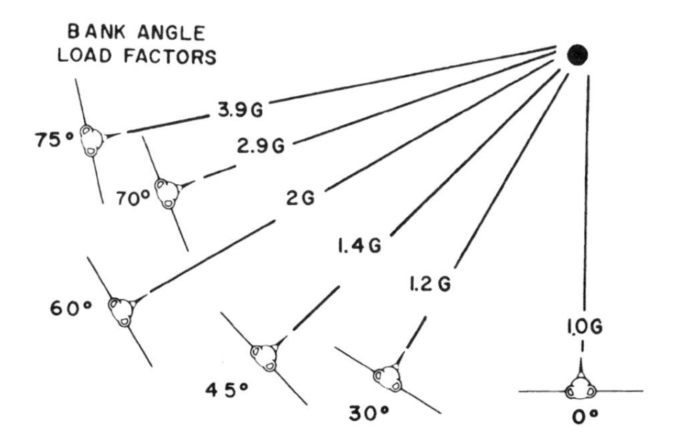

Fig. 14.3. Load factors at various bank angles.

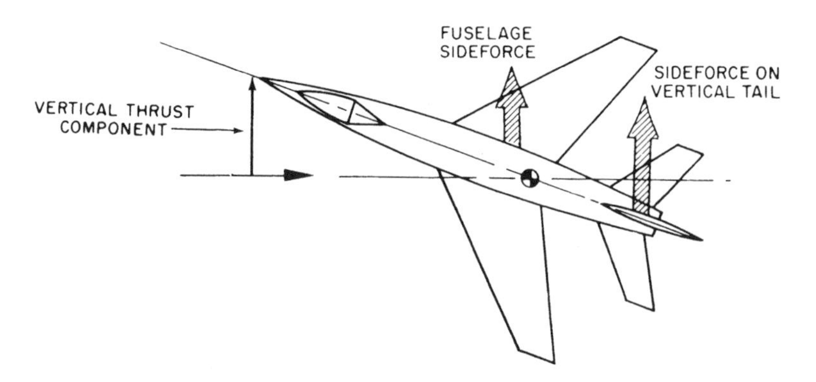

Fig. 14.4. Forces on an aircraft during a 90° roll.

Figure 14.4 shows the forces on an aircraft after it has rolled 90°. One of three possible limiting factors on turning performance is the structural strength limits of the airplane. We have just discussed this limitation. The bank angle determines the load factor and may limit the turn radius.

Effect of a Coordinated Banked Turn on Stall Speed

In basic aerodynamics we learned that the aircraft always develops its slow speed stall at the stall AOA. At this AOA the value of the lift coefficient is a maximum, $C_{L(max)}$. So V_S occurs at $C_{L(max)}$. The basic lift equation 4.1 can be rewritten as

$$V_S = \sqrt{\frac{295L}{C_{L(max)}\sigma S}}$$

and since $L = GW$,

$$V_S = \sqrt{\frac{295GW}{C_{L(max)}\sigma S}}$$

Thus, stall speed depends on the square root of the G loading. All other factors in the equation are constant for the same altitude. Therefore,

$$\frac{V_{S2}}{V_{S1}} = \sqrt{G} \quad \text{or} \quad V_{S2} = V_{S1}\sqrt{G} \tag{14.5}$$

where

V_{S1} = stall speed under 1G flight

V_{s_2} = stall speed under other than $1G$ flight
G = load factor for condition 2

Substituting the value of G from Eq. 14.3 gives

$$\frac{V_{s2}}{V_{s1}} = \sqrt{\frac{1}{\cos\phi}} \tag{14.6}$$

The V–G Diagram (Flight Envelope)

Equation 14.5 shows how the stall speed increases as greater than $1G$ loading is applied to an aircraft. If an aircraft is under a $4G$ load, the stall speed is the square root of 4 or twice the $1G$ stall speed. At zero G the stall speed is zero because no lift is being developed. This equation cannot be applied when an aircraft is developing negative load factors (square root of a negative number is impossible), but an aircraft can be stalled under negative G loading.

Figure 14.5 shows the first construction lines of the V–G diagram (flight envelope), the plot of the stall speed at various G loadings. These curved lines can also be thought of as the number of G's that can be applied to the aircraft before it will stall at a given airspeed. They are called the aerodynamic limits of the aircraft. It is impossible to fly to the left of these curves, because the aircraft is stalled in this region.

Equation 14.3 relates the G loading to the bank angle. Rearranging this equation we see that the bank angle is the angle whose cosine is $1/G$. Mathematically, this is

$$\phi = \arccos\left(\frac{1}{G}\right) \tag{14.7}$$

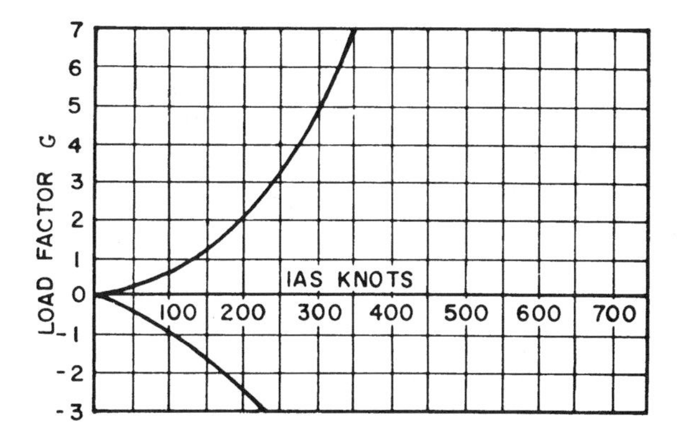

Fig. 14.5. First-stage construction of a V–G diagram.

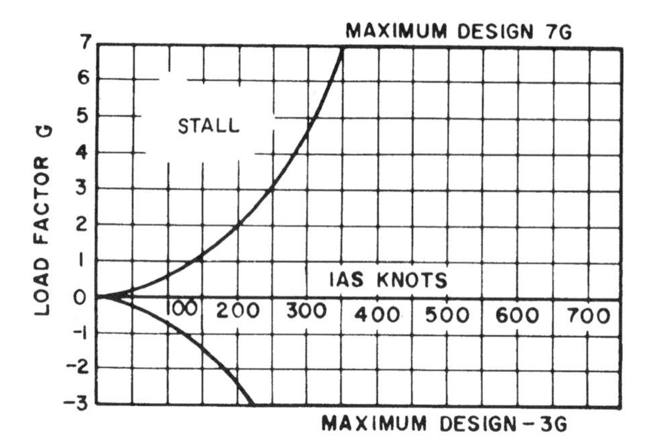

Fig. 14.6. Second-stage construction of V–G diagram.

Table 14.1 shows the bank angles that produce the load factors shown in Fig. 14.5.

The second stage of construction of the V–G diagram consists of drawing horizontal lines at the positive and negative limit load factors, LLFs. These limits are specified for each model aircraft and are shown in Fig. 14.6 as limit load factor $+7G$ and as limit load factor $-3G$.

The structural limit load factors show the maximum G's that may be imposed on the aircraft in flight without damaging the structure. Several important facts about load factors should be understood: First, the limits are for a certain gross weight, and they change if the gross weight differs from that specified. Second, the limits are for symmetrical loading only. Third, the assumption is made that there is no corrosion, metal fatigue or other damage to the aircraft. Fourth, up-gusts and turbulent air can add to pilot imposed load factors.

The aircraft designer designs a certain strength into the airframe. The allowable total strength divided by the weight of the aircraft determines the

Table 14.1. Load Factors at Various Bank Angles

Load Factor, G	Bank Angle, $\theta°$
1	0
2	60
3	70.5
4	75.5
5	78.5
6	80.4
7	81.8

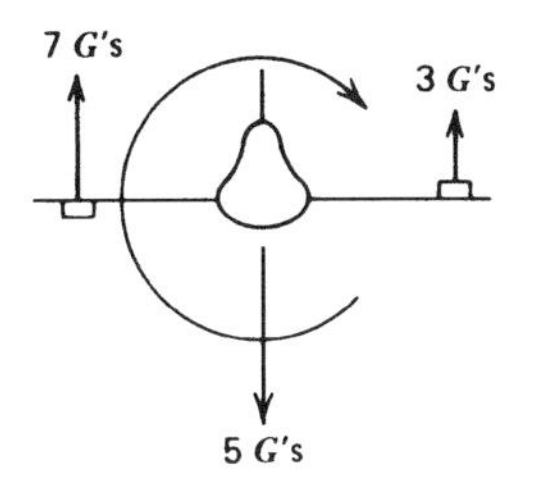

Fig. 14.7. Antisymmetrical loading.

limit load factor. If the design weight is exceeded, then the number of G's that can be tolerated by the airframe must decrease. This is an inverse proportion:

$$\frac{\mathrm{LLF}_2}{\mathrm{LLF}_1} = \frac{W_1}{W_2} \tag{14.8}$$

During unsymmetrical maneuvers, such as a rolling pullout, the wing that is rising will have more lift on it, and thus a greater load factor, than the downgoing wing. This is shown in Fig. 14.7. The accelerometer is located on the center line of the aircraft and will measure the average load factor. In Fig. 14.7, the pilot reads 5G's, but the upgoing wing will actually be subjected to 7G's.

Maneuver Speed

An interesting point on the V–G diagram is the intersection of the aerodynamic limit line and the structural limit line. The aircraft's speed at this point is called the *maneuver speed*, V_A, commonly called *corner speed*. At any speed below the maneuver speed the aircraft cannot be overstressed. It will stall before the limit load factor is reached. Above this speed, however, the aircraft can exceed the limit load factor before it stalls. At the maneuver airspeed the aircraft's limit load factor will be reached at the lowest possible airspeed.

The maneuver airspeed, V_A, is shown in Fig. 14.8. It can be calculated by

$$V_A = V_s\sqrt{\mathrm{LLF}} \tag{14.9}$$

where

V_A = maneuver speed (knots)
V_s = stall speed (knots)
LLF = limit load factor

All missions that an aircraft is designed to perform can be accomplished

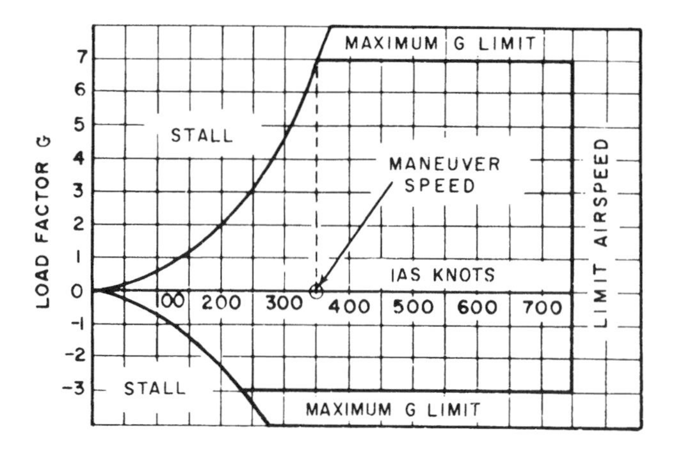

Fig. 14.8. Maneuver (corner) speed.

without exceeding the limit load factors, but on occasion a pilot will cause the aircraft to exceed these load factors. If this happens, the aircraft will probably suffer permanent objectionable deformation, resulting in costly repairs, but it will not necessarily result in failure of the primary structure.

There is another load factor that, if exceeded, may lead to catastrophic failure of the airframe. This load factor is called the ultimate load factor, ULF. Numerically the positive and negative ultimate load factors are 1.5 times the limit load factors. These are shown in Fig. 14.9 and are labeled as structural failure.

The high-speed limit of the V–G diagram is called the never exceed velocity, V_{NW}, or, more commonly, the redline speed. It is shown in Fig. 14.9. If the redline speed is exceeded structural failure may occur.

The pilot must stay inside of the flight envelope or the airplane will be in the structural damage or structural failure area.

Radius of Turn

A body traveling in a circular path is undergoing an acceleration toward the center of rotation. This is called radial acceleration, a_{r}. It is a function of the velocity of the body, V, and the radius, r, of the circle:

$$a_{\mathrm{r}} = \frac{V^2}{r} \tag{14.10}$$

As was shown in Fig. 14.1, the horizontal component of the total lift is the centripetal force that causes the radial acceleration. Also, the reaction force to the centripetal force, called the centrifugal force (CF) is equal in magnitude and

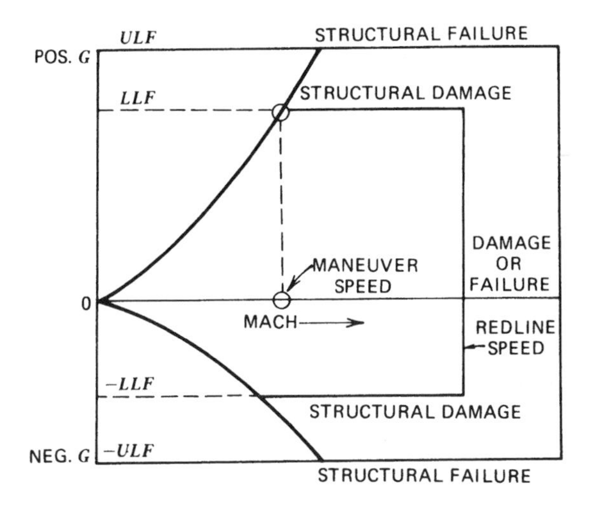

Fig. 14.9. Ultimate load factors.

opposite in direction to the centripetal force. The centripetal force in an automobile making a turn is generated by the friction of the car's tires on the pavement. The force that pulls the driver outward, away from the center of the turn, is the *centrifugal force*.

Since the centrifugal force is equal to the centripetal force in magnitude and results from the radial acceleration, it is, according to Newton's second law, equal to the mass times the radial acceleration:

$$\mathrm{CF} = ma_{\mathrm{r}} = \frac{W}{g}\frac{V^2}{r} \tag{14.11}$$

Equation 14.4 showed that this force is

$$\mathrm{CF} = L \sin \phi$$

Equating CF values gives

$$L \sin \phi = \frac{WV^2}{gr} \tag{14.12}$$

Equation 14.1 then can be rewritten as

$$L \cos \phi = W \tag{14.13}$$

By dividing Eq. 14.12 by Eq. 14.13 we obtain

$$\tan \phi = \frac{V^2}{gr} \quad \text{or} \quad r = \frac{V^2}{g \tan \phi} \tag{14.14}$$

This is the radius of turn equation.

Equation 14.14 was derived in basic units, and thus the velocity is in units of feet per second. If V is measured in knots, then

$$r = \frac{V_k^2}{11.26 \tan \phi} \tag{14.15}$$

where

r = radius of turn (ft)
V_k = velocity (knots TAS)
ϕ = bank angle (degrees)

Rate of Turn

The *rate of turn*, ROT, is primarily used during instrument flight and is, simply, the change in heading of the aircraft per unit of time:

$$\text{ROT} = \frac{g \tan \phi}{V_{\text{fps}}} \quad \text{(radians/sec)}$$

If velocity is in knots and ROT is measured in degrees per second, then

$$\text{ROT} = \frac{1091 \tan \phi}{V_k} \tag{14.16}$$

The importance of bank angle and velocity can be seen in Eqs, 14.15 and 14.16. High bank angles and slower airspeeds produce small turn radii and high ROT. The maneuver speed is the speed at which highest bank angle can be achieved at minimum airspeed. Minimum turn radius and maximum rate of turn will be realized at this speed. Figure 14.10 is a chart that shows turn radii and rates of turn for various bank angles and airspeeds for an aircraft making a coordinated, constant altitude turn.

In our discussion of induced drag in Chapter 5 we saw that the rearward component of the total lift vector was induced drag and that the vertical component was effective lift. These are shown in Fig. 14.11. We see that the effective lift must equal the weight of the plane if constant altitude is to be maintained. We also see that more thrust is required to overcome this increase in induced drag, if airspeed is to be maintained. Thus, it may be that the turn radius may be limited by thrust available instead of structural limitations.

Figure 14.12 shows a thrust-limited aircraft's turn radius plotted against true airspeed. The aerodynamic limits are the same as were described for Fig. 14.8. The thrust limit curve is determined by a combination of bank angle and airspeed.

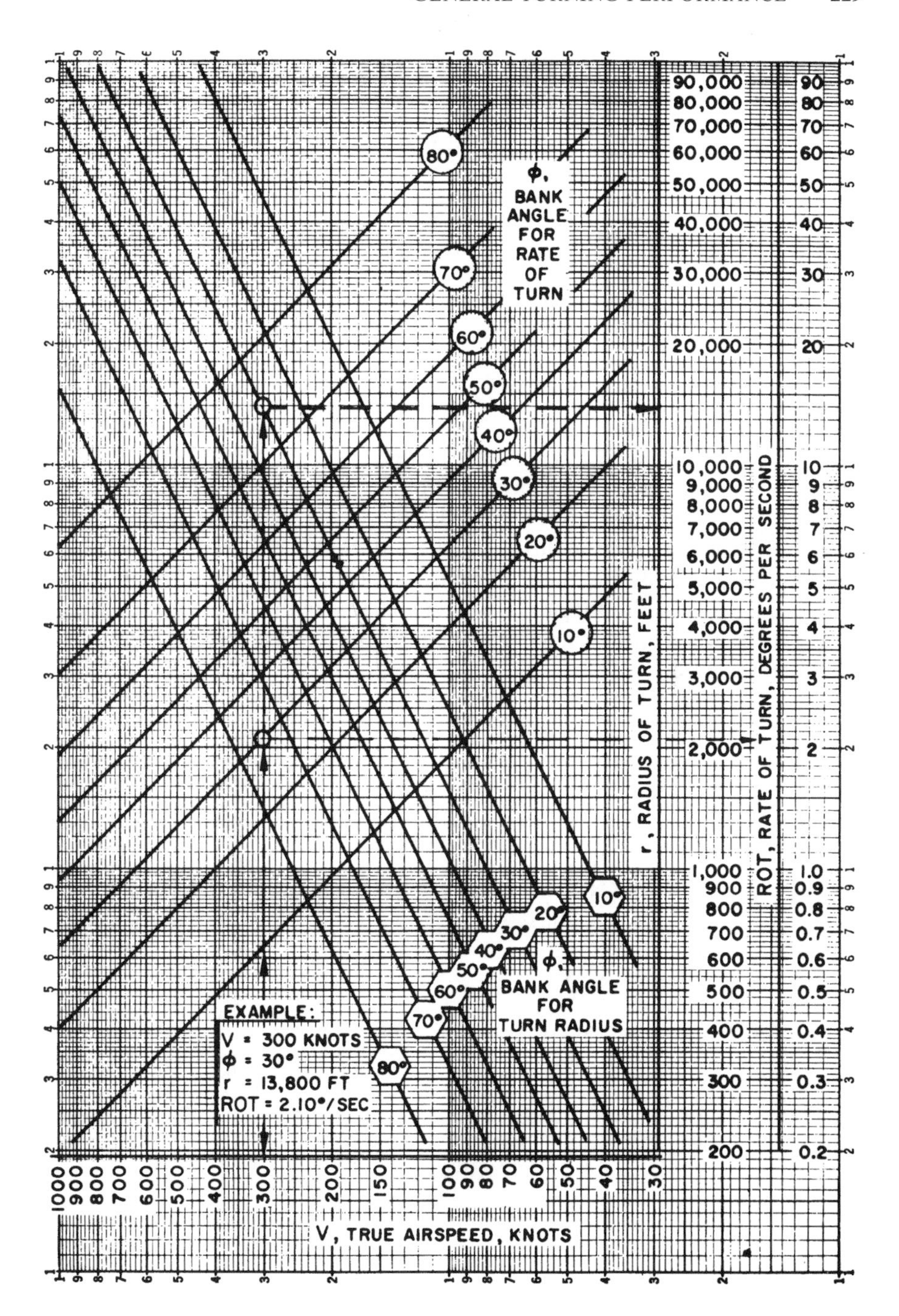

Fig. 14.10. Constant altitude turn performance.

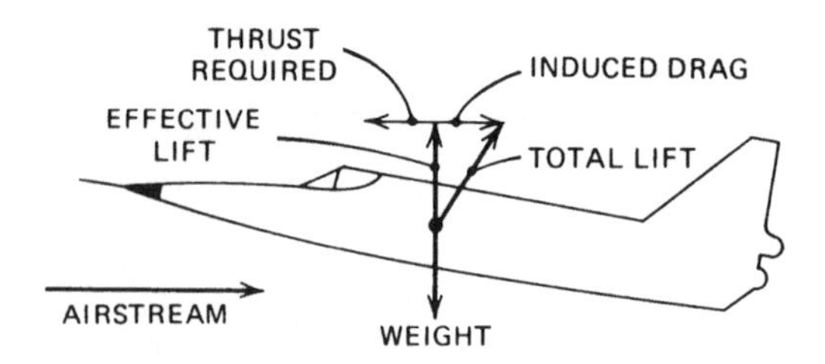

Fig. 14.11. Forces on the complete aircraft.

Vertical Loop

The loop has little tactile value but is taught as a confidence maneuver. In theory a "perfect loop is performed without a change in airspeed, but in practice this is seldom, if ever, achieved. The perfect loop is an exact vertical circle and would require a thrust/weight ratio greater than 1.0 to maintain constant airspeedP during the climb portion of the loop. The airplane would also need large speed braking devices to enable it to maintain constant airspeed during the last half of the maneuver. In addition, extensive throttle "jockeying" would be required.

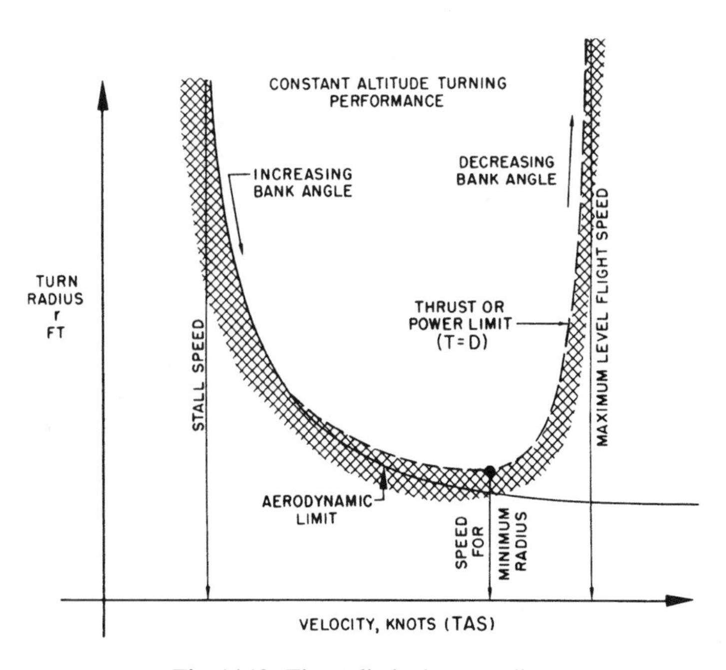

Fig. 14.12. Thrust-limited turn radius.

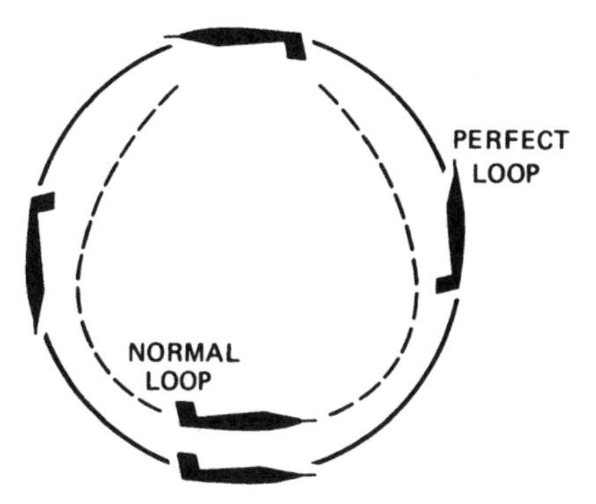

Fig. 14.13. Perfect and normal loop.

The perfect loop and a normal loop are shown in Fig. 14.13. The loop does not seem to fit into either of the two categories of maneuverability (*general turning performance* or *energy maneuverability*). Instead, it is somewhere in between.

The centrifugal force on the airplane depends on the weight, airspeed, and loop radius and can be calculated by use of Eq. 14.11:

$$\text{CF} = \left(\frac{W}{g}\right)\left(\frac{V^2}{r}\right)$$

The centrifugal force, CF, acts radially away from the center of the loop and is constant throughout the entire loop:

$$\text{CF} = \frac{WV^2}{gr}$$

This is shown in Fig. 14.14. The G loading applied by the centrifugal forces alone is

$$G = \frac{\text{CF}}{W} = \frac{V^2}{gr} \tag{14.17}$$

Example An aircraft flying at 296 knots (50 fps) and having a loop radius of 2604 ft experiences a centrifugal force of 3G's. The weight of the aircraft acting toward the earth also affects the G loading. The total G's on the aircraft vary as the aircraft's altitude changes. Maximum G's are $3 + 1 = 4$ and act at the bottom of the loop when the weight and centrifugal force act in the same direction.

At the top of the loop the aircraft is inverted, and the weight acts away from the centrifugal force. G's at the top are $3 - 1 = 2$. At the three and nine o'clock

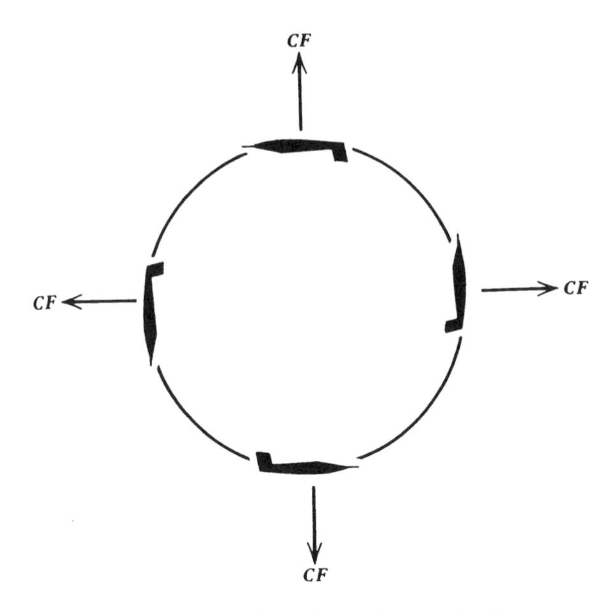

Fig. 14.14. Centrifugal forces in a vertical loop.

positions the weight does not affect the loading of the aircraft and $G = 3$. These load factors are shown in Fig. 14.15.

Actual G's depend on airspeed and loop radius. However, the loss of $2G$ as an aircraft moves from the bottom of the loop to the top and the increase of $2G$s as the aircraft moves from the top to the bottom will hold true no matter what the applied G's are.

EQUATIONS

14.1 $$\cos\phi = \frac{W}{L}$$

14.2 $$G = \frac{L}{W}$$

14.3 $$G = \frac{1}{\cos\phi}$$

14.4 $$\mathrm{CF} = L\sin\phi$$

14.5 $$\frac{V_{s_2}}{V_{s_1}} = \sqrt{G}$$

14.6 $$\frac{V_{s_2}}{V_{s_1}} = \sqrt{\frac{1}{\cos\phi}}$$

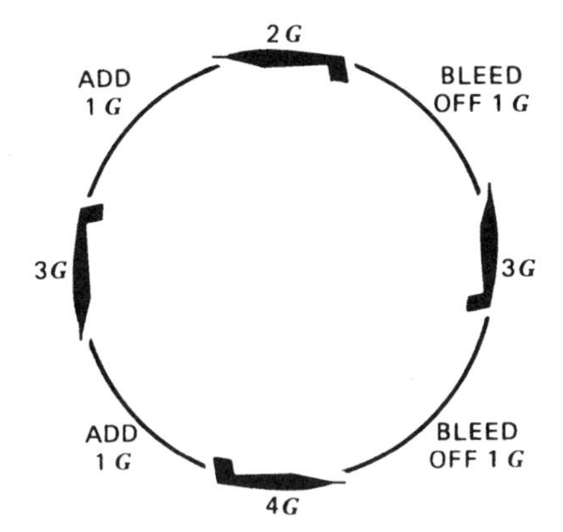

Fig. 14.15. Loading on an example aircraft.

14.7 $\phi = \arccos\left(\frac{1}{G}\right)$

14.8 $\frac{\text{LLF}_2}{\text{LLF}_1} = \frac{W_1}{W_2}$

14.9 $V_A = V_s\sqrt{\text{LLF}}$

14.10 $a_r = \frac{V^2}{r}$

14.11 $\text{CF} = \frac{WV^2}{gr}$

14.12 $L \sin\phi = \frac{WV^2}{gr}$

14.13 $L \cos\phi = W$

14.14 $r = \frac{V^2}{g \tan\phi}$

14.15 $\frac{V_k^2}{11.26 \tan\phi} = r$

14.16 $\text{ROT} = \frac{1091 \tan\phi}{V_k}$

14.17 $G = \frac{V^2}{gr}$

PROBLEMS

1. In a constant altitude, constant airspeed turn, an aircraft is in a state of equilibrium.

a. True

b. False

2. The G forces on an airplane in a constant altitude turn depend on

a. airspeed.

b. type of airplane (jet or prop).

c. bank angle.

d. All of the above

3. An airplane performing an 8-point roll can stop in the 90° position because

a. the wings are not supporting the weight.

b. the fuselage and vertical tail are supporting much of the weight.

c. the pilot pulls the nose above the horizontal, thus some of the thrust helps support the weight.

d. All of the above

4. When an airplane is in a constant altitude bank the stall speed

a. remains the same as in level flight.

b. increases as the square root of the G's.

c. increases as the square root of $1/\cos\phi$.

d. Both (b) and (c)

5. What can we find from the aerodynamic limit line on a V–G diagram?

a. It shows the maximum G's that can be pulled at any airspeed below the corner speed.

b. It shows the stall speed when G's are being pulled.

c. It is impossible to fly to the left of this line because the airplane will stall there.

d. All of the above

6. The limit load factor (LLF) is also called maximum design G.

a. True

b. False

7. If the weight of an airplane, W_2, is increased above the design gross weight, W_1, the limit load factor

a. remains the same.

b. increases by the weight ratio, W_1/W_2.

c. decreases by the weight ratio, W_1/W_2.

d. None of the above

8. If an airplane is flying in symmetrical flight at its maneuver speed

- **a.** it cannot be overstressed.
- **b.** it can make the smallest turn radius.
- **c.** it can make the highest rate of turn.
- **d.** All of the above

9. The constant altitude turn performance chart (Fig. 14.10)

- **a.** can be used for only one particular model of airplane.
- **b.** does not show if the airplane is overstressed.
- **c.** is good for any model airplane.
- **d.** Both (b) and (c)

10. When an airplane makes a level banked turn, the airspeed will decrease (unless throttle is added), because

- **a.** parasite drag is increased the most.
- **b.** induced drag is increased the most.
- **c.** profile drag is increased the most.

11. A twin-engine airplane has an engine failure shortly after takeoff. The pilot tries to turn back toward the field by making a left turn but crashes after making a 180° turn. The wreckage diagram shows that the turn radius was 6000 ft. The airspeed during the turn is estimated to be 200 knots TAS. Calculate the bank angle of the airplane.

12. Using Fig. 14.10 verify the answer to Problem 11.

13. Assuming a level turn, find the *G*'s on the airplane in Problem 11 during the turn.

14. For the airplane in Problem 11, calculate the rate of turn and time to make the 180° turn.

15. Using Fig. 14.10 verify the ROT you calculated in Problem 14.

Super Hornet: Freedom Defender (Courtesy the Boeing Company).

15 Longitudinal Stability and Control

In addition to adequate performance, an aircraft must have satisfactory handling qualities. An aircraft must be able to maintain uniform flight and be able to recover from the effects of disturbing influences, such as gusts. This ability is called the *stability* of an aircraft. Adequate stability is necessary to minimize the workload of the pilot. In some cases, such as helicopter flight, it is necessary to provide artificial stabilization by use of *automatic stabilization equipment* (ASE).

Control is the response of an aircraft to the directions of a pilot. For an aircraft to respond to the controls, its stability must be overcome. Stability and control are often compared to a seesaw, with stability at one end and control at the other. The more stability an aircraft has, the less controllability it has, and vice versa. Modern complex, high-performance aircraft have complicated stability problems that are beyond the scope of this book. Our discussion here is basic and simple. Some simplifying assumptions are made to keep the discussion from getting too involved.

DEFINITIONS

Equilibrium An aircraft is said to be in a state of equilibrium if the sum of all moments and forces at its center of gravity are equal to zero. This means that there is no pitching, yawing, or rolling, nor is any change of velocity taking place.

Static Stability Static stability is the *initial tendency* of an aircraft to move once it has been displaced from its equilibrium position. If it has the tendency to return to its equilibrium position, it is said to have positive static stability.

This is illustrated by the ball in Fig. 15.1a. Point A is the equilibrium position of the ball. If the ball is moved to point B, it has the tendency to return toward point A. It has *positive static stability*. Note that for the ball to have positive static stability it is not important that the ball actually returns to point A, only that it has the tendency to return.

If an aircraft that has been disturbed from its equilibrium position has the initial tendency to move farther away from its equilibrium position, it is said to have *negative static stability*. This is illustrated by the ball in

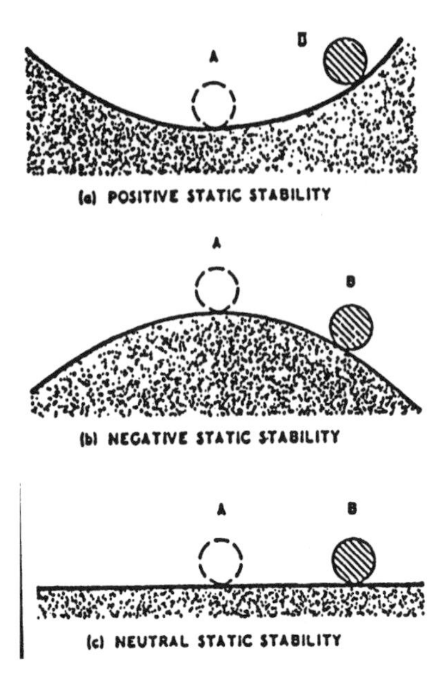

Fig. 15.1. Static stability.

Fig. 15.1b. The ball has been displaced from its equilibrium position at point A to point B. The initial tendency of the ball is to move farther away from point A. The ball has negative static stability.

If an aircraft is disturbed from its equilibrium position and has the tendency neither to return nor to move farther away from its equilibrium position, it is said to have *neutral static stability*. This is illustrated by the ball on a flat surface, as shown in Fig. 15.1c. If the ball is displaced to point B, it does not have a tendency to return to point A, nor does it have a tendency to move away from point A. This is neutral static stability.

For an aircraft to have positive stability, it must first have positive static stability.

Dynamic Stability Dynamic stability is the movement of an aircraft with respect to time. If an aircraft has been disturbed from its equilibrium position and the maximum displacement decreases with time, it is said to have *positive dynamic stability*. If the maximum displacement increases with time, it is said to have *negative dynamic stability*. If the displacement remains constant with time, it is said to have *neutral dynamic stability*.

OSCILLATORY MOTION

Aircraft motions are oscillatory in nature. Therefore, let us consider the

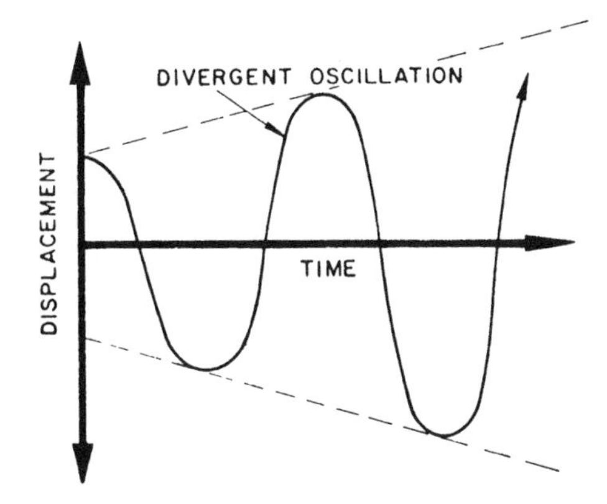

Fig. 15.2. Positive static and negative dynamic stability.

possibilities of these oscillations. As we just mentioned, an aircraft must have positive static stability, but this does not ensure that it will have positive dynamic stability as well.

Consider the graph of aircraft displacement versus time shown in Fig. 15.2. The first reaction to the displacement is to return toward the equilibrium position. This is positive static stability. The subsequent oscillations, however, are divergent, so the dynamic stability is negative. This is a case of positive static and negative dynamic stability, and it is unacceptable.

Another possibility is that of positive static stability and neutral dynamic stability, shown in Fig. 15.3. No damping of the oscillations occurs. This is also unacceptable.

The desired combination of positive static and positive dynamic stability is called a damped oscillation. An airplane will return to its equilibrium condition when this occurs. This is shown in Fig. 15.4.

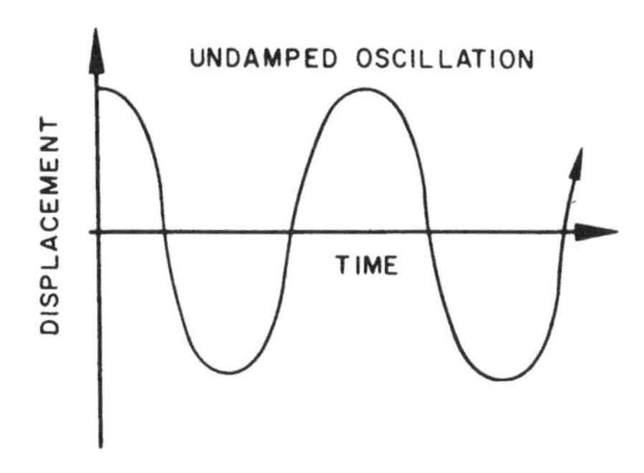

Fig. 15.3. Positive static and neutral dynamic stability.

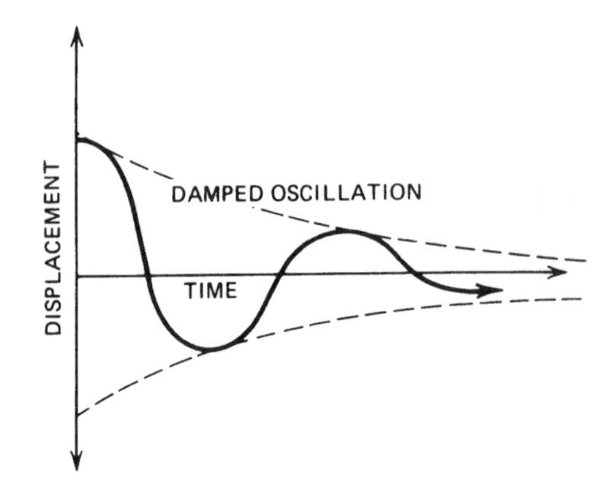

Fig. 15.4. Positive static and positive dynamic stability.

AIRPLANE REFERENCE AXES

To better envision the forces and moments acting on an airplane, we assign three mutually perpendicular reference axes, all intersecting at the CG. These are shown in Fig. 15.5. The longitudinal axis is assigned the letter *X*, the lateral axis is the *Y* axis, and the vertical axis is the *Z* axis. As seen in Fig. 15.6, the positive direction of the axes is determined by positioning the thumb, index finger, and middle finger of the right hand so that they are at right angles to each other. Point the thumb in the forward direction of the *X* axis, and the other fingers point to the positive directions of the *Y* and *Z* axes.

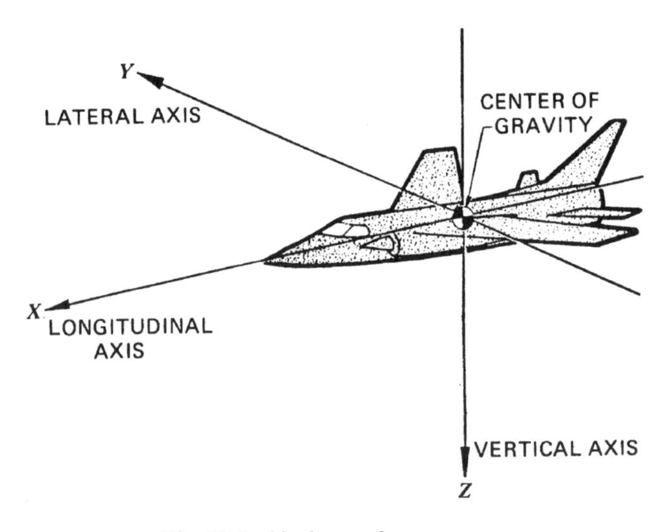

Fig. 15.5. Airplane reference axes.

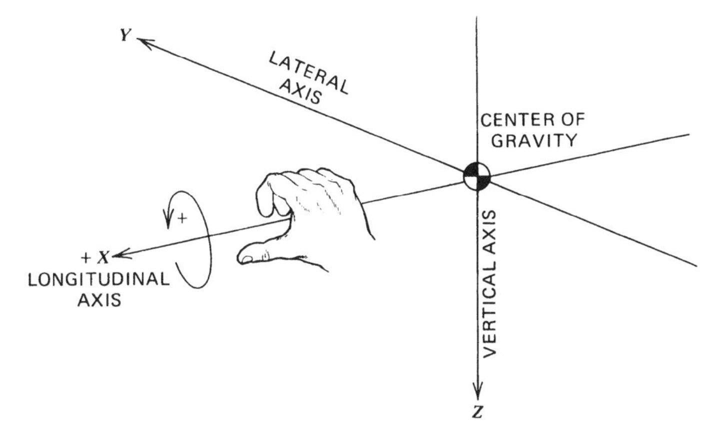

Fig. 15.6. Establishing positive moment direction.

Three moments are possible about the three axes. These are given three identifying letters that occur in sequence in the alphabet. The rolling moment is L (to avoid confusion with lift we will call this L'), the pitching moment is M, and the yawing moment is N. The positive direction of these moments is determined by using the right-hand rule, shown in Fig. 15.6. The right thumb points to the positive axis, and the curvature of the fingers shows the direction of the positive moments. The completed axes and moment illustration is shown in Fig. 15.7.

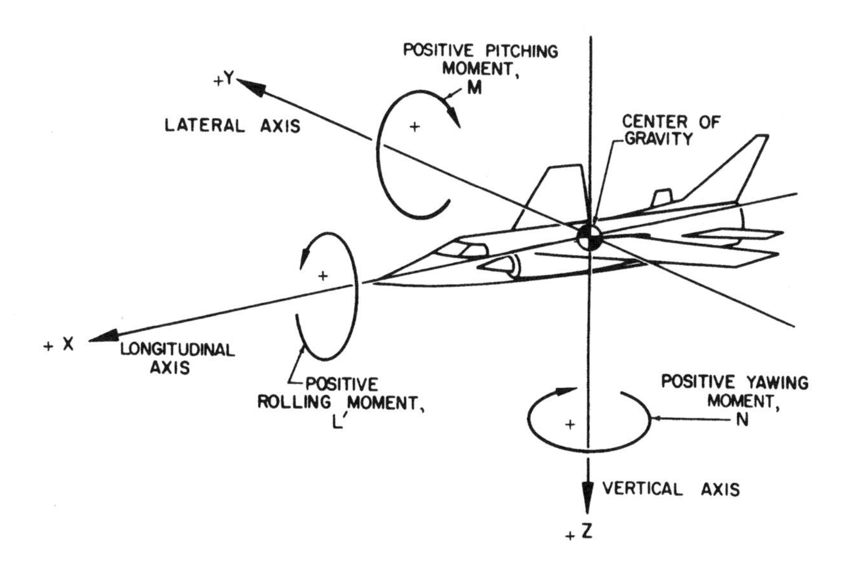

Fig. 15.7. Airplane axes and moment directions.

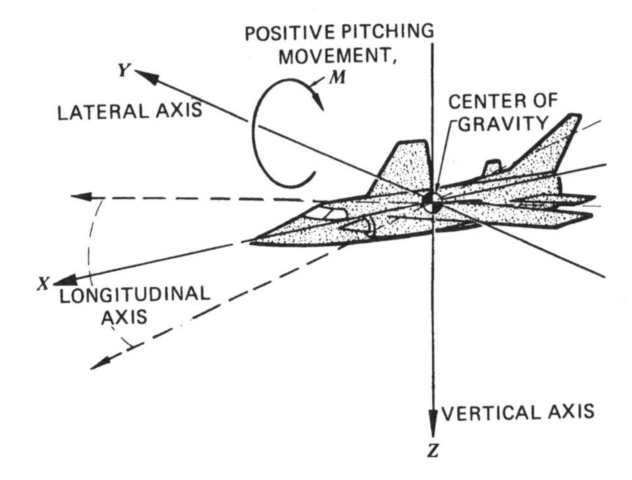

Fig. 15.8. Movement of the longitudinal axis in pitch.

Longitudinal stability and control refer to the behavior of the airplane in pitch, that is, movements of the longitudinal axis about the lateral axis. This movement is illustrated in Fig. 15.8. In the pure pitching case, no rolling or yawing of the aircraft takes place.

STATIC LONGITUDINAL STABILITY

An airplane is said to have positive static longitudinal stability if it tends to maintain a constant AOA in flight once it has been trimmed to that angle. If an airplane is trimmed so that it has zero pitching moments at some AOA and is in a state of equilibrium, and is then disturbed in pitch and tends to return to the trimmed AOA, it is said to have positive static longitudinal stability. An airplane that has negative static longitudinal stability, on the other hand, will continue to pitch away from the trimmed AOA. If an airplane has neutral static longitudinal stability, it will remain at whatever AOA the disturbance has caused.

It is important for an airplane to have positive static longitudinal stability. A stable airplane is easy and safe to fly. It can be trimmed at any desired speed and will tend to maintain that speed. An airplane with negative static longitudinal stability will be impossible to trim and will require the pilot's attention at all times.

The Pitching Moment Equation

The pitching moment about the aircraft CG is

$$M_{\mathrm{CG}} = C_{\mathrm{M(CG)}} qSc \tag{15.1}$$

where

M_{CG} = pitching moment about the CG (ft-lb)
$C_{M(CG)}$ = coefficient of pitching moment about CG
q = dynamic pressure (psf)
S = wing area (ft^2)
c = mean aerodynamic chord, MAC (ft)

Rearranging Eq. 15.1 gives

$$C_{M(CG)} = \frac{M_{CG}}{qSc} \tag{15.2}$$

Because the values of q, S, and c are always positive, it follows that for a nose-up (+) pitching moment, the value of $C_{M(CG)}$ must also be positive; for a nose-down (−) pitching moment, the value of $C_{M(CG)}$ must be negative.

Graphic Representation of Static Longitudinal Stability

A plot of the variation of $C_{M(CG)}$ at different values of C_L (different AOAs) for an airplane with positive static longitudinal stability is shown in Fig. 15.9. The trim point is the value of C_L where the aircraft has no pitching moment. At all values of C_L above the trim point, such as point y, the aircraft will have a nose-down (−) pitching moment. If the aircraft is disturbed by an up-gust, the AOA (and C_L) will be increased. For stability, a nose-down pitching moment is required, so this is a stabilizing condition. If the airplane is disturbed by a down-gust, the AOA and value of C_L will be reduced. This is represented by point X in Fig. 15.9.

The value of $C_{M(CG)}$ at point X is positive, and a nose-up pitching moment results. This is what is required for static stability. Thus, a negative slope on

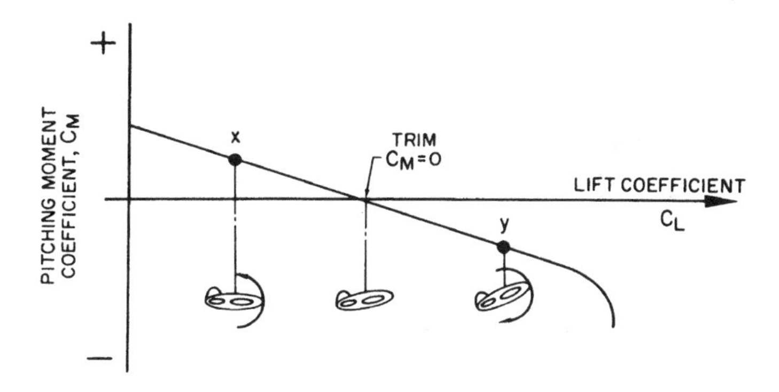

Fig. 15.9. Positive static longitudinal stability.

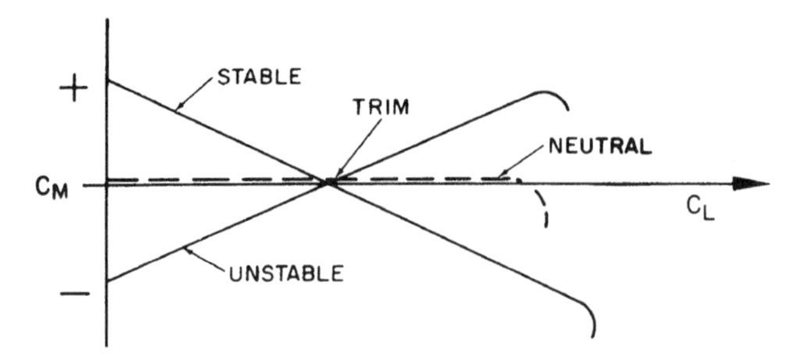

Fig. 15.10. Types of static longitudinal stability.

this graph represents an aircraft with positive static stability. Conversely, a positive slope would indicate an unstable aircraft, and a zero slope would represent an aircraft with neutral stability. These are shown in Fig. 15.10.

There are different degrees of stability. Some aircraft tend to return to their equilibrium positions faster than others. Again, using the analogy of the ball in the curved container, the degree of stability can be illustrated as shown in Fig. 15.11. Degrees of stability are shown on the $C_{M(CG)}$–C_L plot by the slope of the curve. Steeper slope of the stable curve shows more stability, and steeper slope of the unstable curve shows more instability.

An airplane can be stable at lower angles of attack but may be unstable at higher angles of attack. This would indicate a pitch-up problem at high AOA. This is shown in Fig. 15.12.

Contribution of Aircraft Components to Pitch Stability

Wings The static stability contribution of the wings depends on the relative position of the aerodynamic center AC and the center of gravity CG of the airplane. Consider a tailless (flying wing) airplane with a symmetrical airfoil section as shown in Fig. 15.13. We will consider two possibilities of CG and AC location:

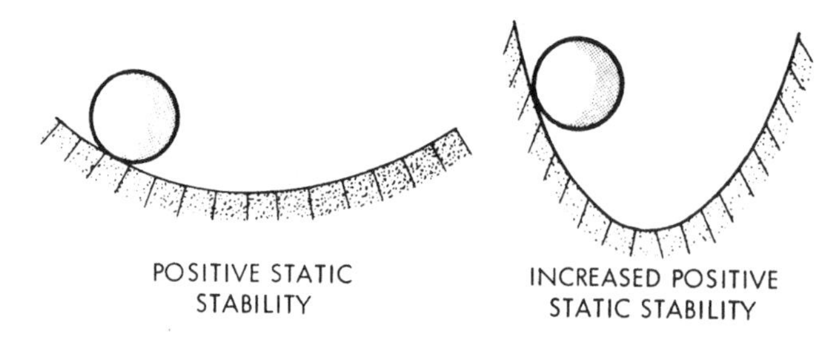

Fig. 15.11. Degrees of positive static stability.

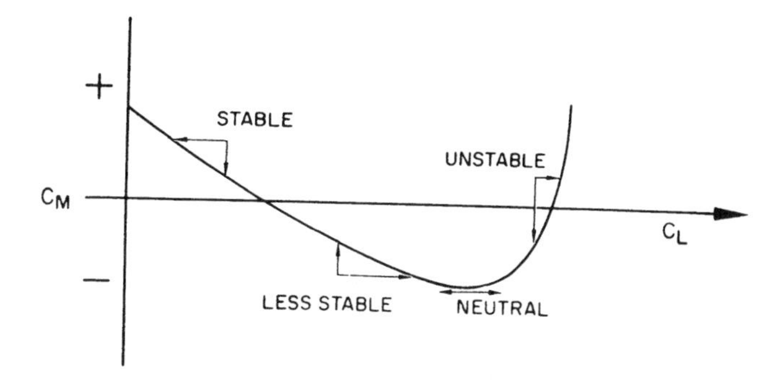

Fig. 15.12. Aircraft static longitudinal stability.

1. CG is located forward of the AC (Fig. 15.13a).
2. CG is located behind the AC (Fig. 15.13b).

If the airplane in Fig. 15.13a experiences an up-gust, the AOA will be increased and the lift at the AC will increase. The airplane will then rotate about the CG, and this nose-down moment will tend to rotate the airplane and return it to its equilibrium AOA. This is the stable condition.

If the airplane in Fig. 15.13b experiences an up-gust, the increase in lift at the AC will rotate the airplane about its CG and create a nose-up pitching moment and rotate the airplane away from its equilibrium position. This is the unstable condition. The CG must be ahead of the AC for a stable flying wing.

The airplane in Fig. 15.13a is stable in pitch, but it is not in equilibrium. The conditions for equilibrium require that there be no unbalanced forces or unbalanced moments acting on the airplane. The first requirement is easily satisfied by adjusting the AOA so that lift equals weight. The second requirement, however, has not been met. The forces of lift and weight create a nose-down pitching moment, which must be canceled out by an opposing nose-up pitching moment.

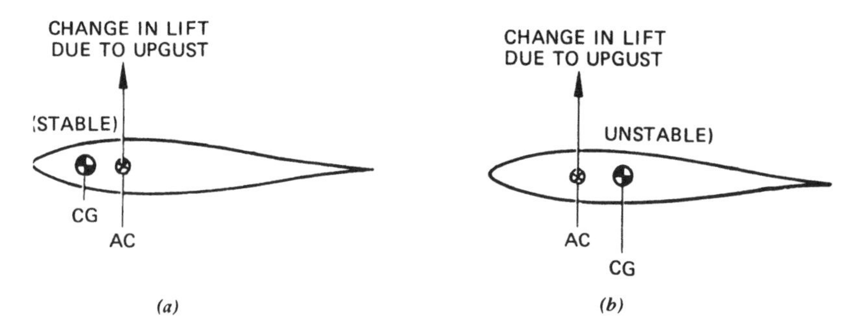

Fig. 15.13. Effect of CG and AC location on static longitudinal stability.

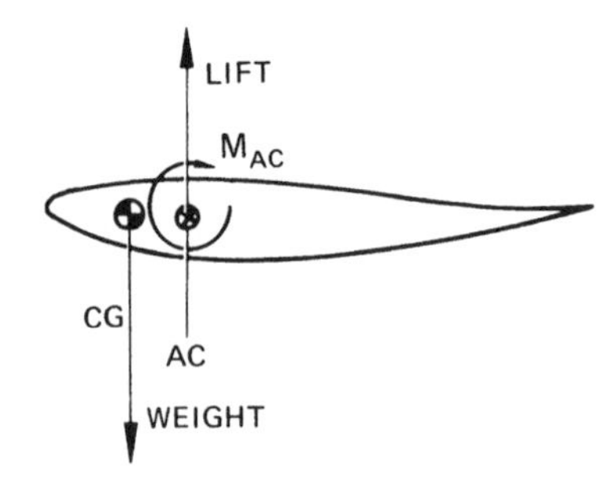

Fig. 15.14. Static longitudinally stable flying wing in equilibrium.

In our discussion of pressure distribution on airfoils in Chapter 3, we saw that the forces acting on the top and bottom surfaces of a symmetrical airfoil were located in the same position along the chord and that no pitching moment resulted from their location (see Fig. 3.9). The pressure location on cambered airfoils, on the other hand, was found not to be located at the same chordwise position, and so a nose-down pitching moment resulted from positive camber (Fig. 3.10). A negatively cambered airfoil produces a nose-up pitching moment, which is what we need to cancel the nose-down moment caused by the lift and weight vectors. This is shown in Fig. 15.14. Delta-wing airplanes use reflexed (negatively cambered) trailing edges to create the nose-up pitching moments required for equilibrium.

The overall contribution of the wings of an aircraft to the static longitudinal stability depends on the location of the AC and the CG of the aircraft. If the AC is behind the CG, the contribution is stabilizing. If the AC is ahead of the CG, the contribution is de-stabilizing. The trend in recent years is to locate the wings farther back on the fuselage, thus increasing the aircraft's pitch stability.

There is one disadvantage in doing this. Consider the vertical forces acting on the aircraft if the AC is located behind the CG as shown in Fig. 15.15.

It is assumed that this airplane has a symmetrical airfoil so it has no pitching moment due to the pressure distribution on the airfoil. For balance,

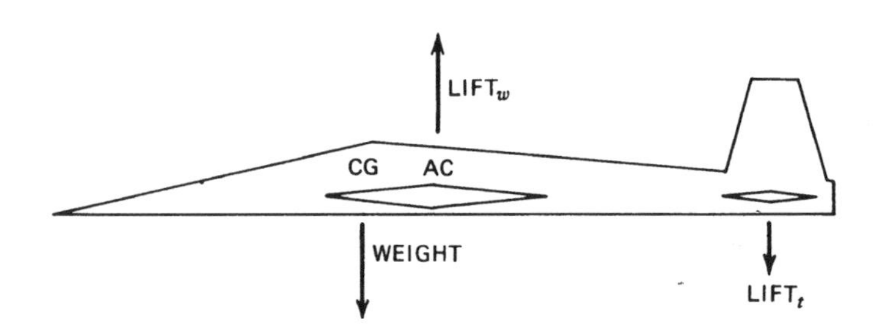

Fig. 15.15. Airplane with static longitudinal stability.

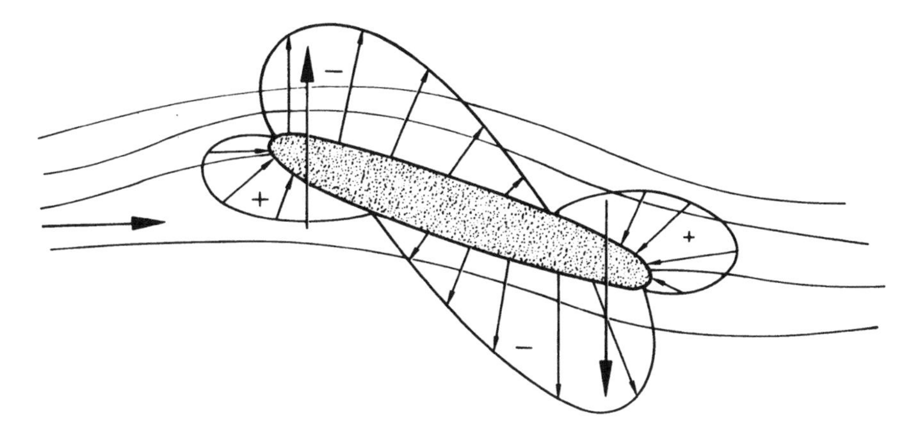

Fig. 15.16. Pressure distribution about a body of revolution.

the airplane must have a download on the tail ($LIFT_T$). The lift on the wing ($LIFT_W$) must equal the weight of the aircraft plus the lift on the tail. Thus, it must operate at a higher AOA than if the lift on the tail acted upward. Higher angles of attack produce more drag. One of the newer concepts is to reduce the static stability by moving the CG backward, and thus reduce the drag. This may require automatic stabilization devices.

Fuselage A streamlined fuselage has a pressure distribution similar to that of a body of revolution when placed in an airstream. The pressure distribution about such a body is shown in Fig. 15.16. It can be seen that no net lift is developed by the pressure distribution, but a nose-up pitching moment is developed by an up-gust, so the pitching moment is destabilizing. Thus, the fuselage is a destabilizing component.

Engine Nacelles The direction of the airflow through a propeller disk or through a turbojet engine is not changed if the axis of the engine is in line with the aircraft's flight path. If the engine axis is at an angle with the relative wind, however, the airflow is turned so that it flows in the direction of the engine axis. When this happens, a side force is developed on the propeller shaft or on the side of the jet engine intake, in accordance with Newton's third law. This is shown on the aircraft in Fig. 15.17.

The axes of the aircraft are at a positive AOA to the relative wind and the resulting force creates a nose-up pitching moment. This is a destabilizing moment. If the engines are mounted so that the propellers or jet intakes are behind the CG and the axis of the engine makes a positive angle with the relative wind, the resulting upward side force produces a nose-down moment. This is a stabilizing moment. Propellers or jet engine intakes located forward of the CG are destabilizing components, and propellers or jet engine intakes located behind the CG are stabilizing components.

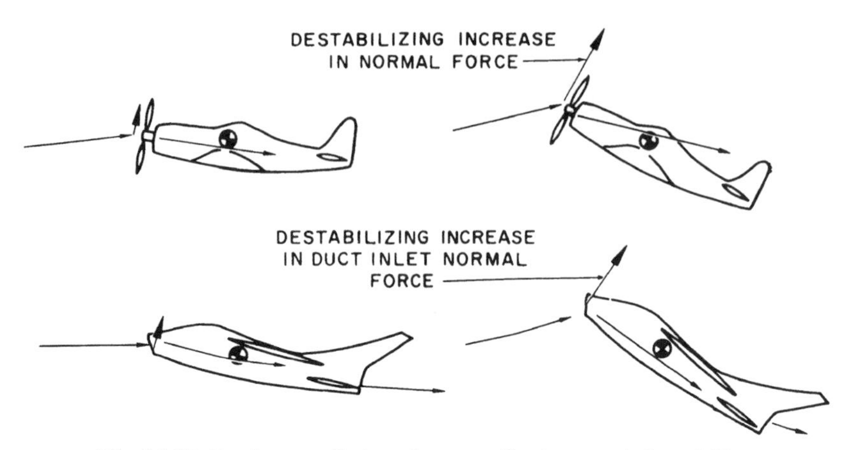

Fig. 15.17. Engine nacelle location contribution to pitch stability.

Horizontal Stabilizer As the name implies, the horizontal stabilizer is a strongly stabilizing influence on static longitudinal stability. It is usually a symmetrical airfoil because it must produce both upward and downward airloads. The contribution of the horizontal stabilizer to the pitch stability of the aircraft can be seen in Fig. 15.18. Figure 15.18a shows that if an up-gust causes the aircraft to pitch up, then an upward lift is developed by the horizontal tail. This creates a nose-down moment, which is stabilizing. In Fig. 15.18b the opposite effect is achieved when the aircraft is pitched downward by a down-gust. A nose-up moment is developed in this case.

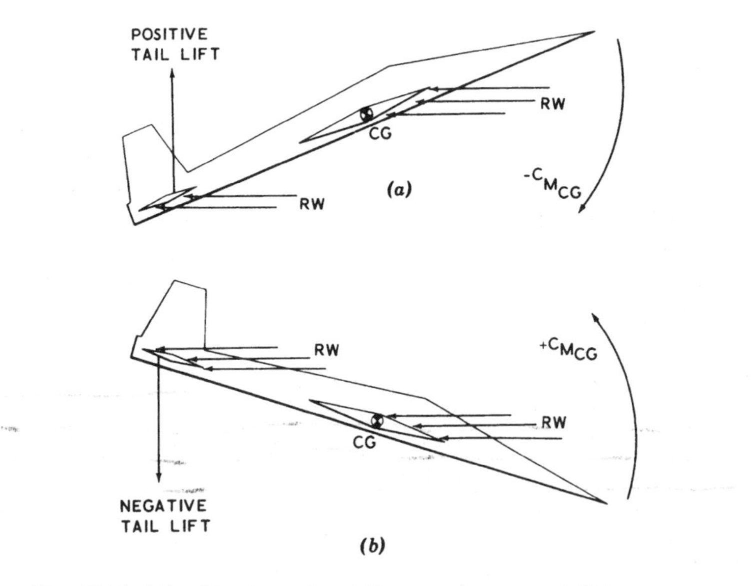

Fig. 15.18. Lift of horizontal stabilizer produces a stabilizing moment.

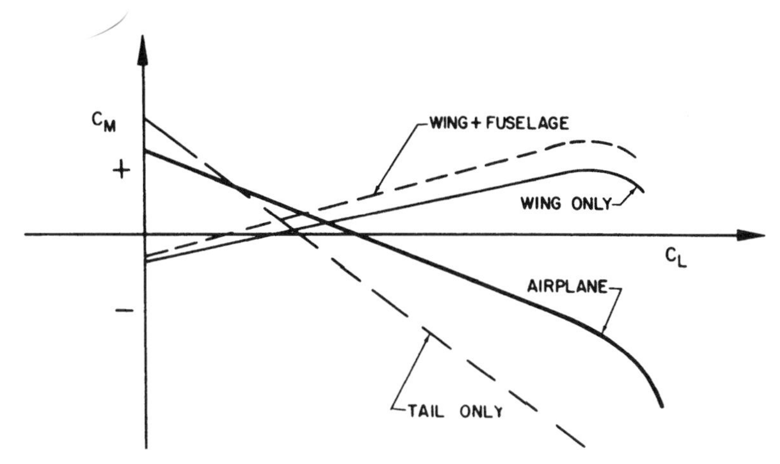

Fig. 15.19. Typical buildup of aircraft components.

The degree of stability produced by the tail is determined by the tail size and the moment arm to the aircraft's CG. The tail area (ft^2) multiplied by the moment arm (ft) equals the dimension of ft^3, called the "tail volume." It is an indication of the stabilizing effectiveness of the horizontal stabilizer. An increase in either the size of the surface or the distance between the CG of the airplane and the AC of the stabilizer will increase the tail volume and thus the stability of the aircraft.

Total Airplane Figure 15.19 shows a typical buildup of wing with its AC forward of the CG, the wing plus the fuselage, the horizontal tail alone, and the total airplane. The horizontal stabilizer has enough stability to overcome the negative stability of the wing and fuselage combined. Stability of the engine location is not shown in this figure.

Effect of CG Position Varying the CG position has a great effect on the longitudinal static stability of an airplane as shown in Fig. 15.20. Forward positioning of the CG results in increased pitch stability, as was explained earlier and illustrated in Fig. 15.13. This is shown by the slope of the C_M–C_L curves in Fig. 15.20. At 40% MAC the slope is zero, and the aircraft has neutral stability at this point. Moving the CG behind this point results in an unstable condition.

Stick-Fixed–Stick-Free Stability If the elevators (or other control surfaces) are allowed to float free, they will be moved from their neutral positions by outside disturbances, and the total stabilizing surface will be reduced. The stability of an aircraft with "free-floating" control surfaces is known as *stick-free stability*. The stability of an aircraft with the controls held in a fixed position will be increased. This is known as *stick-fixed stability*.

Figure 15.21 shows the C_M–C_L curves for both conditions. The slope of the stick-fixed condition is greater; hence, the stability is greater for the stick-fixed case. Unless the airplane is equipped with irreversible powered controls, better stability will be realized if the controls are held in the neutral position.

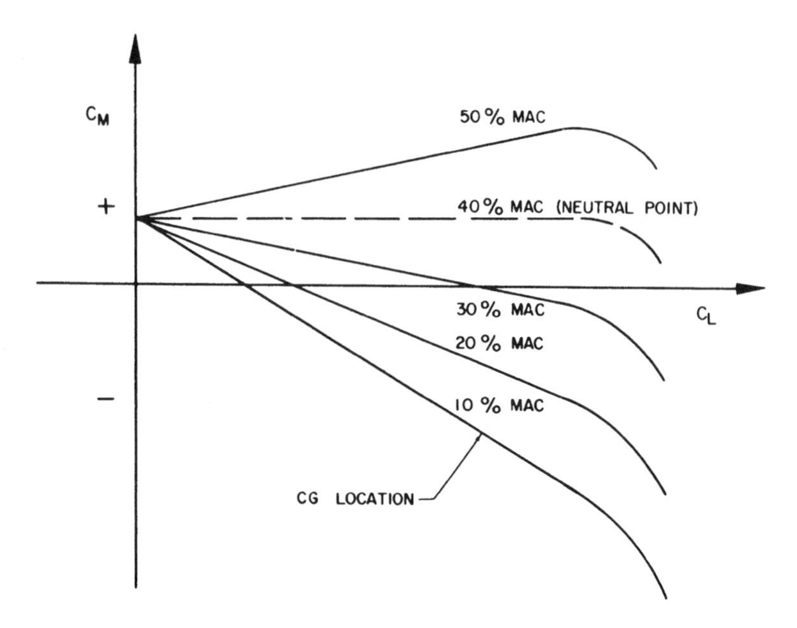

Fig. 15.20. Effect of CG location on static longitudinal stability.

DYNAMIC LONGITUDINAL STABILITY

Stability considerations that we have discussed up to now have dealt with static stability. *Dynamic stability* involves the response of an aircraft to disturbances over a period of time. Dynamic stability exists when the amplitude of these disturbances dampens out with time.

The reaction of an aircraft to disturbances differs, depending on whether the controls are in the stick-free or the stick-fixed configuration. First, let us consider the stick-free or reversible-controls aircraft. Most light general avi-

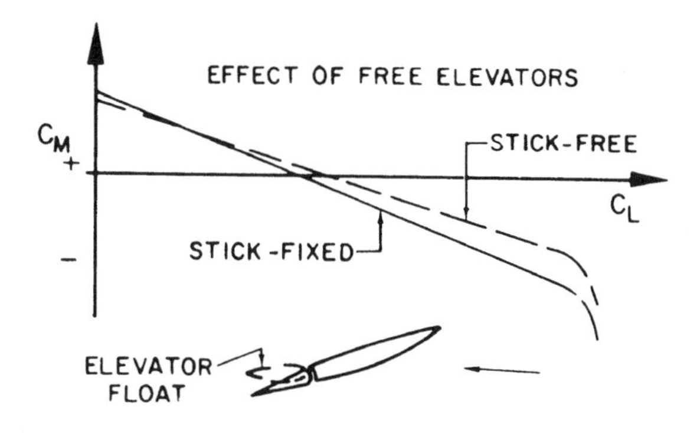

Fig. 15.21. Stick free–stick fixed stability.

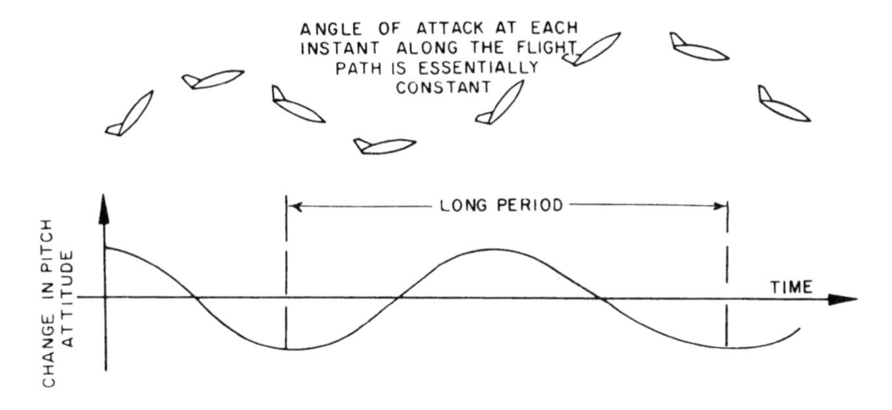

Fig. 15.22. Phugoid longitudinal dynamic mode.

ation aircraft are in this category. If a control surface is capable of being manually moved from outside the aircraft, it is in this category.

Two types of dynamic oscillations are possible. One is the long period with poor damping oscillation, called the phugoid mode. This is shown in Fig. 15.22. With phugoid oscillation, the airspeed, pitch, and altitude of the airplane vary widely, but the AOA remains nearly constant. The motion is so slow that the effects of inertia forces and damping forces are very low. The whole phugoid can be thought of as a slow interchange between kinetic and potential energy.

The second mode is a short-period mode as shown in Fig. 15.23. The short-period mode in the stick-free configuration is caused by the elevator flapping about its hinge line. A typical flapping mode may have a period of 0.3 to 1.5 sec and has heavy damping of the oscillations. Recovery from this type of pitching can be accomplished by releasing the controls or, more rapidly, by holding the controls in their neutral positions. This has the effect of putting the airplane into the stick-fixed configuration.

Pilots must be careful not to try to dampen out the oscillations. Pilot

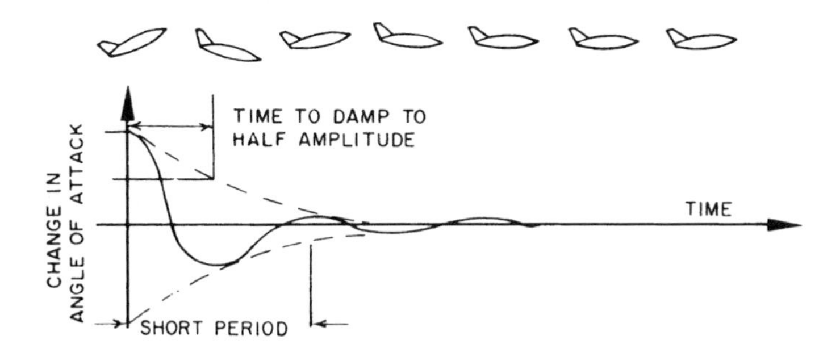

Fig. 15.23. Short-period dynamic mode.

reaction time is close to the natural period of the oscillations and inadvertent reinforcement of the pitching moment may result. While this may only result in a rough ride for a relatively slow light airplane, it can prove disastrous for a high-performance jet aircraft. If this reinforcement occurs in a high-performance jet, it is called a *pilot-induced oscillation* (PIO). A PIO can destroy the aircraft in a few seconds.

A similar dynamic stability problem, called *collective bounce*, exists in helicopter flight. If an up-gust is encountered and the helicopter is forced upward, inertia forces the pilot downward in the seat. If the pilot has hold of the collective stick, this will also be forced downward, and the AOA of all the rotors will be decreased, causing the helicopter to descend. Again, the inertia forces cause the pilot to be forced upward, the collective stick is raised, and the process is repeated.

PITCHING TENDENCIES IN A STALL

Low-Tailed Aircraft

The forces in the pitching plane are shown in Fig. 15.24. Assume that the aircraft is trimmed in straight and level flight and that the aircraft's CG is forward of the AC (as shown). A downward balancing force on the tail is required. In case of an aft CG location, an upward force on the tail is required. If the speed is reduced, a gentle up-deflection of the elevator is required to maintain altitude, and the downward load on the tail is increased. The static stability of the aircraft causes a nose-down pitching tendency, which has to be resisted by further up elevator to keep the nose from dropping. The low-set tail soon becomes engulfed in the turbulent, low-energy air from the wing wake. This reduces the efficiency of the tail.

At the stall, two distinct things happen. First, the airplane responds to the traditional nose-down pitching tendency at the stall, and the whole airplane responds with a nose-down pitch. Second, at the moment of stall, the wing wake passes straight aft and goes above the low-set tail. This leaves the tail in undisturbed, high-energy air, and it now is at a high positive AOA, causing an

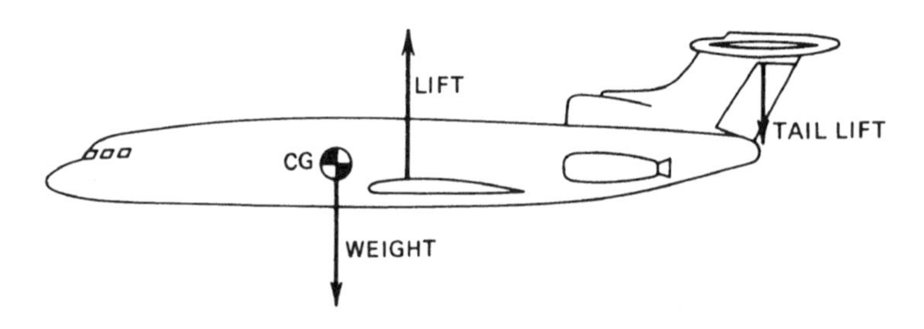

Fig. 15.24. Forces on a pitching plane.

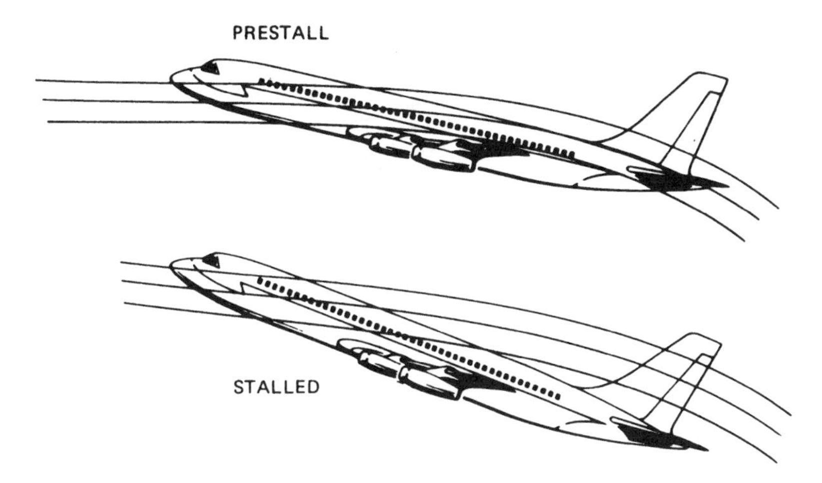

Fig. 15.25. Wing wake influences on a low-tail aircraft.

upward lift on the tail. This lift increases the nose-down pitching tendency (Fig. 15.25).

T-Tail Aircraft

Again assume that an aircraft is trimmed in straight and level flight. As the aircraft is slowed toward a stall, the handling characteristics are much the same as for the low-tailed aircraft, except that the high tail remains clear of the wing wake and retains its effectiveness. Continued speed reduction is, therefore, more efficient.

At the stall two distinct things happen. First, the swept-wing, high-tail airplane tends to suffer a marked nose-up pitch after the stall (this is explained in detail later). Second, the wing wake, which has now become low-energy turbulent air, passes straight aft and immerses the T tail, which is now in the right position to catch it. This greatly reduces the tail effectiveness and makes it incapable of combating the nose-up pitch and so the airplane continues to pitch up. The great reduction in lift and the increase in drag cause a rapidly increasing descent path. Thus, the angle of attack is increased and the pitch-up problem is worsened. The airplane is well on its way to extreme angles of attack and deep stall. This is shown in Fig. 15.26.

Explanation of Nose-Up Pitch Following Stall in Swept-Wing Aircraft

Three things affect the pitching tendency:

1. *Wing Section Characteristics.* The section characteristics of most wings are such as to cause a nose-down pitch, providing that the whole wing

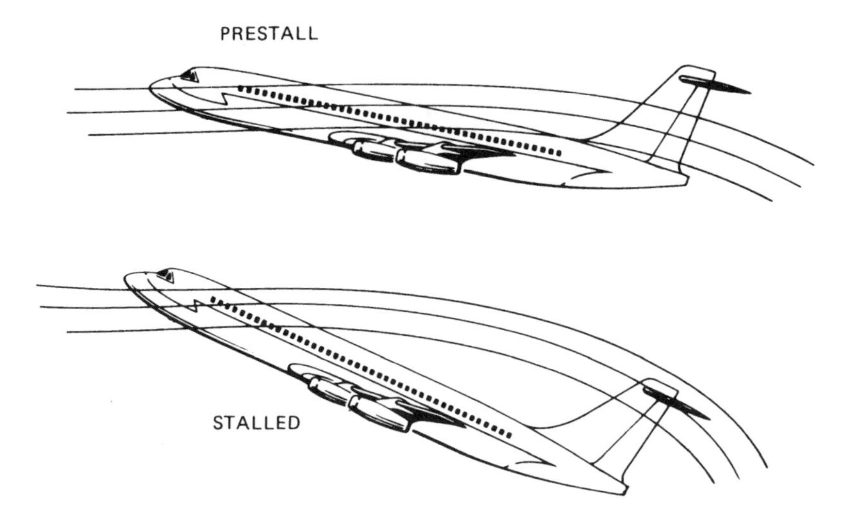

Fig. 15.26. Wing wake influences on a T-tail aircraft.

could be stalled at the same instant (spanwise). The pressure distribution is rather flat in normal flight, as shown in Fig. 15.27. As the stall is approached the pressure distribution changes to an increasingly leading-edge peaked pattern because of the enormous low-pressure area developed by the nose profile. At the stall this peak collapses, the pressure distribution pattern changes, and the result is usually a nose-down pitch.

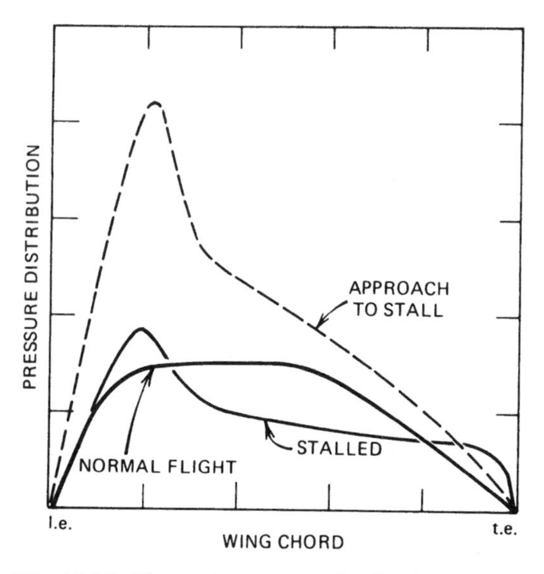

Fig. 15.27. Change in pressure distribution at stall.

2. *Wing Sweep.* In practice the whole wing does not stall at the same instant. Swept and tapered wings tend to stall at the tips first because of the high wing loading at the tips. The boundary layer outflow also resulting from sweep slows the airflow, reduces the lift near the tips, and worsens the situation. This loss of lift outboard (and therefore aft) causes the center of pressure to move forward, which augments the pitch-up tendency.
3. *The Fuselage.* On a modern aircraft the forward fuselage overhangs the wing by a greater extent than on the older straight-wing models. As we have seen, the fuselage is a destabilizing factor, and an even greater one with the longer overhang. Thus, the fuselage contributes to the nose-up pitching tendency as AOA exceeds its stall value.

LONGITUDINAL CONTROL

To control an airplane it is necessary to overcome its stability. As we have seen, the farther forward the CG of the airplane is, the more static pitch stability the airplane has. However, since stability and control oppose each other, the forward CG results in lower controllability. One of the pitch control requirements is that sufficient elevator forces are available to overcome the stability of forward CG location.

Takeoff and landing maneuvers are critical as far as control forces are concerned. During takeoff, an aircraft must be able to rotate to takeoff attitude. The requirement here is that the aircraft be able to attain takeoff attitude at $0.9V_s$. Unlike the airborne case, the aircraft rotates about its main wheels, rather than its CG, during takeoff. The forces on the aircraft are shown in Fig. 15.28.

In many aircraft little or no lift is developed until the aircraft is rotated to takeoff attitude. The angle of incidence of the wing is selected to produce minimum drag, instead of producing lift, as the aircraft accelerates during the takeoff run. The nose-down moments that the elevator must overcome to rotate the aircraft are (1) the moment caused by the thrust line being above the main wheels, (2) the moment caused by the CG being ahead of the main wheels, and (3) the nose-down moment of either a cambered wing or takeoff flaps (if used).

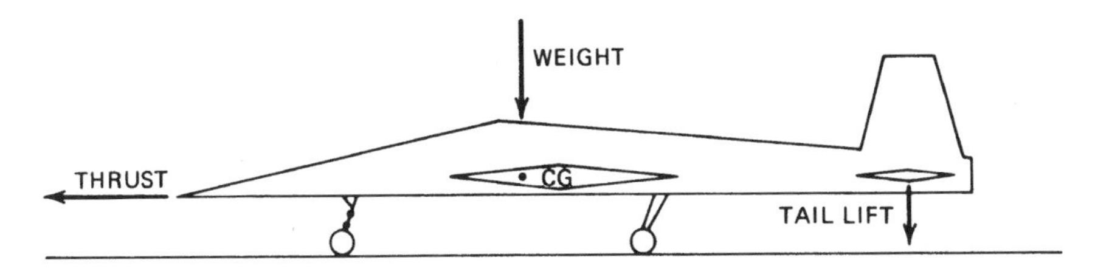

Fig. 15.28. Forces producing moments during takeoff.

An effect that is not shown on the drawing is the nose-down pitching moment caused by ground effect, which reduces the downwash over the horizontal tail of a low-tailed airplane. The elevator must be able to produce a nose-up pitching moment that will overcome all the nose-down pitching moments mentioned above.

Landing requirements include flaring the aircraft prior to touchdown and overcoming the nose-down moment caused by the airplane entering ground effect. If the airplane fulfills the takeoff requirements satisfactorily, it will usually have enough elevator control for landings.

SYMBOLS

$C_{M(CG)}$	Pitching moment coefficient (dimensionless)
L'	Rolling moment (ft-lb)
M_{CG}	Pitching moment about CG
N	Yawing moment
PIO	Pilot-induced oscillation
X	Longitudinal axis of aircraft
Y	Lateral axis of aircraft
Z	Vertical axis of aircraft

EQUATIONS

15.1 $M_{CG} = C_{M(CG)} qSc$

15.2 $C_{M(CG)} = \dfrac{M_{CG}}{qSc}$

PROBLEMS

1. The more stability an airplane has, the easier it is to control.
   a. True
   b. False
2. Static stability of an airplane is
   a. the ability to return to its equilibrium position once it has been disturbed.
   b. the long time reaction to a disturbance.
   c. the initial tendency to move toward the equilibrium position after a disturbance.
   d. None of the above
3. Dynamic stability of an airplane is ______ to a disturbance.
   a. the immediate reaction
   b. the long time reaction

c. oscillatory in nature.
d. Both (b) and (c)

4. A damped oscillation is one that
   a. shows positive static stability.
   b. shows positive dynamic stability.
   c. Both (a) and (b)
   d. Neither (a) nor (b)

5. A ball in a flower vase as compared to a ball in a soup bowl illustrates
   a. increased dynamic stability.
   b. decreased dynamic stability.
   c. increased static and dynamic stability.
   d. decreased static and dynamic stability.

6. Weight and balance of airplanes is important because
   a. the CG affects the stability and control.
   b. if the CG is moved forward the static pitch stability is increased.
   c. if the CG is too far forward, the plane may not respond to elevator commands.
   d. All of the above

7. The farther back the wings are located, the more static pitch stability an airplane has.
   a. True
   b. False

8. Placing the jet engines at the rear of an airplane
   a. increases the static pitch stability.
   b. decreases the static pitch stability.
   c. increases the pitch control.
   d. decreases the pitch control.
   e. Both (a) and (d)
   f. Both (b) and (c)

9. Phugoid oscillation is dangerous because pilot reaction can lead to PIO.
   a. True
   b. False

10. The most critical condition for pitch control is that the elevators must be able to
    a. rotate the airplane for takeoff.
    b. flare the airplane for landing.
    c. overcome a forward CG location.
    d. All of the above.

MD-90-30 (Courtesy of the Boeing Company).

16 Directional and Lateral Stability and Control

DIRECTIONAL STABILITY AND CONTROL

Directional stability and control refers to the behavior of the airplane in yaw, that is, movements of the longitudinal axis when it is rotated about the vertical axis. This rotation is caused by yawing moments and is illustrated in Fig. 16.1. In the pure yawing case, there is no pitching or rolling of the aircraft.

In this section we discuss static directional stability and directional control. *Dynamic directional stability*, however, is coupled with dynamic roll stability, and these coupled effects are discussed later in the chapter. *Sideslip angle*, β (beta), is the angle between the relative wind and the airplane's longitudinal axis. When the relative wind is from the right, the sideslip is positive in value. Figure 16.2 shows a positive sideslip angle.

STATIC DIRECTIONAL STABILITY

An airplane is said to have *positive directional stability* if it is trimmed for nonsideslip flight, is then subjected to a sideslip condition, and reacts by turning into the new relative wind so that the sideslip angle is reduced to zero. If the airplane reacts to the sideslip by turning away from the new relative wind, thus increasing the sideslip angle, the airplane has *negative directional stability*. If no reaction results from sideslip, the airplane is said to have *neutral directional stability*.

Figure 16.2a shows an airplane with *negative static directional stability*, while Fig. 16.2b shows an airplane with positive static directional stability. With negative static directional stability, an airplane tends to turn away from the relative wind. This leads to sideward flight that is often called "swapping ends," a clearly unacceptable condition.

The Yawing Moment Equation

The yawing moment about the aircraft CG is

$$N_{CG} = C_{N(CG)} qSb \tag{16.1}$$

where

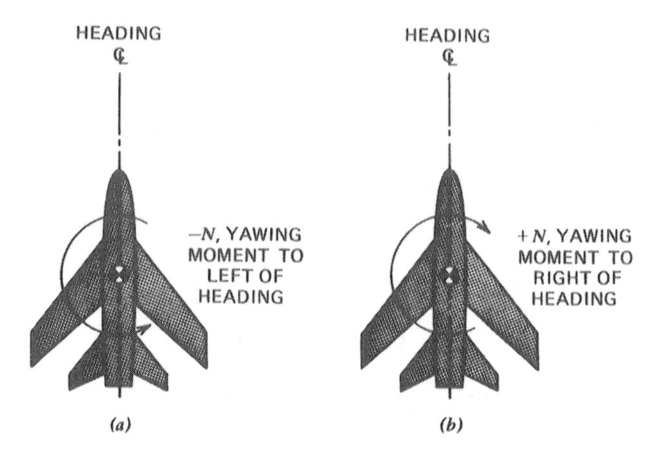

Fig. 16.1. (a) Negative yawing moment, (b) positive yawing moment.

N_{CG} = yawing moment about CG (ft-lb)
$C_{N(CG)}$ = coefficient of yawing moment about CG
q = dynamic pressure (psf)
S = wing area (ft^2)
b = wing span (ft)

Rearranging Eq. 16.1 gives

$$C_{N(CG)} = \frac{N_{CG}}{qSb} \tag{16.2}$$

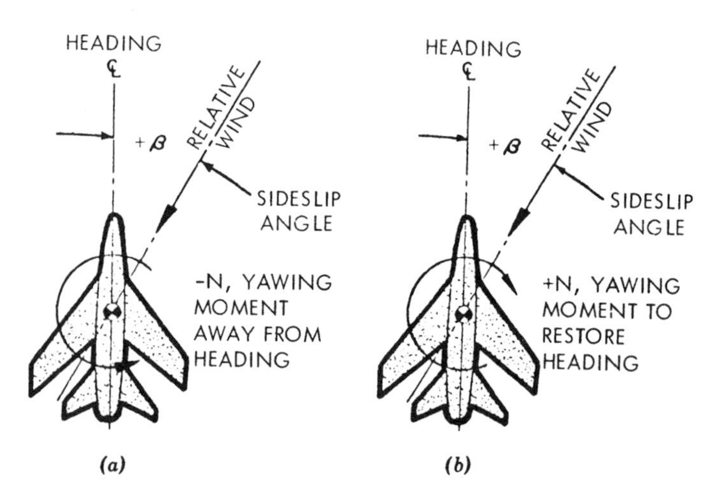

Fig. 16.2. (a) Unstable, (b) stable in yaw.

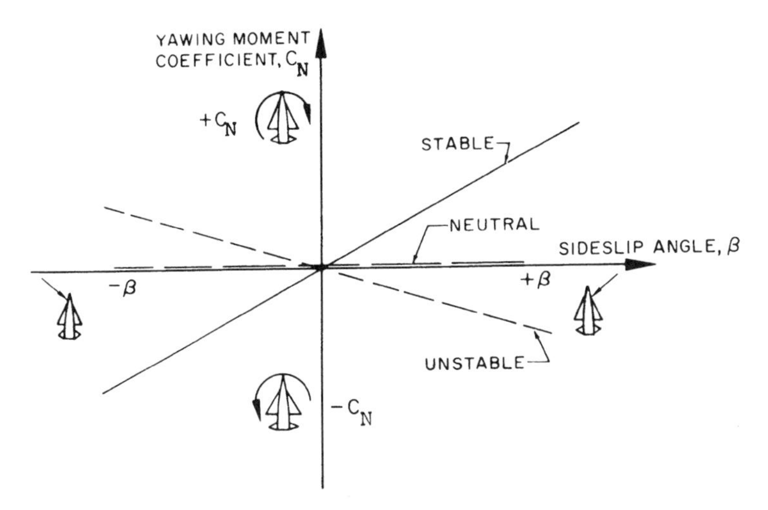

Fig. 16.3. Static directional stability.

Because the values of q, S, and b are always positive, it follows that for a nose-right (+) yawing moment, the value of the yawing coefficient must also be positive. Similarly, nose-left (−) yawing moments require the value of the yawing coefficient to be negative.

Graphic Representation of Static Directional Stability

A plot of the variation of $C_{N(CG)}$ at different sideslip angles for an aircraft with positive static directional stability is shown in Fig. 16.3. It has a positive slope. This is exactly opposite to the slope for static pitch stability (Fig. 15.10).

The trim point is where there is no yawing moment. This is where there is no sideslip angle, at the intersection of the two axes. Assume that an airplane is at the trim point and experiences a right (+) sideslip. If the airplane has positive directional stability, it will develop a nose-right (+) yaw coefficient, which will yaw the airplane into the new relative wind. Thus, a positive slope shows positive stability.

Again, as in the case of pitch stability, the degree of the slope is an indication of the degree of the stability. Steeper slope means increased stability (or instability). It is not unusual for an airplane to be stable at small sideslip angles but unstable at high sideslip angles. This is shown in Fig. 16.4. The unstable condition must not exist at angles of sideslip encountered in ordinary flight conditions.

Contribution of Aircraft Components to Yaw Stability

Wings The wing contribution to positive static directional stability is small, but increases with the amount of sweepback. This is shown in Fig. 16.5. In Fig.

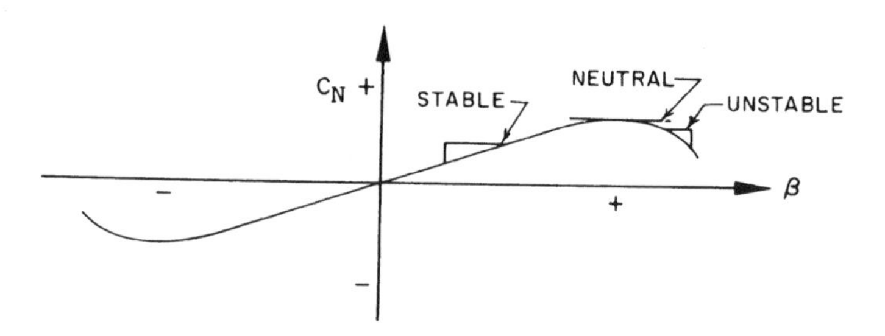

Fig. 16.4. Static directional stability at high sideslip angles.

16.5, the sideslip is to the right, and the right wing produces more drag than the left wing. This is a stabilizing influence. The right wing also produces more lift, but this is a roll factor, which is discussed later in this chapter.

Fuselage In our discussion of pitch stability we discovered that the fuselage contribution was a destabilizing factor. A similar effect occurs with sideslip. The center of pressure is located near the quarter length of the fuselage, which is ahead of the CG, and is destabilizing.

Another factor that contributes to the directional instability of the fuselage is that if the side area of the fuselage forward of the CG is greater than that behind it, the relative wind hitting this area creates a destabilizing yawing moment. Figure 16.6 shows the fuselage effect.

Engine Nacelles The contribution of the propeller or jet engine inlets to directional stability of an aircraft depends on their fore and aft location with respect to the aircraft's CG. If the aircraft is sideslipped, the relative wind must be changed to a direction parallel to the axis of the engine. This is similar to the discussion of pitch stability (see Fig. 15.17). A sideward force is developed when the direction of the airflow is changed, and this will be a destabilizing

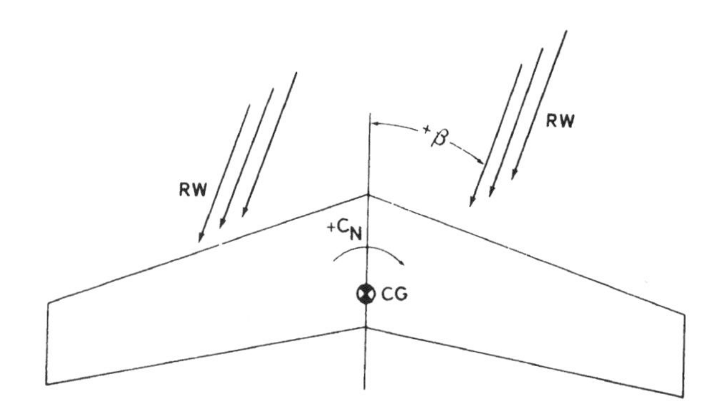

Fig. 16.5. Effect of wing sweepback on directional stability.

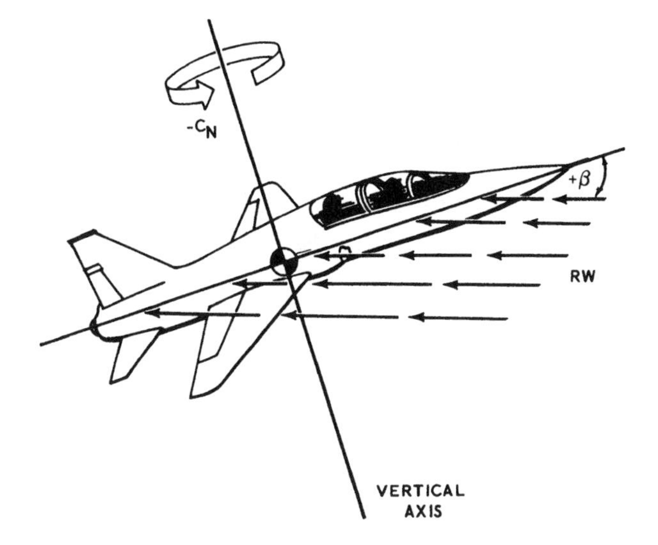

Fig. 16.6. Directional instability of fuselage.

factor if the propeller or engine inlet is forward of the CG. Aft engines are stabilizing in both pitch and yaw.

Vertical Tail As the name implies, the vertical stabilizer is a strongly stabilizing factor. It is located behind the CG, and a sideslip creates a horizontal tail force on the aerodynamic shape of the tail. This produces a stabilizing yawing moment about the CG, as shown in Fig. 16.7.

Tail stability can be enhanced by the addition of a dorsal fin (extension of the tail below the fuselage), which does not increase the parasite drag as much as increasing the size of the vertical tail would. Another advantage of the dorsal fin is that it delays the tendency of the vertical tail to stall at high sideslip angles. Figure 16.8 shows both the reduction of parasite

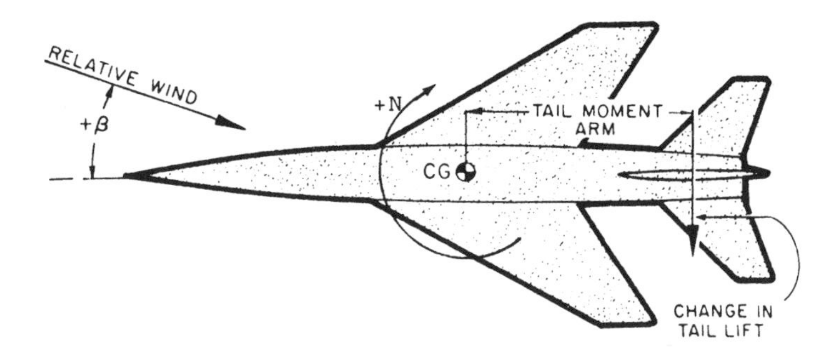

Fig. 16.7. Vertical tail is stabilizing in yaw.

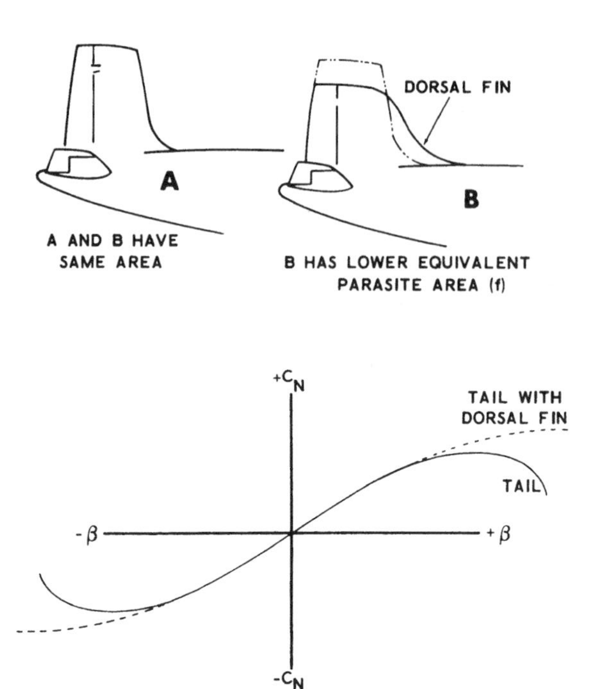

Fig. 16.8. Dorsal fin decreases drag and increases stability.

drag (sketches) and the increased yaw stability (graph) resulting from the addition of a dorsal fin.

Total Airplane Figure 16.9 shows a typical buildup of the fuselage, vertical tail, and dorsal fin as they affect static directional stability. Wing and engine effects are not shown. Wing effect is small, and engine effect depends on location, as discussed earlier.

Rudder-Fixed–Rudder-Free Stability Fixing the rudder in the neutral position will prevent "rudder float" and will effectively increase the vertical tail area and thus increase the directional static stability. This is similar to the stick-fixed stability that we discussed earlier. For aircraft with conventional, reversible controls, increased directional stability will result if the pilot keeps both feet on the pedals and holds the rudder in the neutral position. Figure 16.10 shows this.

Effect of High Angle of Attack If the vertical tail is engulfed in stalled air from the wings at high angles of attack, it will not be effective in developing sideward forces, and static directional stability will deteriorate. This is shown in Fig. 16.11. The decay in stability will have a strong effect on the ability to recover from spins and unusual attitudes.

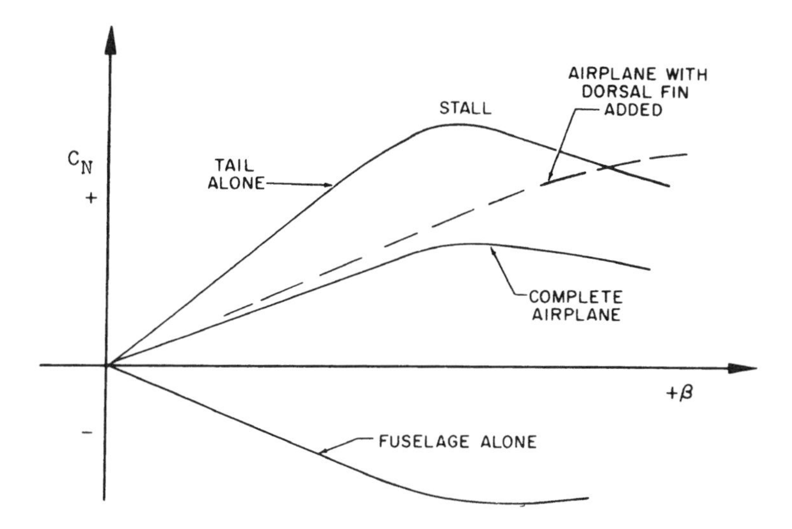

Fig. 16.9. Typical buildup of component effects on static directional stability.

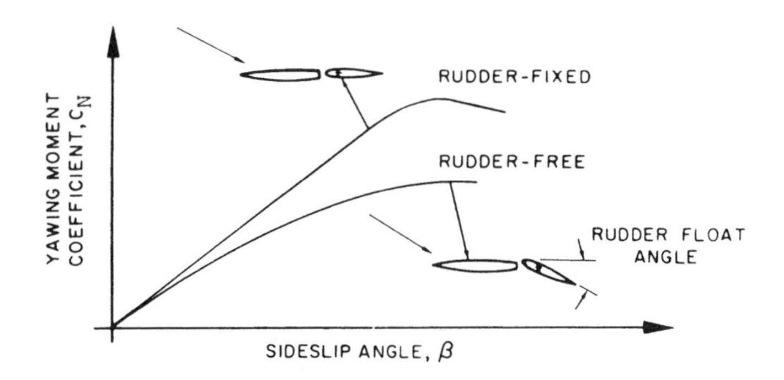

Fig. 16.10. Rudder-fixed–rudder-free yaw stability.

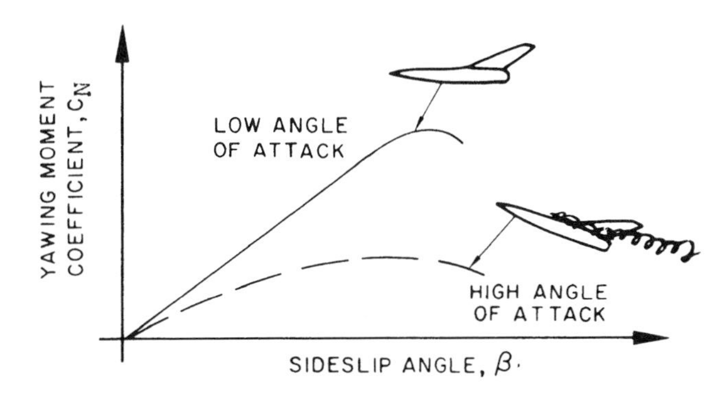

Fig. 16.11. Loss of directional stability at high AOA.

DIRECTIONAL CONTROL

There are five conditions of flight that can be critical to the directional control forces exerted by the rudder. The type of airplane and its mission will determine which of these is the most important.

1. *Spin Recovery* This was discussed in Chapter 11.
2. *Adverse Yaw* This is a coupled dynamic stability problem involving both directional and roll stability. It is discussed later in this chapter in the section on coupled effects.
3. *Slipstream Rotation* This effect is predominant in single-engine propeller aircraft. It may be critical at high-power and low-airspeed combinations, thus it is important during takeoff and landing operations. The slipstream from the propeller rotates about the fuselage of U.S. aircraft as shown in Fig. 16.12. If it strikes the left side of the vertical stabilizer it will cause a nose-left yawing moment that must be overcome with rudder force to maintain directional control.
4. *Crosswind Takeoff and Landing* The aircraft rudder forces must be great enough to maintain a track down the runway when a crosswind exists. In effect the airplane is sideslipping with respect to the wind. The positive static yaw stability of the airplane will try to turn it into the crosswind. This is undesirable during takeoffs and landings, and the rudder must be able to overcome this tendency.

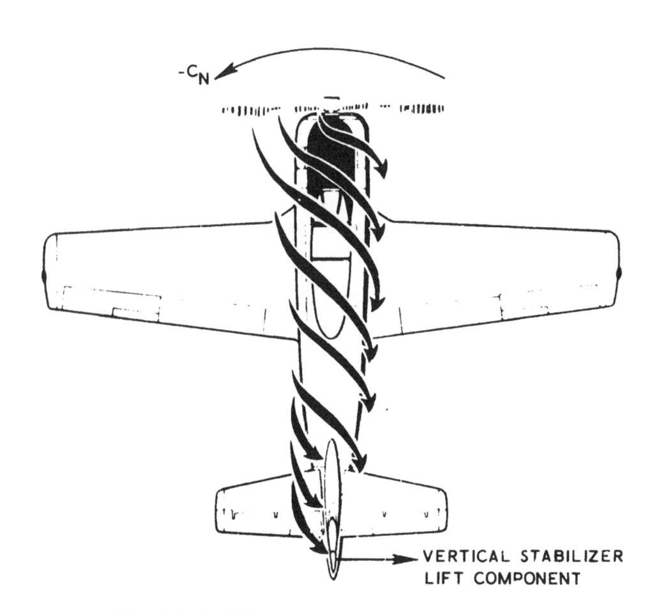

Fig. 16.12. Slipstream rotation causes yaw.

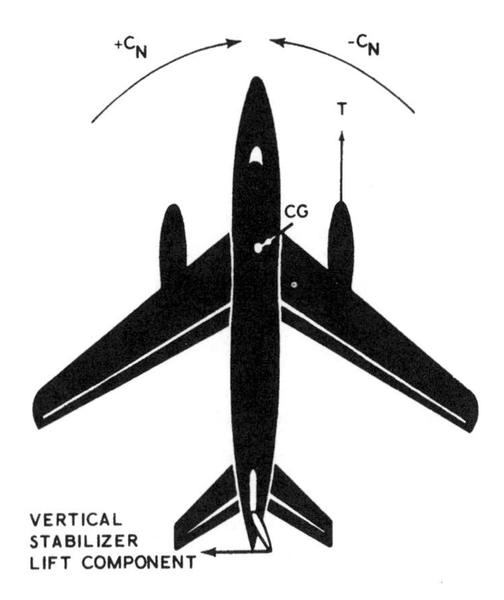

Fig. 16.13. Yawing moment due to asymmetrical thrust.

5. *Asymmetrical Thrust* Multiengined airplanes have an additional control requirement that single-engine airplanes do not have. If an engine malfunctions, the aircraft must be able to maintain its heading. The airplane in Fig. 16.13 shows this requirement. The left engine is assumed to have lost thrust, and a nose-left yawing moment has resulted. To maintain the original heading an equal and opposite yawing moment must be developed by the rudder and vertical stabilizer combination.

   Note that even when the moments are canceled out, there will be an unbalance in sideward forces acting on the airplane. This will cause a sideslip to the left unless the right wing is dropped slightly. The minimum airspeed where the rudder can produce the necessary yawing moment to maintain heading, if an engine fails, is called the in-flight minimum control speed, V_{MCA}. The usual requirement is that V_{MCA} is not greater than $1.2V_s$.

LATERAL STABILITY AND CONTROL

Lateral stability and control refer to the behavior of the airplane in roll, that is, movement of the lateral axis when it is rotated about the longitudinal axis. Roll results when a rolling moment (L') acts on the airplane. This rolling moment can be caused either by the pilot activating the ailerons (or spoilers) or by the aircraft being subjected to a sideslip angle. From the stability standpoint, we are more interested in the reaction of the airplane to a sideslip angle.

In the next section we discuss static lateral stability and lateral control. Dynamic lateral stability is coupled with dynamic directional stability, and they are discussed later in this chapter.

STATIC LATERAL STABILITY

Figure 16.14 shows an airplane that is sideslipping to the right. This sideslip causes a rolling moment to develop on the aircraft. The right wing will drop and the airplane will sideslip even farther to the right. This is an unstable condition.

For static stability in roll we need a wings leveling rolling moment to be developed if a wing drops and sideslip results. Figure 16.15 shows the three possibilities. In Fig. 16.15a the airplane is sideslipping to its right (the viewer's left, since this view is looking aft) and is experiencing a + sideslip angle. To level the wings and have positive static roll stability, a left-wing-down (−) rolling moment is required. In Fig. 16.15b there is no rolling moment developed when a right sideslip is experienced, so this airplane has neutral static roll stability. Figure 16.15c shows an unstable airplane (the same situation shown in Fig. 16.14).

The Rolling Moment Equation

The rolling moment equation about the aircraft CG is

$$L'_{\text{CG}} = C_{L'(\text{CG})} qSb \tag{16.3}$$

where

L'_{CG} = rolling moment about the CG (ft-lb)

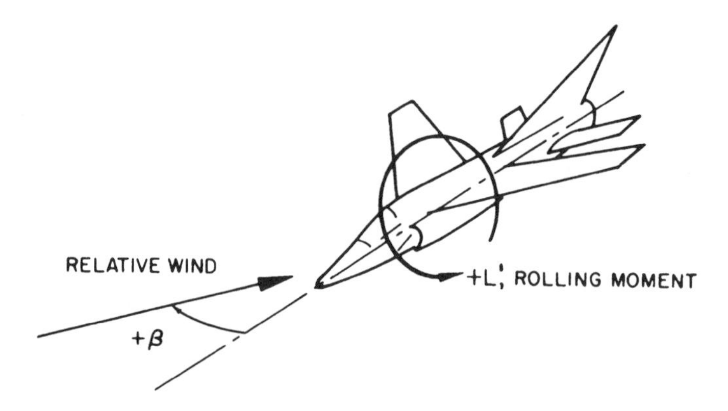

Fig. 16.14. Rolling moment caused by sideslip.

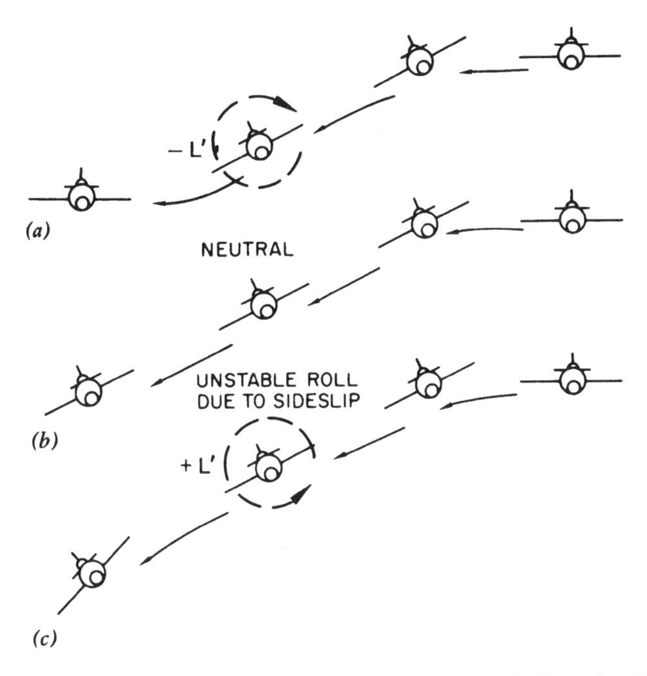

Fig. 16.15. (a) Stable, (b) neutral, and (c) unstable static lateral stability.

$C_{L'(\mathrm{CG})}$ = coefficient of rolling moment about CG
q = dynamic pressure (psf)
S = wing area (ft^2)
b = wing span (ft)

Rearranging Eq. 16.3 gives

$$C_{L'(\mathrm{CG})} = \frac{L'_{\mathrm{CG}}}{qSb} \tag{16.4}$$

For a right-wing-down, +, rolling moment, $C_{L'}$ must be positive. For a left-wing-down, −, rolling moment, $C_{L'}$ must be negative.

Graphic Representation of Static Lateral Stability

A plot of the variation of $C_{L'(\mathrm{CCG})}$ at different sideslip angles for an aircraft with positive static lateral stability is shown in Fig. 16.16. The plot has a negative slope. This slope is the same as that for static longitudinal stability, but it is opposite to that for static directional stability (see Figs. 15.9 and 16.3). The trim point is the point where there is no rolling moment. This occurs where there is no sideslip angle, at the intersection of the two axes.

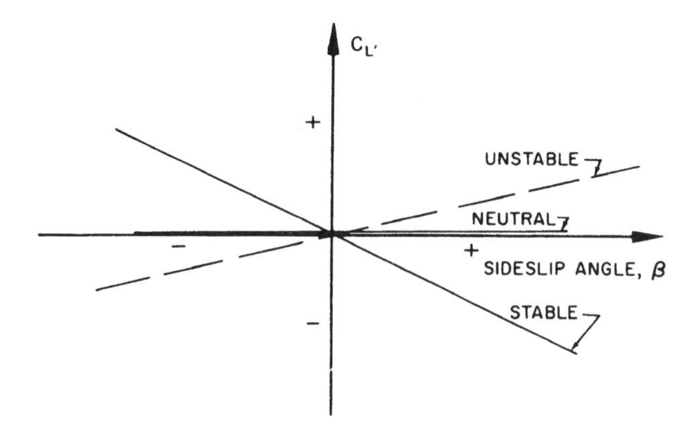

Fig. 16.16. Static lateral stability.

Assume that an airplane is trimmed for zero roll and experiences a right, +, sideslip. If the airplane has positive lateral stability it will develop a negative rolling moment coefficient, $-C_{L'}$, which raises the right wing. This is stabilizing. Thus, a negative slope shows positive stability. The degree of the slope is an indication of the degree of the stability (or instability) just as it was for both pitch stability and yaw stability.

Contributions of Aircraft Components to Roll Stability

Wing Dihedral Dihedral is defined as the spanwise inclination of a wing or horizontal stabilizer to the horizontal, or to a plane equivalent to the horizontal. Upward inclination is positive dihedral and downward inclination is negative dihedral (commonly called anhedral). Dihedral is shown in Fig. 16.17.

It is difficult to illustrate in a two-dimensional drawing the relative wind as it hits the aircraft in a sideslip. Figure 16.18 is an attempt to do this. The airplane is in a sideslip to the right. The sideslip angle is positive as defined previously. The AOA of the right wing is greater than that of the left wing because of the dihedral of the wings. Therefore, more lift, L is produced on the

Static Lateral Stability

Fig. 16.17. Dihedral angle.

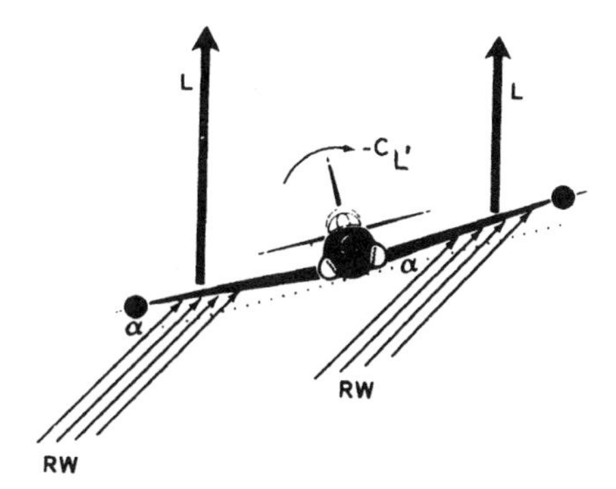

Fig. 16.18. Dihedral producing static lateral stability.

right wing, and the negative coefficient of rolling moment, $-C_{L'}$, is generated and a stabilizing rolling moment results.

Vertical Wing Position A high wing position places the CG of an airplane below the center of pressure of the wing. This results in a pendulum effect that is equivalent to 1 to 3° of positive effective dihedral. Conversely, a low-mounted wing with the CG of the airplane above the center of pressure of the wing will be unstable and is equivalent to 1 to 3° of negative effective dihedral. This is apparent in observing the larger dihedral angles of low-wing airplanes.

Wing Sweepback The contribution of swept wings in producing positive static directional stability was discussed previously (see Fig. 16.5). We saw that if a swept-wing airplane was sideslipped to the right, then the wing on the right had more drag than the wing on the left. The right wing also has more lift than the left wing. The relative wind hits the right wing at a more favorable angle. This is shown in Fig. 16.19. This is stabilizing, much in the same way that dihedral is stabilizing; in fact, it is called the *dihedral effect of sweepback*.

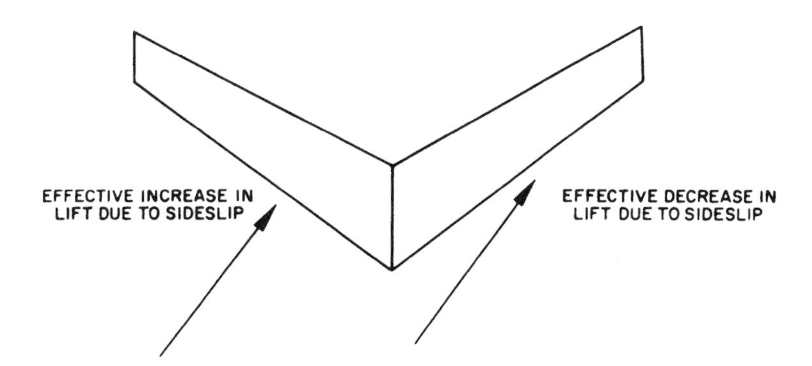

Fig. 16.19. Dihedral effect of sweepback.

Fig. 16.20. Vertical tail effect on lateral stability.

Vertical Tail Side forces on the vertical tail will be a stabilizing influence since the tail is above the CG. The effect of the vertical tail on positive static lateral stability is shown in Fig. 16.20.

Complete Airplane The total airplane must have a positive lateral stability. Some components may have negative stability, but this must be overcome by the stabilizing moments of other components, and the entire aircraft must be laterally stable.

LATERAL CONTROL

The lateral control of an airplane is accomplished by providing differential lift on the wings. This is usually obtained by some type of ailerons or spoilers. Delta-wing aircraft often combine the ailerons and elevators into a single control unit called an elevon or ailevator. Both left and right surfaces act together when elevator action is called for. They act in opposition to each other when roll motion is required. A combination of pitch and roll response is also possible. Pilot inputs are similar to those for conventional controls.

Critical lateral control conditions are takeoffs and landings in crosswind situations. Military fighters and acrobatic airplanes may need high roll rates and therefore may desire less lateral stability and more lateral control than transport aircraft.

DYNAMIC DIRECTIONAL AND LATERAL COUPLED EFFECTS

In our discussion of static directional and lateral stability, we saw that the static stability depended on the aircraft's reaction to an imposed sideslip angle. Both yawing and rolling of an aircraft produce sideslip. Conversely, sideslip produces both yawing and rolling moments. These two moments interact and result in coupled effects that determine the dynamic stability of an aircraft in yaw and roll. Several effects of coupled yaw and roll stability are described below.

Roll Due to Yawing

The normal way to produce rolling moments is by use of ailerons. However, yawing can also produce roll. If the pilot applies right rudder, the aircraft yaws to the right. In rotating about the CG, to the right, the left wing moves faster than the right wing. The left wing then develops more lift, and the aircraft rolls to the right.

This ties into our previous discussion on static roll stability. The yaw to the right creates a negative sideslip angle (relative wind from the left). This causes a positive rolling moment (right wing down), and the airplane rolls to the right.

Adverse Yaw

Normally an airplane will yaw in the same direction that it is rolled. However, it is possible that an airplane will yaw in the direction opposite to the roll. This can lead to a loss of control. It is called adverse yaw.

First, let us examine the lift forces on each wing when an airplane is rolled. Figure 16.21 shows the rear view of an airplane that is making an aileron roll to the left. The effective relative wind on the upgoing (right) wing is the vector resultant of the free stream relative wind and the downward relative wind, due to the wing moving upward. The lift vector is perpendicular to the effective relative wind and is thus tilted backward. This is similar to the discussion of induced drag in Chapter 5 (see Fig. 5.10). The effective relative wind on the downgoing (left) wing is the vector resultant of the free stream relative wind and the upward relative wind, due to the wing moving downward. The lift vector is perpendicular to the effective relative wind and, for this wing, is tilted forward.

The rearward direction of the lift vector on the upgoing wing and the forward direction of the lift vector on the downgoing wing both oppose the yaw of the airplane in the intended direction of turn. In this example they try to turn the airplane to the right, hence the term adverse yaw. Adverse yaw effects are amplified at high angles of attack. This is shown in Fig. 16.22. The drag of the upgoing wing (Fig. 16.22a) is much greater than on the downgoing wing (Fig. 16.22b), thus additional adverse yaw tendency exists. To

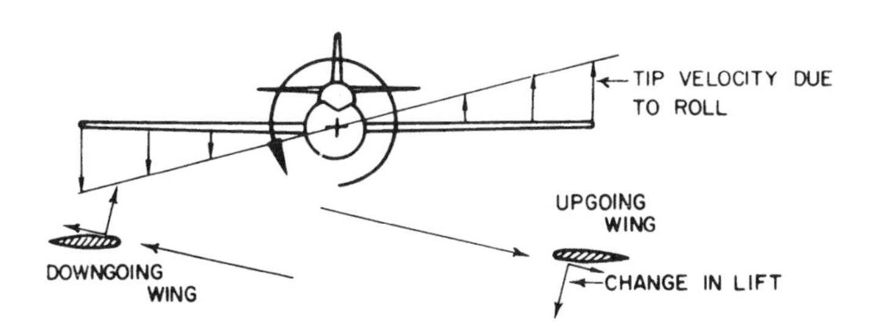

Fig. 16.21. Adverse yaw.

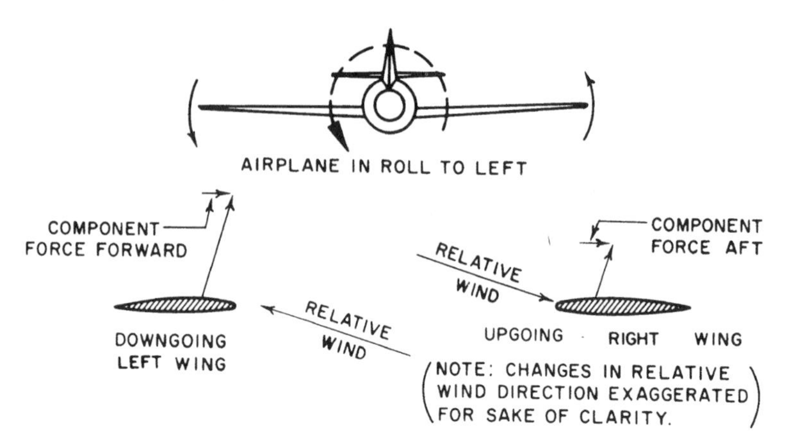

Fig. 16.22. High AOA: (a) upgoing wing; (b) downgoing wing.

decrease adverse yaw tendencies, neutralize the ailerons, once bank has been established.

Types of Motion Resulting From Coupled Effects

Three types of airplane motion can result from the interaction of yaw and roll:

1. *Spiral divergence* results when the static directional stability is great in comparison to the static lateral stability (dihedral effect). If a wing is lowered, the directional stability is greater than the roll stability and the aircraft will not sideslip readily. Thus, the dihedral effect is weak and the wing will not raise to the level position. The airplane tends to enter an ever-tightening spiral dive commonly called a graveyard spiral.
2. *Directional divergence* results from a negative directional stability. The airplane develops sideslip after being disturbed either in roll or yaw and develops a yawing moment that causes the airplane to yaw farther in the same direction. Once the yawing motion has started it will continue until the airplane is broadside to the relative wind. Obviously, this condition cannot be tolerated, and design to prevent it is a prime factor in dynamic stability.
3. *Dutch roll* is a coupled directional and lateral oscillation that occurs when the dihedral effect is very strong in comparison to the directional stability. If a stable airplane is in a sideslip to the right, for instance, it will yaw to the right. At the same time, the right wing will develop more lift and the airplane will roll to the left. If not controlled, the upgoing right wing will cause sideslip to the left. The whole process will then be repeated on the left side of the airplane. When static directional stability is strong, the Dutch roll is heavily damped, but this does lead to spiral divergence. A

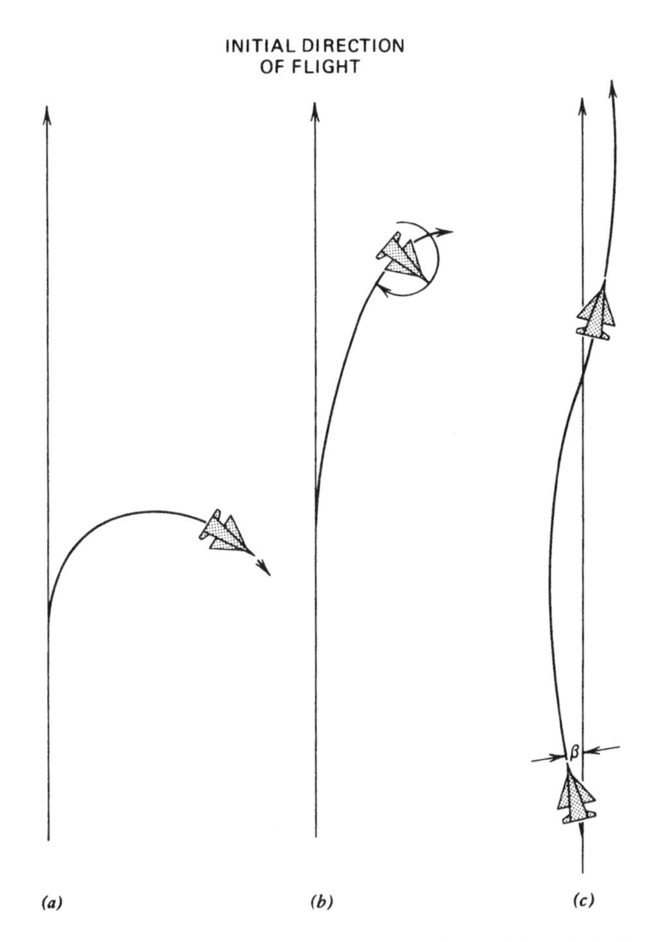

Fig. 16.23. Flight paths due to coupled dynamic effects: (a) spiral divergence, (b) directional divergence, (c) Dutch roll.

small amount of spiral divergence is tolerated because it is greatly preferred to Dutch roll.

Three possible flight paths due to coupled dynamic directional and lateral stability are shown in Fig. 16.23.

SYMBOLS

$C_{L'(\mathrm{CG})}$ Rolling moment coefficient about CG

$C_{N(\mathrm{CG})}$ Yawing moment coefficient about CG

L'_{CG} Rolling moment about CG (ft-lb)
N_{CG} Yawing moment about CG
β (beta) Sideslip angle (degrees)

EQUATIONS

16.1 $N_{CG} = C_{N(CG)} qSb$

16.2 $C_{N(CG)} = \dfrac{N_{CG}}{qSb}$

16.3 $L'_{CG} = C_{L'(CG)} qSb$

16.4 $C_{L'(CG)} = \dfrac{L'_{CG}}{qSb}$

PROBLEMS

1. For static directional stability, if an airplane sideslips to the left, the plane must
   - **a.** keep its old heading.
   - **b.** yaw to the left.
   - **c.** yaw to the right.
2. The wings of an airplane contribute to static directional stability
   - **a.** if they are forward of the CG.
   - **b.** if they are behind the CG.
   - **c.** Either of the above
3. The fuselage is unstable in both static pitch and static yaw stability.
   - **a.** True
   - **b.** False
4. Engine nacelles with propellers or jet inlets behind the airplane's CG are stable in both static pitch and static yaw stability.
   - **a.** True
   - **b.** False
5. Pilots can increase the stability of airplanes without powered controls by
   - **a.** resting their feet on the rudders, even though the plane is trimmed.
   - **b.** holding the stick (or yoke) in the neutral position.
   - **c.** Both of the above
   - **d.** Neither (a) nor (b)

6. Operating an airplane at a high AOA will result in a loss of yaw stability.
   a. True
   b. False
7. Which of these vertical wing locations will require the most wing dihedral for static roll stability?
   a. High wing
   b. Mid wing
   c. Low wing
8. Swept wings affect static roll stability more than straight wings by
   a. increasing the stability.
   b. decreasing the stability.
   c. They do not change the stability at all.
9. If the pilot of a stable airplane applies right rudder, the airplane will
   a. roll to the left and yaw to the left.
   b. roll to the right and yaw to the right.
   c. roll to the left and yaw to the right.
   d. roll to the right and yaw to the left.
10. Of the three types of coupled effects discussed, which is the most dangerous?
   a. Spiral divergence
   b. Directional divergence
   c. Dutch roll

17 High-Speed Flight

Flight speeds have been arbitrarily named as follows:

Subsonic Aircraft speeds where the airflow around the aircraft is completely below the speed of sound (about Mach 0.7 or less).

Transonic Aircraft speeds where the airflow around the aircraft is partially subsonic and partially supersonic (from about Mach 0.7 to Mach 1.3).

Supersonic Aircraft speeds where the airflow around the aircraft is completely above the speed of sound but below hypersonic airspeed (from about Mach 1.3 to Mach 5.0).

Hypersonic Aircraft speeds above Mach 5.0.

In this chapter we discuss the airflow as the aircraft approaches the speed of sound, transonic flight, and supersonic flight. Hypersonic flight is not discussed.

In subsonic flight, the density change in the airflow is so small that it can be neglected in the flow equations without appreciable error. The airflow at these lower speeds can be compared to the flow of water and is called *incompressible flow*. At high speeds, however, density changes take place in the airstream that are significant. Thus, this type of airflow is called *compressible flow*. Transonic, supersonic, and hypersonic flight all involve compressible flow.

THE SPEED OF SOUND

The speed of sound is an important factor in the study of high-speed flight. Small pressure disturbances are caused by all parts of an aircraft as it moves through the air. These disturbances move outward from their source through the air at the speed of sound. A two-dimensional analogy is that of the ripples on a pond that result when a stone is thrown in the water.

The speed of sound in air is a function of temperature alone:

$$a = a_0\sqrt{\theta} \tag{17.1}$$

where

a = speed of sound
a_0 = speed of sound at sea level, standard day
θ = temperature ratio, T/T_0

Because the aircraft's speed in relation to the speed of sound is so important in high speed-flight, airspeeds are usually measured as Mach number (named after the Austrian physicist Ernst Mach). Mach number is the aircraft's true airspeed divided by the speed of sound:

$$M = \frac{V}{a} \tag{17.2}$$

where

M = Mach number
V = true airspeed (knots)
a = speed of sound (knots)

When an aircraft is flying below the speed of sound, the pressure disturbances will be moving faster than the airplane, and those disturbances that travel ahead of the aircraft influence the approaching airflow. This "pressure warning" can be observed in a smoke wind tunnel as it causes the upwash well ahead of the wing. This is shown in Fig. 17.1a.

If the aircraft is flying at a speed greater than the speed of sound, the airflow ahead of the aircraft is not influenced by the pressure field of the aircraft, since the speed of the pressure disturbances is less than the speed of the aircraft. The airflow ahead of the wing will have no warning of the approach of the wing and will not change its direction ahead of the wing's leading edge. This is shown in Fig. 17.1b.

HIGH SUBSONIC FLIGHT

Compressibility effects are not limited to aircraft that fly supersonically. A high subsonic velocity will produce local supersonic flow on top of the wings, fuselage, and other parts of the aircraft. Helicopters often experience compressibility effects on rotor tips. This can occur even when the helicopter is in a hover. The most important compressibility effect that occurs in subsonic aircraft is the formation of normal shock waves on the aircraft's wings or rotor blades.

Normal Shock Wave Formation on Wings

Airflow increases in velocity as it passes over a wing surface, so the local Mach number on top of the wing is always greater than the flight Mach number. For example, if an aircraft is flying at $M = 0.50$, the local velocity might be $M = 0.78$ at the thickest point of the wing (depending on the thickness and/or camber of the airfoil). This is shown in Fig. 17.2.

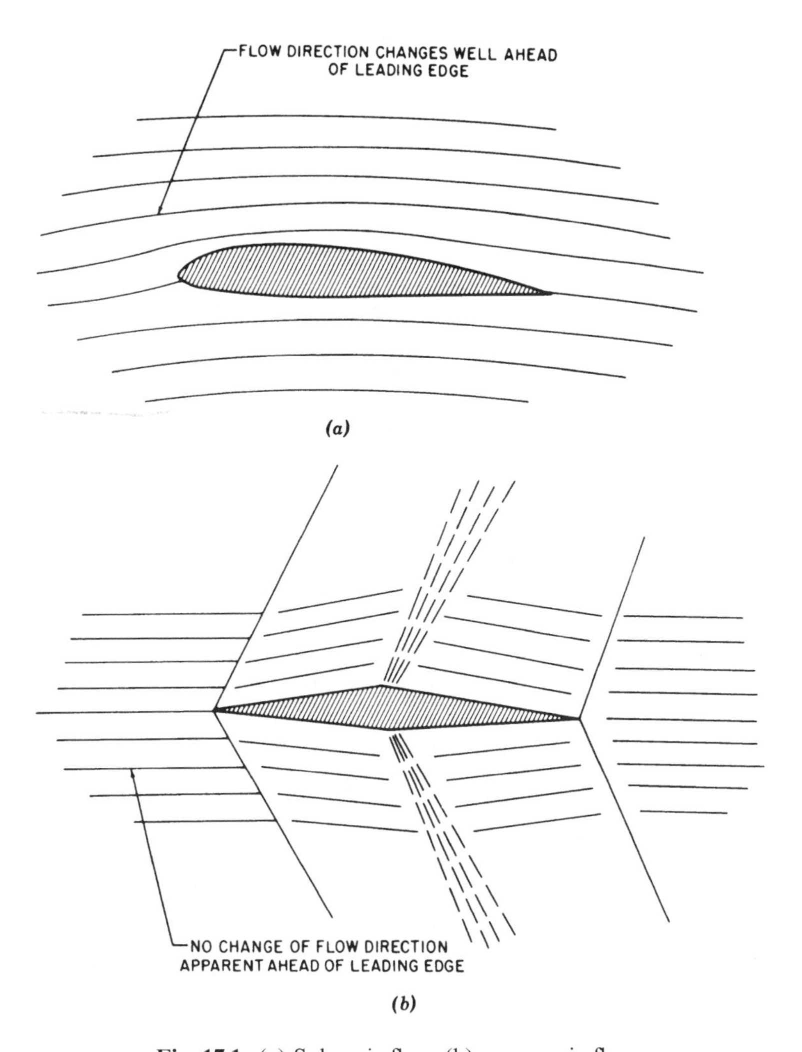

Fig. 17.1. (a) Subsonic flow, (b) supersonic flow.

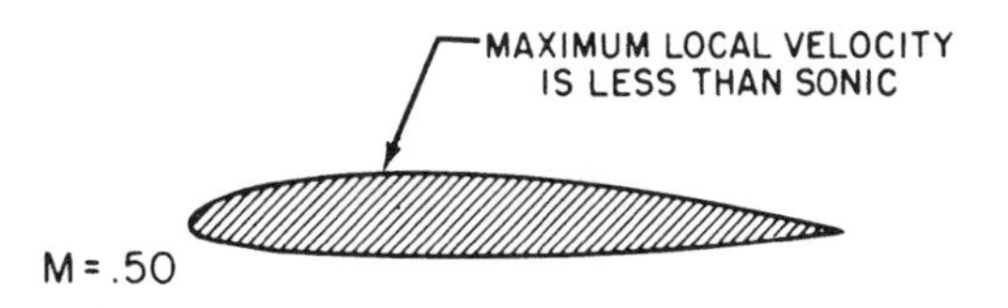

Fig. 17.2. Subsonic flow about a wing section.

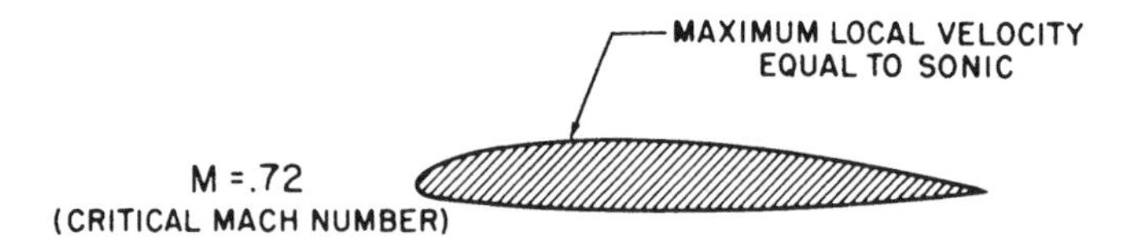

Fig. 17.3. Critical Mach number.

If the flight Mach number increases, the local velocity on top of the wing also increases. At some flight Mach number the local maximum velocity reaches sonic speed, $M = 1.0$. This flight Mach number is called the critical Mach number, M_{crit}. For the example shown in Fig. 17.3, M_{crit} is 0.72.

Once M_{crit} is exceeded, the aircraft is flying in the transonic speed range. Supersonic airflow exists in the area of maximum thickness on top of the wing, but subsonic flow exists elsewhere. All pressure disturbances behind this sonic flow cannot be propagated forward because they run into sonic velocities traveling rearward. A normal shock wave is formed as shown in Fig. 17.4. This shock wave is present where the air slows from supersonic to subsonic. As the air flows through the normal shock wave it undergoes a rapid compression. The compression decreases the kinetic energy of the airstream and converts it into a pressure and temperature increase behind the shock wave. The heat rise behind the shock wave is either radiated to the atmosphere or absorbed by the wing surface, but in either case it is lost. The lost energy must be continuously supplied by the engines. This creates a type of drag known as wave drag.

The increase in static pressure caused by the normal shock wave has the same effect as the adverse pressure gradient that was discussed in basic aerodynamics. It slows the air in the boundary layer down to the point where faster moving air from outside the boundary layer rushes in and reverse flow occurs. This process is at the "possible separation" area in Fig. 17.4.

DESIGN FEATURES FOR HIGH SUBSONIC FLIGHT

Subsonic jet aircraft can increase airspeed without encountering shock wave problems if the critical Mach number can be raised. Several design features

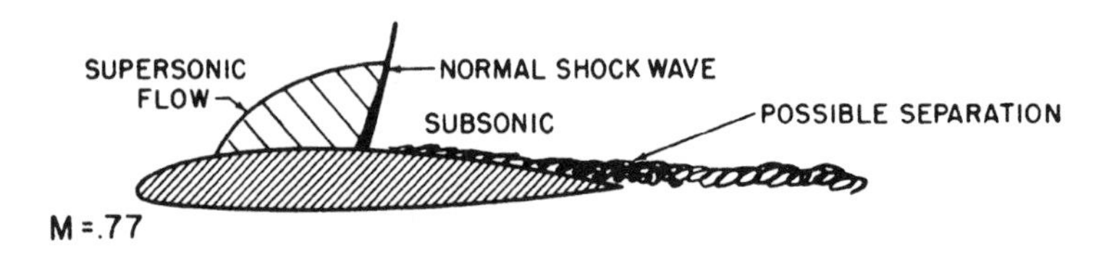

Fig. 17.4. Normal shock wave formation on wing.

help accomplish this:

- Thin airfoil sections
- Airfoil sections that have good high subsonic Mach number characteristics
- Sweepback
- Vortex generators

Thin Airfoil Sections

The thinner the airfoil, the less the air is speeded up in passing over the top surface; thus, the airspeed can be higher before the air reaches M_{crit}. One disadvantage of thin wings is that they do not have high values of $C_{L(max)}$, so higher takeoff and landing speeds are required. Another disadvantage is that it is difficult to design the structural strength and rigidity required in a thin wing. Finally, there is less room for fuel tanks in a thin wing as compared to a thicker wing of the same planform.

High-Speed Subsonic Airfoils

The laminar flow airfoil discussed in Chapter 5 was the first high-speed airfoil. Moving the maximum thickness backward from about 25% C to 40–50% C did reduce drag and increase the critical Mach number, but one disadvantage still existed for this type of airfoil. The shock wave that develops on a laminar airfoil occurs in the adverse pressure gradient region. As the airflow is slowing up in this region it has a tendency to separate (stall) easily.

The *supercritical airfoil* shown in Fig. 17.5 was designed to correct this deficiency. It is shaped so that the normal shock will occur where the upper surface pressure gradient is favorable or zero. At that point the boundary layer is able to encounter the pressure increase across the shock wave without separating. The surface curvature on the top surface of the supercritical airfoil is less than a conventional laminar airfoil, so the local Mach number will be less than that for the laminar flow airfoil. If M_{crit} is exceeded, the airflow will be supersonic nearer the leading edge on the supercritical airfoil. However, the airflow will remain at a much lower Mach number, and the supersonic flow will be terminated by a gradual deceleration. Therefore, no shock wave will result and drag will be less.

Sweepback

One of the most common methods of increasing critical Mach number is to sweep the wings backward (and more recently, to sweep them forward). Instead of presenting the complicated conventional discussion of this, a simpler explanation is presented here.

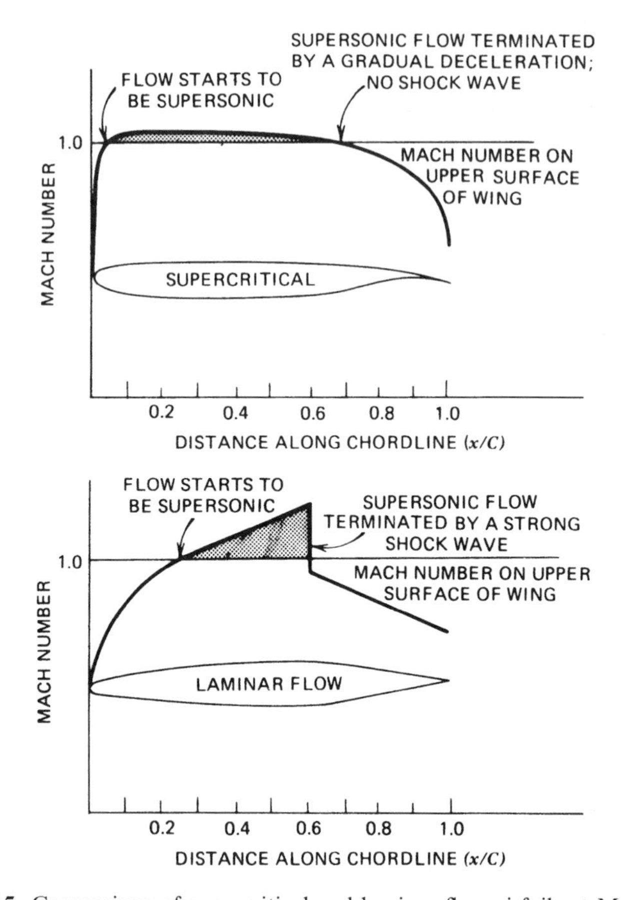

Fig. 17.5. Comparison of supercritical and laminar flow airfoils at Mach 0.75.

We learned earlier that thin wings will increase the M_{crit}. By sweeping the wings the effective chord (parallel to the aircraft's longitudinal axis) has been lengthened, but the wing thickness has not been changed. Thus, the ratio of thickness/chord has been reduced. In effect, the wing thickness has been reduced and a higher M_{crit} results.

The coefficient of lift curve is affected by sweep in two ways. First, the value of $C_{L(max)}$ is reduced, so takeoff and landing speeds are increased. Second, the curve flattens and has no sharp stall AOA point. Figure 17.6 shows these.

The lack of a definite stall point has caused some complacency in pilots who are making a transition from straight-wing to swept-wing aircraft. While it is true that the swept-wing aircraft does not show a clean stall, it is also true that sweeping the wings decreases the aspect ratio, and this means a large increase in induced drag coefficient. This is particularly dangerous during takeoffs and landings, as was emphasized earlier. Wing sweep is also important in directional and lateral stability as discussed in Chapter 16. Wingtip stall is increased

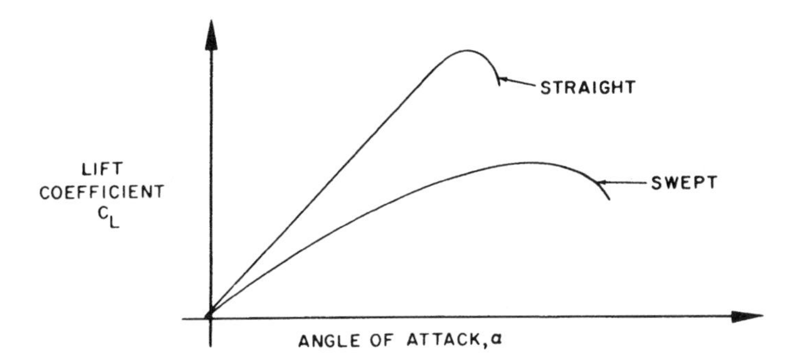

Fig. 17.6. Effect of wing sweep on a C_L-α curve.

by swept-back wings as was presented in Chapter 11. Sweeping the wings forward will eliminate this problem.

Vortex Generators

Use of vortex generators to invigorate the boundary layer and thus delay separation in the low-speed region of flight was discussed in Chapter 4. The same general principles apply to the high-speed region of flight. Shock-induced separation occurs because the boundary layer does not have enough energy to overcome the adverse pressure gradient through a shock wave. Vortex generators mix higher kinetic energy air from outside the boundary layer with the slower air in the boundary layer and delay separation.

If a normal shock wave does develop, vortex generators are effective in breaking it up. The additional drag of the generators is small in comparison to the wave drag that they help dissipate. Figure 17.7 shows the action of vortex generators.

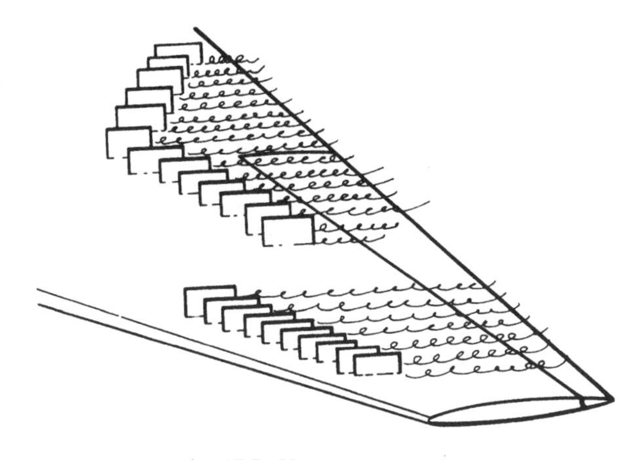

Fig. 17.7. Vortex generators.

TRANSONIC FLIGHT

Early attempts to "break the sound barrier" were not very successful because of problems encountered in the transonic region. When these problems were better understood, several important design changes were made, and today supersonic aircraft have little difficulty in flying through this speed region. Subsonic aircraft do not have these design changes and will experience difficulties if the critical Mach number is exceeded. This discussion also applies to subsonic aircraft and helicopters, if they venture into the transonic region of flight, as well as to supersonic aircraft.

Force Divergence

At airspeeds of about 5% above M_{crit} the normal shock wave on the top of the wing causes the boundary layer to separate from the wing. This causes a change in the aerodynamic force coefficients, both C_L and C_D. This airspeed is called the *force divergence Mach number.*

Figure 17.8 shows the C_D curve (for a constant value of C_L) plotted against Mach number. When the force divergence Mach number is reached, the value of C_D rapidly increases. This velocity is also known as the *drag divergence* or *drag rise Mach number.* The drag increase is caused by the energy loss across the normal shock wave and the boundary layer separation and is called wave drag. Note that even though the value of C_D again decreases above Mach 1, that the total drag continues to increase with airspeed. The rate of increase will be less, however, above Mach 1.

In addition to the drag increase behind the normal shock wave, the airflow separation also results in a loss of lift. Figure 17.9 shows a sudden lowering of the coefficient of lift at the force divergence Mach number. Shock-induced separation creates a local stall situation similar to the low-speed stall, except that it occurs behind the normal shock wave rather than at the trailing edge of the wing. This causes the center of pressure to be shifted forward and

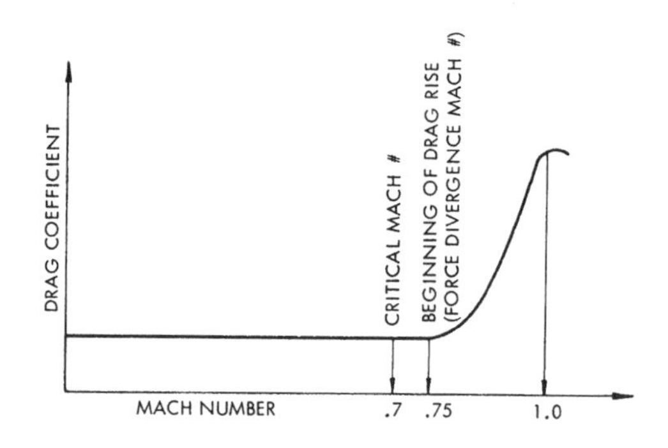

Fig. 17.8. Force divergence effect on C_D.

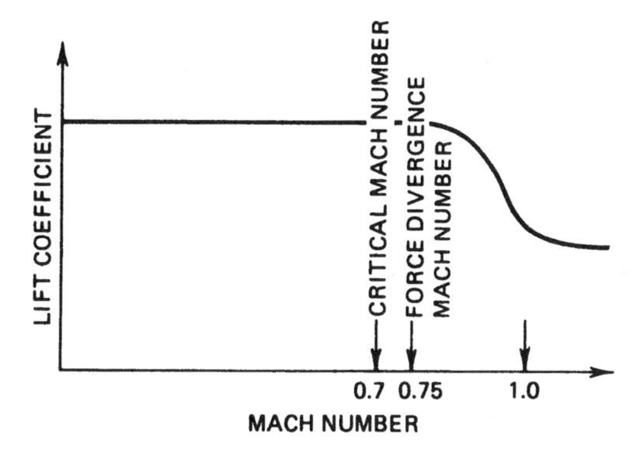

Fig. 17.9. Force divergence effect on C_L.

produces a nose-up pitching moment. If one wing develops a normal shock before the other, a rolling moment can be produced toward the wing with the earlier shock. If a wing drops, the aircraft will tend to yaw in that direction, and a condition similar to Dutch roll may develop.

Some of the effects of reaching force divergence Mach number are

- An increase in C_D for a given value of C_L
- A decrease in C_L for a given AOA
- A change in pitching moment as the aerodynamic center shifts

Tuckunder

The downwash behind the wing will be decreased when airflow separation takes place. As a result the horizontal stabilizer AOA is effectively increased and develops more lift. This is one of the factors that causes the aircraft to pitch nose down. It is called *tuckunder*. If the flight Mach number is increased beyond the force divergence Mach number, the normal shock wave on top of the wing increases in intensity and moves backward. A second normal shock wave then appears on the bottom of the wing. This is shown in Fig. 17.10. A

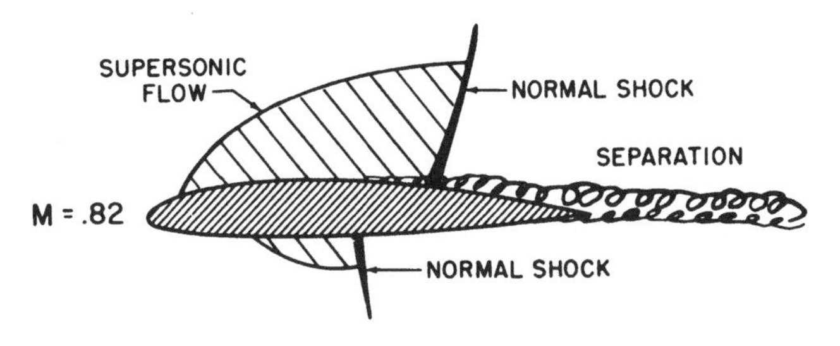

Fig. 17.10. Normal shock wave on bottom of wing.

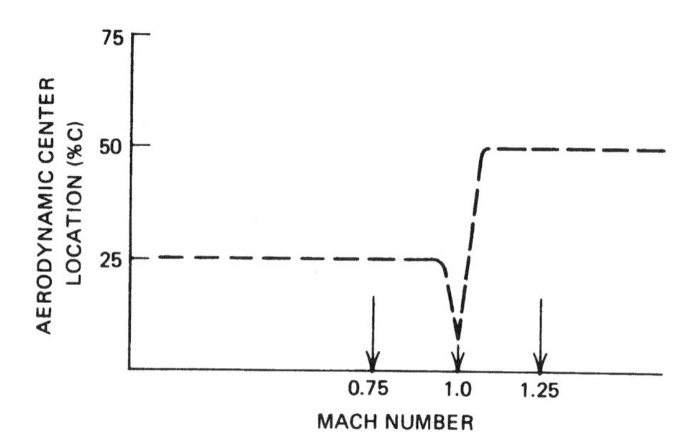

Fig. 17.11. Aerodynamic center location shift.

second factor influencing tuckunder is the movement of the shock waves toward the rear of the wing. As the top shock moves rearward the separation point also moves rearward and so does the center of pressure, thus adding to the tuckunder tendency. A third factor is that the aerodynamic center moves from the quarter chord point to the 50% chord point when the aircraft reaches supersonic flight.

All aircraft flying supersonically suffer a nose-down pitching moment. The shift of the aerodynamic center is not a smooth movement. Sharp leading edges make the shift smoother, but in many aircraft the aerodynamic center moves forward before the eventual rearward shift. This is shown in Fig. 17.11. This movement can cause a pitch-up moment in the early stages of transonic flight, but eventually it will cause tuckunder as the speed increases.

Buffet

The violently turbulent separated air behind a normal shock wave often produces buffeting of the aircraft. This is caused by the airflow hitting the horizontal stabilizers. The T-tail configuration helps eliminate this problem by putting the horizontal tail above the wing wake. Flight in the transonic range is undesirable, so supersonic aircraft fly through this region as rapidly as possible.

Control Surface Buzz

Boundary layer separation acting on control surfaces often causes rapid oscillations called buzz. This can cause metal fatigue problems to hinge fittings and other parts of the control surfaces.

Control Effectiveness

Shock-induced separation can reduce the effectiveness of control surfaces in two ways. First, if the surface is located in a region of stalled air, it is operating

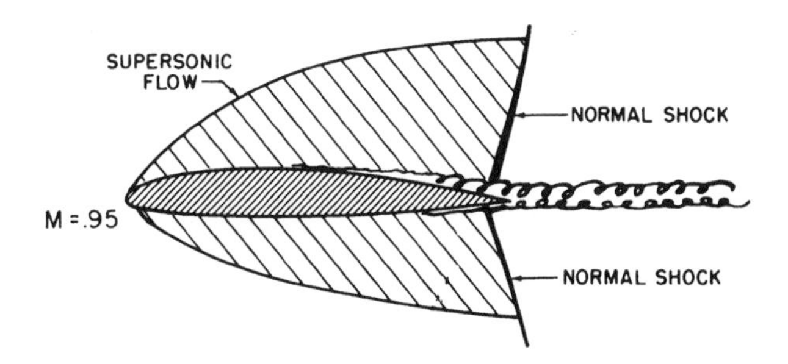

Fig. 17.12. Normal shock waves move to trailing edge.

in an "aerodynamically dead" air mass and cannot produce effective aerodynamic forces. Second, control surfaces are effective if they change the airflow around the entire wing or stabilizer. When there is a shock wave ahead of the control surface, deflection of that surface cannot influence the airflow ahead of the shock wave, so what little aerodynamic force is developed is restricted to the control surface and area behind the shock wave.

SUPERSONIC FLIGHT

As the flight Mach number increases in the transonic speed range, the upper and lower normal shock waves increase in size and strength and move to the trailing edge of the wing, as shown in Fig. 17.12.

When the flight Mach number is 1.0, the entire airfoil is supersonic, except the leading-edge stagnation area. The shock waves at the trailing edge are still normal shock waves, so the velocity behind them is still subsonic. A new shock wave now appears ahead of the airfoil. It is a compression wave just like the normal shock wave, and because the air at the leading edge is subsonic, the new wave, called a bow wave, must also be a normal shock wave, at least in the area directly ahead of the leading edge.

Figure 17.13 shows a bow wave ahead of the leading edge. Fully supersonic flight exists when the bow wave becomes attached to the leading edge of the wings, nose of the aircraft, tail sections, and all other parts of the aircraft.

The bow wave may not attach easily if the leading edges are too blunt. Sharp leading edges are used on supersonic aircraft, but it still is necessary for the flight Mach number to be greater than 1.0 for all the airflow to be supersonic. Thus, the supersonic range starts at about Mach 1.2 to 1.3, depending on individual aircraft design. The normal waves at the trailing edges also are deflected backward and now are called oblique shock waves.

Oblique Shock Waves

An oblique shock wave, like a normal shock wave, is a compression wave. It differs from the normal shock because the airflow direction changes as air flows

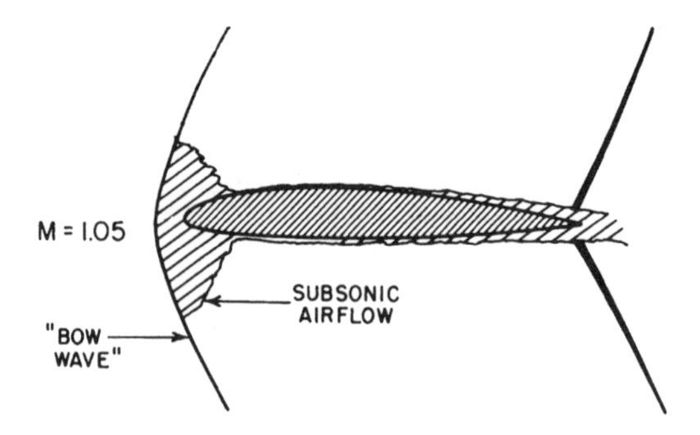

Fig. 17.13. Unattached bow wave at transonic speed.

through it and the airflow slows down but remains supersonic. To see how an oblique shock wave forms, consider a wedge-shaped object placed in a supersonic flow as shown in Fig. 17.14.

The airflow next to the horizontal surface must change direction when the wedge is encountered. It does this and assumes the direction of the arrow V_2. This sudden change in direction, as the air particles turn into the path of other particles, V_1, traveling in the horizontal direction, produces a shock wave at an oblique angle to the horizontal. The angle that the wave makes with the horizontal is called the wave angle. This is shown in Fig. 17.14.

The wave angle depends on the Mach number of the approaching flow and the angle of the wedge. It is the sum of the Mach wave angle and the wedge angle. The Mach wave angle is the angle whose sine is $1/M$ (arc sin $1/M$):

$$\sin \mu = \frac{1}{M} \tag{17.3}$$

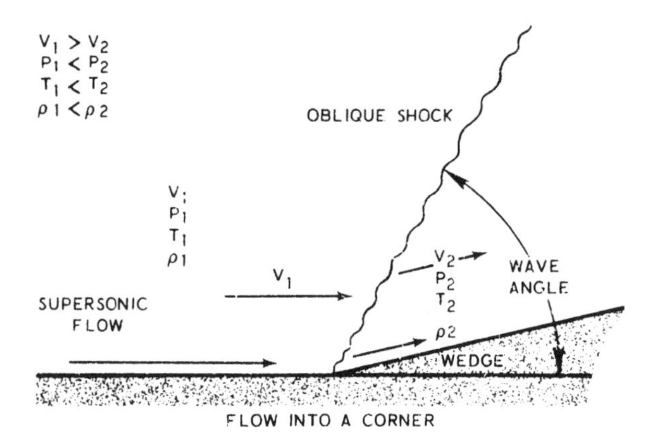

Fig. 17.14. Formation of an oblique shock wave.

For example, if $M = 2.0$ and the wedge angle is 10°, then

$$\text{Wave angle} = \arcsin 0.5 + 10^\circ = 30^\circ + 10^\circ = 40^\circ$$

The oblique shock wave is weaker than the normal shock wave, but it still is a shock wave and is wasteful of energy. As air passes through an oblique shock wave, its density, pressure, and temperature all rise and its velocity decreases. The principal difference is that the air remains supersonic behind an oblique shock wave, but it is always slowed to subsonic behind a normal shock wave.

Expansion Waves

The third kind of wave that occurs in supersonic flow is the expansion wave. Unlike the normal and oblique shock waves, the expansion wave is not a shock wave. In fact, the expansion wave is just the opposite of a compression wave. Density, pressure, and temperature all decrease as the air flows through an expansion wave. Air velocity increases through this wave and no energy is lost. Expansion waves form where the airstream turns around a convex corner.

This is shown in Fig. 17.15. In flowing around the convex corner, the airstream makes an infinite number of minute direction changes until it has turned the corner and the flow direction is parallel to the downstream surface. The expansion wave takes place through a fan-shaped area as shown by the dashed lines in Fig. 17.15.

Shock waves, on the other hand, take place through a very narrow area. As the static pressure decreases through an expansion wave, lift is produced in this region. Subsonic flow could not negotiate a convex corner because the flow would separate from the wing. Supersonic flow, however, has a higher energy

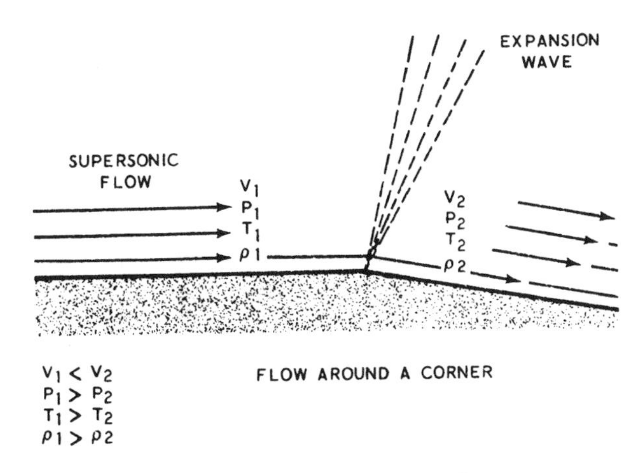

Fig. 17.15. Formation of an expansion wave.

Type of wave formation	Oblique shock wave	Normal shock wave	Expansion wave
	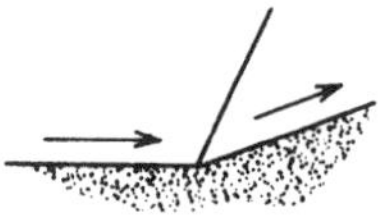	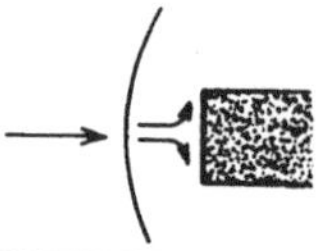	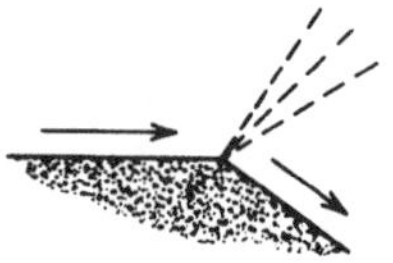
Flow direction change:	"Flow into a corner," turned to flow parallel to the surface	No change	"Flow around a corner," turned to flow parallel to the surface
Effect on velocity and Mach number:	Decreased but still supersonic	Decreased to subsonic	Increased to higher supersonic
Effect on static pressure and density:	Increase	Great increase	Decrease
Effect on energy or total pressure:	Decrease	Great decrease	No change (no shock)
Effect on temperature:	Increase	Great increase	Decrease

Fig. 17.16. Summary of supersonic wave characteristics.

level and can turn these sharp corners by means of an expansion wave, without separation.

Characteristics of the waves are summarized in Fig. 17.16.

Aerodynamic Forces in Supersonic Flight

Aerodynamic forces are produced by differential pressures acting on the surfaces of a wing in an airstream. This is true whether the airstream is subsonic or supersonic. In supersonic flight sudden increases in pressure result when the airflow passes through shock waves. Expansion waves produce less rapid changes in pressure. The pressure drops when passing through an expansion wave. An expansion wave at a sharp corner produces a fairly rapid pressure drop, while an expansion over a curved surface will produce a gradual pressure drop.

Supersonic Airfoils

One type of supersonic airfoil is the double-wedge airfoil shown in Fig. 17.17. This airfoil cannot be used subsonically because the air will become separated when it reaches the convex angles at the top and bottom of the wedges. It does have use in supersonic missiles that are boosted through the subsonic flight region.

Figure 17.17a shows the wave pattern about a double-wedge airfoil at zero AOA and thus zero lift in supersonic flight. Oblique shock waves, of the same strength, form at the top and the bottom of the leading and trailing edges.

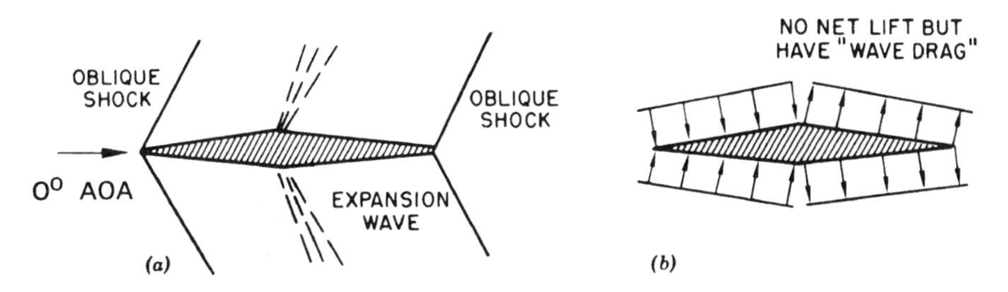

Fig. 17.17. Double-wedge airfoil in supersonic airflow: (a) wave pattern, (b) pressure distribution.

Behind these shocks the pressure rises but no net lift results. At the sharp corners at the 50% chord point expansion waves form, of the same strength. Lower pressures exist behind the expansion waves, but again the pressures on the top and bottom are equal in value and cancel each other. This is shown in Fig. 17.17b. The rearward components of the pressure arrows represent "wave drag" that exists even when no lift is being developed.

Consider the double-wedge airfoil at a small positive AOA, as shown in Fig. 17.18a. The oblique shock waves are now not equal in strength. The shock wave at the top of the leading edge does not have to turn as high an angle as it did at zero AOA and is now weaker. The pressure rise through the top shock is now lessened.

The reverse is true for the bottom leading-edge shock. It is stronger and the pressure rise is greater than at zero AOA. The expansion waves at the midpoint corners are not of equal strength at a positive AOA. The top expansion is now greater, because the angle to be turned is greater than it was at zero AOA. The pressure decrease is also greater on the top. The opposite reasoning applies to the bottom expansion wave, and the pressure decrease behind the bottom expansion wave is lessened. Net lift is now being developed. The pressure distribution is shown in Fig. 17.18b.

To avoid the subsonic problems of the double-wedge airfoil, another supersonic airfoil section is used. This is called the circular arc or biconvex

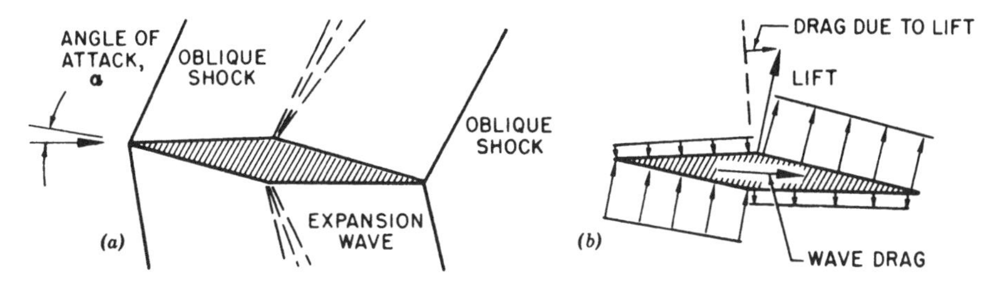

Fig. 17.18. Double-wedge airfoil developing lift: (a) wave pattern, (b) pressure distribution.

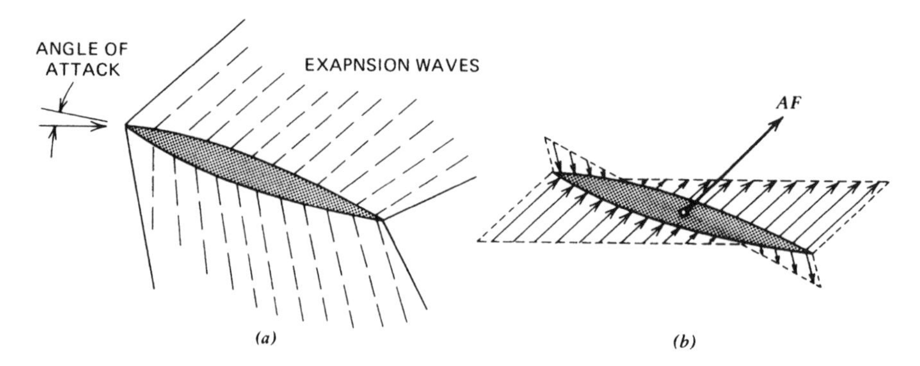

Fig. 17.19. Circular arc airfoil in supersonic flow: (a) wave pattern, (b) pressure distribution.

airfoil. It uses two arcs of circles to define its shape. This airfoil develops shock waves at the leading and trailing edges, but because there are no sharp angles at the midchord point, no concentrated expansion wave occurs at that point. Instead, a continuous expansion wave, from the leading edge to the trailing edge, forms on the top and bottom airfoil surfaces. A symmetrical circular arc airfoil produces no net lift at zero AOA because the waves are symmetrical. But at positive AOA the shock wave on the top leading edge is weaker and the top expansion wave is stronger than the corresponding waves on the bottom of the airfoil. Thus, positive lift is created as shown in Fig. 17.19. The center of pressure and the aerodynamic center are both at the same position for both the double-wedge and the circular arc airfoils. This position is at the 50% chord point. Even though the circular arc airfoil can be flown at subsonic speeds, it does not develop a high value of $C_{L(\max)}$ and thus takeoff and landing speeds are high.

Wing Planform

Earlier in this chapter we discussed the use of sweepback as a means of increasing M_{crit} for subsonic aircraft. This same principle has other advantages for supersonic aircraft as well. In addition to increasing the force divergence Mach number, sweepback also reduces the amount of C_D increase. This is shown in Fig. 17.20. The increase in drag coefficient is both delayed and reduced in value by sweepback.

It is also interesting to compare straight wings (0° sweep) with swept wings. Note that the straight-wing and the swept-wing curves cross at some Mach number greater than 1.0. For the 45° wing they cross at about $M = 1.4$. This means that at this speed both wings would have the same drag and at higher speeds the straight wing actually has less drag. Therefore, for high supersonic aircraft the straight-wing configuration may be desirable. Of course, at lower speeds than the crossover point, the advantage is with the swept-wing aircraft.

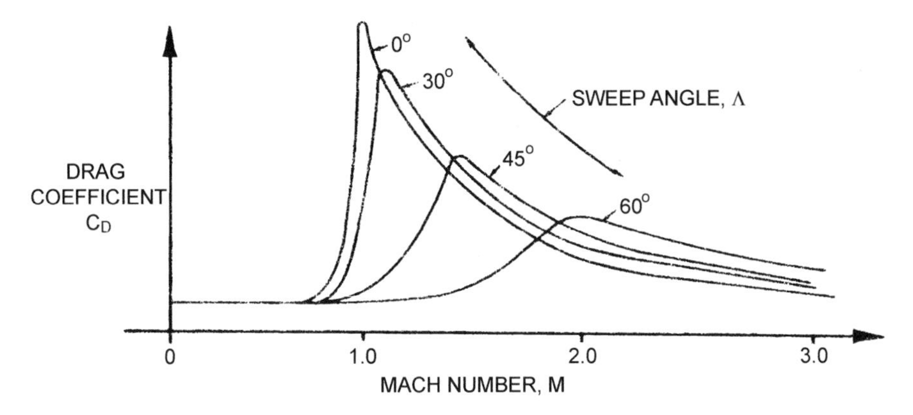

Fig. 17.20. Effect of wing sweep on C_D.

In our discussion of the formation of oblique shock waves we limited our analysis of Mach wave angle to two-dimensional flow. In reality, the pressure disturbances radiate outward in all directions, like expanding spheres. Thus, instead of a Mach wave we have a Mach cone. This is shown in Fig. 17.21.

Another advantage of sweep is that it places the wing within the Mach wave (cone) of the aircraft. This allows subsonic airfoil sections to be used for supersonic flight. Using subsonic airfoils is desirable because they develop higher $C_{L(\max)}$ values and thus reduce takeoff and landing speeds.

Consider a Mach wave that is generated at the wing root as in Fig. 17.22. The air is supersonic ahead of the Mach wave and it is also supersonic in the direction of flight behind the Mach wave. The component of air behind the Mach wave that is perpendicular to the Mach wave will, however, be subsonic. If the wing is swept back so that its leading edge is behind the Mach wave, it will be flying in a subsonic flow and thus can use a subsonic airfoil section.

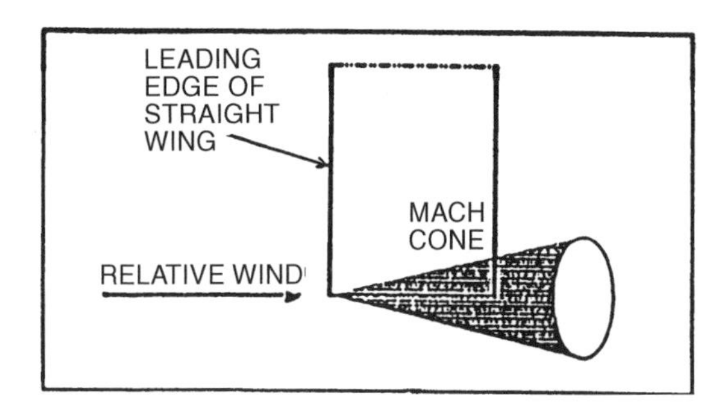

Fig. 17.21. Mach cone.

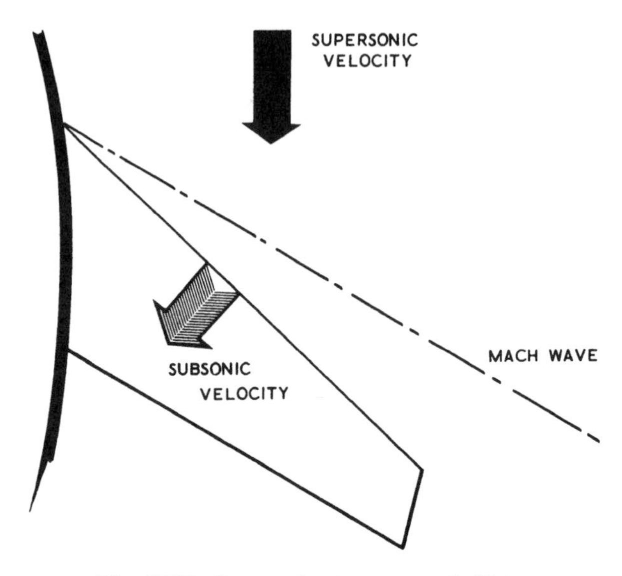

Fig. 17.22. Swept wing in supersonic flight.

Area Rule Drag Reduction

Experiments in wind tunnels showed that bodies with sudden changes in cross-sectional area produced more drag than similar bodies with gradual changes in area. This principle was applied to aircraft. A large increase in the cross-sectional area occurs where the wings attach to the fuselage. By reducing the fuselage area at this point, the total area was enlarged gradually and drag was reduced. At the trailing edge of the wing the area of the fuselage was again increased to compensate for the loss of wing area. This results in the "coke bottle' shape of the fuselage. This method is no longer used, because it is not cost-effective.

Use of All Movable Control Surfaces

Early attempts to break the sound barrier were often disastrous because the aircraft did not have the design features found in more modern aircraft. To attain the required speed, the aircraft had to be put into a steep dive. When the aerodynamic center shifted to the 50% C location, the aircraft's center of gravity was ahead of the aerodynamic center. This created a stable dive condition that could not be overcome with the elevators, and often the aircraft continued in the dive until it hit the earth. The elevators were constructed in the conventional manner, as shown in Fig. 17.23. The control forces were limited to the small elevator itself and could not influence the flow about the horizontal stabilizer. The amount of force that the pilot could exert was also limited, because no power assist controls were available.

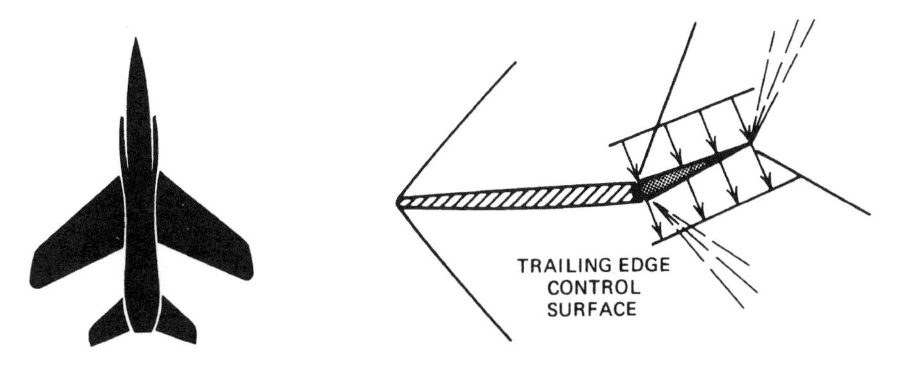

Fig. 17.23. Subsonic control surface.

As the state of the art progressed, all moving control surfaces, as shown in Fig. 17.24, were developed, as were fully powered, irreversible control systems. Much more control was then available, and pitch control was no longer a problem.

Supersonic Engine Inlets

The compressor blades of a turbine engine can be made to operate either subsonically or supersonically, but not both. The necessity for slow-speed flight, as well as supersonic flight, rules out the use of supersonic flow in the compressor. Therefore, the flow must be slowed to subsonic when the aircraft is flying supersonically. This can be done only by the airflow passing through a normal shock wave.

The Mach number behind a normal shock wave is approximately $1/M$, where M is the Mach number ahead of the shock. If the flight Mach is only slightly above 1.0 this is no problem, and a normal shock inlet such as shown in Fig. 17.25 can be used.

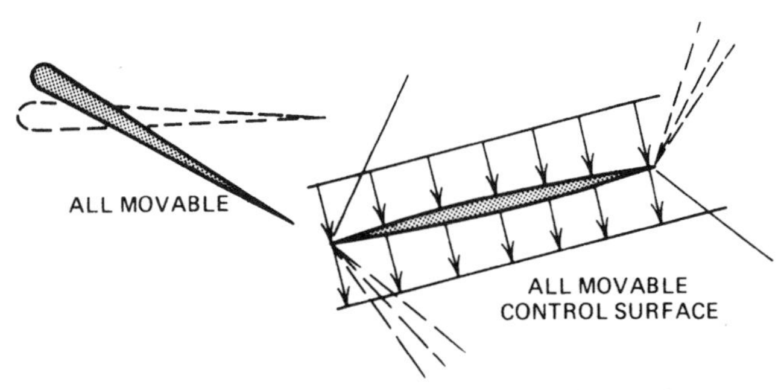

Fig. 17.24. Supersonic control surface.

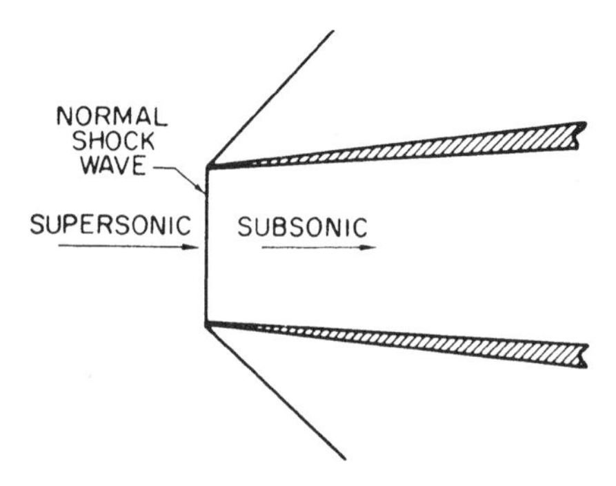

Fig. 17.25. Normal shock engine inlet.

As the flight Mach number is increased, the reduction in airspeed in the normal shock wave becomes greater, the intensity of the shock increases, and the energy loss in the shock wave is increased. To reduce this energy loss an oblique shock inlet is used. This type of inlet may use a "spike" or a "ramp," which cause oblique shock waves to form. When air passes through an oblique shock wave its velocity is reduced but remains supersonic. The energy lost in passing through an oblique shock is much less than that lost in passing through a normal shock. Often the air is slowed by a series of oblique shocks, each slowing the air until the air velocity is only slightly supersonic; then the air is passed through a weak normal shock.

Single and multiple oblique shocks formed by spike-type inlets are shown in Fig. 17.26. The Mach cone angle varies with the flight Mach number, so for proper position of the shock waves, the spike (or ramp) must be adjustable in flight.

Aerodynamic Heating

The temperature of the air in the stagnation area of the leading edge of the wing, at the nose of the aircraft, at the leading edges of the stabilizers, and at

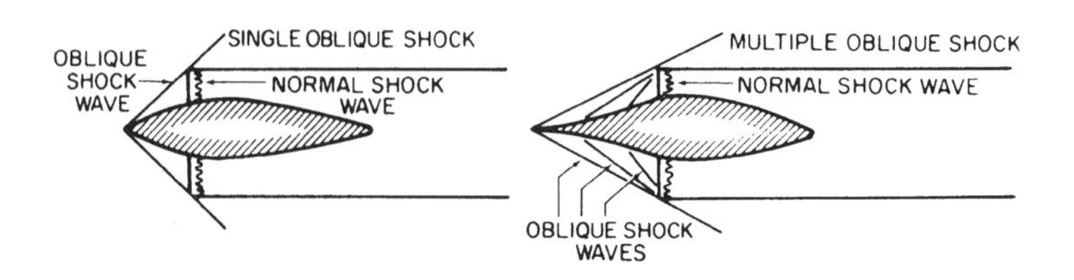

Fig. 17.26. "Spike" oblique shock engine inlets.

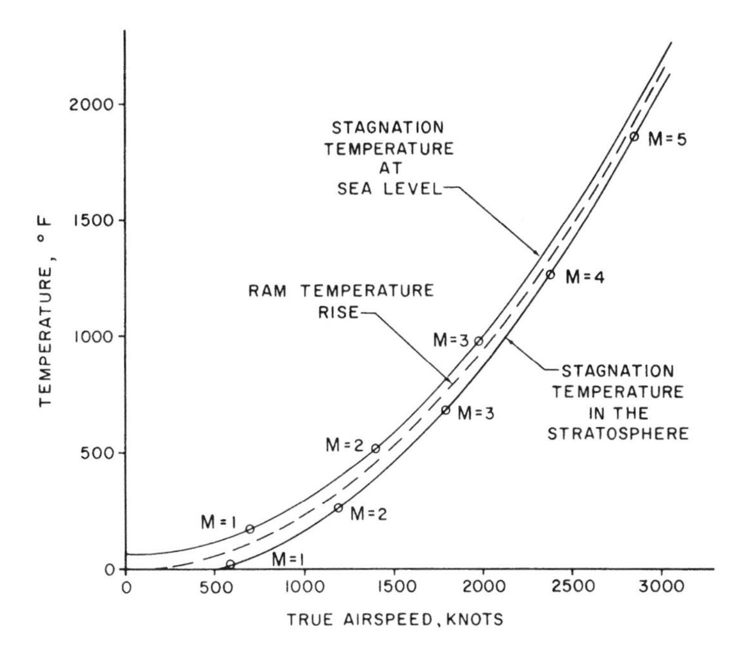

Fig. 17.27. Stagnation temperatures.

other stagnation points of an aircraft is shown in Fig. 17.27. The effect of aerodynamic heating on the tensile strength of structural alloys is shown in Fig. 17.28. These figures help illustrate the vast difference in problems of structural strength in designing a Mach 3 supersonic transport aircraft as compared to a Mach 2 SST. The stagnation temperature in the stratosphere is about 250°F for the Mach 2 aircraft and about 700°F for the Mach 3 SST. Comparing the tensile strength of aluminum alloy at these temperatures shows that at 250°F about 80% of the room temperature tensile strength remains, but at 700°F less than 10% of the room-temperature strength remains. Clearly, it is not feasible to make leading edges out of aluminum alloy for a Mach 3 aircraft.

SYMBOLS

a	speed of sound (knots)
a_0	speed of sound at sea level, standard day
M	Mach number (dimensionless)
θ (theta)	Temperature ratio $= T/T_0$
μ (mu)	Mach wave angle (degrees)

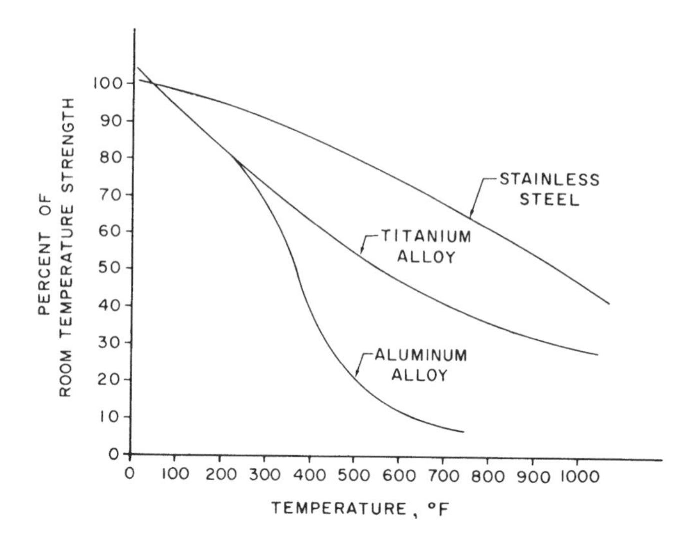

Fig. 17.28. Effect of temperature on tensile strength of metals after half-hour exposure.

EQUATIONS

17.1 $a = a_0\sqrt{\theta}$

17.2 $M = \dfrac{V}{a}$

17.3 $\sin\mu = \dfrac{1}{M}$

PROBLEMS

1. The speed of sound is an important factor in high-speed flight because
   a. M_{crit} occurs at $M = 1$.
   b. the pressure waves generated by the plane move at sonic speed.
   c. shock waves form when local air velocities are supersonic.
   d. Both (b) and (c)
   e. None of the above

2. The speed of sound depends on the air density, pressure, and temperature.
   a. True
   b. False

3. Critical Mach number, M_{crit}, is the aircraft's speed when

a. it goes supersonic.
b. the airflow first reaches sonic speed.
c. shock waves form.
d. Both (b) and (c)
e. None of the above

4. After passing through a normal shock wave, airflow is
   a. subsonic.
   b. not changed in direction.
   c. heated up.
   d. increased in pressure and density.
   e. All of the above.

5. Airflow passing through an oblique shock wave remains supersonic but changes in direction.
   a. True
   b. False

6. Transonic flight problems may include
   a. force divergence.
   b. increase in C_D.
   c. decrease in C_L.
   d. tuckunder.
   e. buffet.
   f. control surface buzz.
   g. loss of control effectiveness.
   h. All of the above
   i. All except (a)

7. Above about Mach 2, the straight-wing aircraft has a lower drag than a 60° swept-wing aircraft.
   a. True
   b. False

8. Mach wave angle changes with airspeed as follows:
   a. It decreases as airspeed increases.
   b. It increases as airspeed increases.
   c. It remains the same with airspeed changes.

9. All supersonic aircraft have circular arc or biconvex supersonic airfoils.
   a. True
   b. False

10. Airflow passing through an expansion wave
   a. speeds up.
   b. increases the energy of the airstream.
   c. decreases the temperature of the air.
   d. Both (a) and (c)

Answers to Problems

CHAPTER 1

1. m = 500 slugs; 2. a = 12 fps^2; 3. Drag = 282 lb., Lift = 102.6 lb;
4. Tan = 0.75, Distance = 5 mi; 5. s = 8 ft, F = 30 lb;
6. Time = 16.7 sec; 7. Takeoff roll = 1667 ft; 8. a = 5 fps^2; 9. V_t = 450 fps;
10. PE = 80×10^6 ft-lb, KE = 10×10^6 ft-lb, TE = 90×10^6 ft-lb,
11. HP = 5000 hp; 12. $F_B = 1260$ lb

CHAPTER 2

1. c; 2. c; 3. a; 4. a; 5. d; 6. d; 7. c; 8. b; 9. a; 10. b;
11. $d = 0.929$, PA = 2000 ft, $q = 1.078$, $s = 0.8618$, DA = 5000 ft;
12. $H = 2062.8$ psf, $P_2 = 1931$ psf, $q_2 = 131$ psf, $A_3 = 12.5 ft^2$, $P_3 = 2041$ psf, $q_3 = 21$ psf; 13. $q = 116.8$ psf; 14. EAS = 337.5 knots;
15. TAS = 495.5 knots

CHAPTER 3

1. d; 2. c; 3. d; 4. d; 5. c; 6. a; 7. d; 8. d; 9. b; 10.d

CHAPTER 4

1. a; 2. c; 3. a; 4. c; 5. d; 6. d; 7. c; 8. b; 9. d;
10. a. combination, b. energy add, c. combination, d. camber chg;
11. AOA = 4°; 12. V_s = 134 knots; 13. V_s = 116 knots; 14. $R_g = 10.3 \times 10^6$

CHAPTER 5

1. d; 2. c; 3. d; 4. d; 5. b; 6. d; 7. d; 8. c; 9. b; 10.b; 11. $D_{min} = 1020$ lb;
12. V = 172 knots; 13. $D_i = D_p = 510$ lb

14.

V_2	(V_2/V_1)	D_p	$(V_1/V_2)^2$	D_i	D_t
125	0.530	270	1.89	963	1233
150	0.763	389	1.31	669	1058
172	1.0	510	1.0	510	1020
200	1.35	690	0.74	377	1067
300	3.04	1562	0.33	168	1720
400	5.42	2766	0.184	94	2860

CHAPTER 6

1. d; 2. d; 3. a; 4. a; 5. a; 6. a; 7. d; 8. c; 9. c; 10. b;
11. $T = 3000$ lb, Efficiency = 77%; 12. $C_t = 1.5$, $T = 4100$ lb, $FF_0 = 6150$ lb/hr;
13. $FF_{trop} = 1845$ lb/hr;
14. a. 500 knots, b. 240 knots, c. 0.327 (19°), d. 10,940 fpm, e. 240 knots,
f. 300 knots, g. 350 knots

CHAPTER 7

1. d; 2. c; 3. a; 4. b; 5. d; 6. a; 7. d; 8. c; 9. b; 10. d;
11. $V_{12} = 325$ knots, $V_8 = 285$ knots;
12. $SR_{12} = 0.2708$ nmi/lb, $SR_8 = 0.3455$ nmi/lb;
13. $SR_8 = 0.3125$ nmi/lb, SR improvement in Problem 13 = 14.15.4%,
SR improvement in Problem 12 = 27.6%, conclusion: reduce throttle as fuel is burned; 14. $SR_0 = 0.2973$, $SR_{20} = 0.4108$, SR improvement = 38%

CHAPTER 8

1. a; 2. d; 3. c; 4. b; 5. d; 6. b; 7. b; 8. d; 9. c; 10. b;
11. 474 HP, 388 HP, 540 HP, 657 HP, 1588 HP, 3520 HP;
12. a. 340 knots, b. 0.542 (32.8°), c. 0.48 (29.2°), d. 4125 fpm, e. 4950 fpm,
f. 100 knots, g. 140 knots, h. 160 knots

CHAPTER 9

1. c; 2. a; 3. b; 4. c; 5. c; 6. c; 7. c; 8. b; 9. b; 10. a;
11. $V_{30} = 160$ knots, $V_{20} = 130$ knots; 12. $SR_{30} = 0.3765$ nmi/lb,

$SR_{20} = 0.5778$ nmi/lb;
13. $SR_{20} = 0.4471$ nmi/lb, Improvement = 18.7%, Improvement in Problem 12 = 53.5%, conclusion: reduce throttle as fuel is burned;
14. $SR_0 = SR = 0.58$ nmi/lb, conclusion: no change in SR with altitude (considering airframe only)

CHAPTER 10

1. c; 2. b; 3. c; 4. a; 5. d; 6. b; 7. a

CHAPTER 11

1. d; 2. b; 3. d; 4. b; 5. b; 6. a; 7. c; 8. c; 9. d; 10. d; 11. RC = 506.5 fpm;
12. RC = −658.5 fpm

CHAPTER 12

1. d; 2. c; 3. c; 4. d; 5. a; 6. d; 7. d; 8. a; 9. a; 10. $A = 8\ \text{fps}^2$;
11. V = 126.5 fps (75 knots); 12. $S = 2576$ ft; 13. $S = 5796$ ft; 14. $S = 2087$ ft;
15. $S = 4025$ ft

CHAPTER 13

1. d; 2. b; 3. d; 4. d; 5. d; 6. a; 7. a; 8. c;; 9. a; 10. a; 11. $A = -8\ \text{fps}^2$;
12. $S = 1785$ ft; 13. $S = 2901$ ft; 14. $S = 2160$ ft; 15. $S = 2231$ ft

CHAPTER 14

1. b; 2. c; 3. d; 4. d; 5. d; 6. a; 7. c; 8. d; 9. d; 10. b; 11. 30.6°; 12. Verified;
13. 1.162G; 14. ROT = 3.23°/sec; 15. Verified

CHAPTER 15

1. b; 2. c; 3. d; 4. c; 5. c; 6. d; 7. a; 8; 8. c; 9. b; 10. d

CHAPTER 16

1. b; 2. c; 3. a; 4. a; 5. c; 6. a; 7. c; 8. a; 9. b; 10. b

CHAPTER 17

1. d; 2. b; 3. b; 4. e; 5. a; 6. h; 7. a; 8. a; 9. b; 10. d

References

1. Abbott, I. H., and A. E. VonDoenhoff, *Theory of Wing Sections*, Dover, New York, 1959.
2. Anderson, J. D., Jr., *Introduction to Flight*, McGraw-Hill, New York, 1978.
3. Bent, R. D., and J. L. McKinley, *Aircraft Powerplants*, 4th ed., Gregg Div., McGraw-Hill, New York, 1978.
4. Carrol, R. L., *The Aerodynamics of Powered Flight*, Wiley, New York, 1960.
5. Dalton, S., *The Miracle of Flight*, McGraw-Hill, New York, 1977.
6. Dole, C. E., *Flight Theory and Aerodynamics*, Wiley, New York, 1981.
7. ———, *Flight Theory for Pilots*, IAP, Casper, WY, 1993.
8. ———, *Mathematics and Physics for Aviation Personnel*, IAP, Casper, WY, 1991.
9. ———, *Fundamentals of Aircraft Material Factors*, IAP, Casper, WY, 1991.
10. Domasch, D. O., S. S. Sherby, and T. F. Connolly, *Airplane Aerodynamics*, 4th ed., Pitman, New York, 1967.
11. Dwinnell, J. H., *Principles of Aerodynamics*, McGraw-Hill, New York, 1949.
12. Etkin, B., *Dynamics of Atmospheric Flight*, Wiley, New York, 1972.
13. Hoerner, S. F., *Fluid-Dynamic Drag* (published by author), 1965.
14. Hoerner, S. F., and H. V. Borst, *Fluid-Dynamic Lift* (published by Mrs. L. A. Hoerner), 1967.
15. Kuethe, S., *Foundations of Aerodynamics*, 3rd ed., Wiley, New York, 1976.
16. Martynov, A. K., *Practical Aerodynamics*, Macmillan, New York, 1965.
17. McCormick, B. W., *Aerodynamics, Astronautics, and Flight Mechanics*, Wiley, New York, 1979.
18. ———, *Aerodynamics, Astronautics, and Flight Mechanics*, 2nd ed., Wiley, New York, 1995.
19. Montgomery, J. R., *Sikorsky Helicopter Flight Theory for Pilots and Mechanics*, Sikorsky Aircraft, Stratford, CT, 1964.
20. Nicolai, L. M., *Design of Airlift Vehicles*, University of Dayton, Dayton, OH, 1974.
21. Perkins, C. D., and R. E. Hage, *Airplane Performance, Stability and Control*, Wiley, New York, 1949.
22. Roed, A., *Flight Safety Aerodynamics*, Aerotech, Kungsangen, Sweden, 1972.
23. Roland, H. E., and J. F. Detwiler, *Fundamentals of Fixed and Rotary Wing Aerodynamics*, ISSM, University of Southern California, Los Angeles, 1967.
24. Saunders, G. H., *Dynamics of Helicopter Flight*, Wiley, New York, 1975.
25. Seckel, E., *Stability and Control of Airplanes and Helicopters*, Academic Press, New York, 1964.

26. Shapiro, A. H., *The Dynamics and Thermodynamics of Compressible Fluid Flow*, Vols. I and II, Ronald Press, New York, 1953.
27. Sutton, O. G., *The Science of Flight*, Pelican Books (A209), New York, 1955.

U.S. GOVERNMENT PUBLICATIONS

28. Anonymous, *Aerodynamics*, .S. Army Material Command, Washington, 1965.
29. Anomymous, *Aeropdynamics for Pilots*, Air Training Command, USAF, ATC Manual 51-3, 1979.
30. Anonymous, *Academics for Basic*, Air Training Command, USAF, ATC P-V4A-A-AB-SW, 1975.
31. Anonymous, Applied AQerodynamics, Air Training Command, USAF, ATC P-V4A-A-AA-SW, 1977.
32. Anonymous, *Interceptor*, Air Defense Command, USAF, ADCPI 62-1 through 62-16, June 1964–April 1966.
33. Anonymous, *Low Level Wind Shear*, Federal Aviation Administration, AC No. 00-50, 1976.
34. Anonymous, *Pilot Windshear Guide*, Federal Aviation Administration, AC No. 00-54, 1988.
35. Anonymous, *Rotary Wing Flight*, U.S. Army Field Manual, 1-51, 1979.
36. Anonymous, *Wake Turbulence*, Air Training Command, USAF, Pamphlet 51-12, 1976.
37. Hurt, H. H., Jr., *Aerodynamics for Naval Aviators*, NAVWEPS 00-80T-80, Government Printing Office, 1965.

PERIODICALS

38. Anonymous, “Hydroplaning, Fun or Disaster,” *Aerospace Saqfety*, USAF, AFISC, May 1964.
39. Anonymous, “Low Altitude Wind Shear,” *Aerospace Saqfety*, USAF, AFISC, Sept. 1976.
40. Eggleston, B., and D. L. Jones, “The Design of Lifting Supercritical Airfoils Using a Numerical Optimization Method,” *Canadian Aeronautics and Space Journal*, Vol. 23, No. 3, May/June 1977.
41. Haines, P., and J. Luers, “Aerodynamic Penalties of Heavy Rain on Landing Aircraft,” *Journal of Aircraft*, Vol. 20, No. 2, Feb. 1983.
42. Kocivar, B., “Super STOL,” *Popular Science*, Feb. 1976.
43. Luers J., and P. Haines, “Heavy Rain Influence on Airplane Accidents,” *Journal of Aircraft*, Vol. 20, No. 2, Feb. 1983.

Index

Wave drag, 289, 293
Weight, 2
Wing planform:
definitions, 63
supersonic, 294